Steel Rolling
Principle, Process & Application

Steel Rolling
Principle, Process & Application

N.K. Gupta

*Former HOD of Roll Pass Design
Bureau and Operations of Merchant Mill
Bhilai Steel Plant, Bhilai*

*Managing Consultant
Roll Pass Design Solutions
www.rollpassdesignsolution.com*

CRC Press is an imprint of the
Taylor & Francis Group, an **informa** business

First published 2021
by CRC Press
2 Park Square, Milton Park, Abingdon, Oxon, OX14 4RN

and by CRC Press
6000 Broken Sound Parkway NW, Suite 300, Boca Raton, FL 33487-2742

© 2021, Manakin Press Pvt. Ltd.

CRC Press is an imprint of Informa UK Limited

The right of N.K. Gupta to be identified as author of this work has been asserted by him in accordance with sections 77 and 78 of the Copyright, Designs and Patents Act 1988.

Reasonable efforts have been made to publish reliable data and information, but the author and publisher cannot assume responsibility for the validity of all materials or the consequences of their use. The authors and publishers have attempted to trace the copyright holders of all material reproduced in this publication and apologize to copyright holders if permission to publish in this form has not been obtained. If any copyright material has not been acknowledged please write and let us know so we may rectify in any future reprint.

All rights reserved. No part of this book may be reprinted or reproduced or utilised in any form or by any electronic, mechanical, or other means, now known or hereafter invented, including photocopying and recording, or in any information storage or retrieval system, without permission in writing from the publishers.

For permission to photocopy or use material electronically from this work, access www.copyright.com or contact the Copyright Clearance Center, Inc. (CCC), 222 Rosewood Drive, Danvers, MA 01923, 978-750-8400. For works that are not available on CCC please contact mpkbookspermissions@tandf.co.uk

Trademark notice: Product or corporate names may be trademarks or registered trademarks, and are used only for identification and explanation without intent to infringe.

Print edition not for sale in South Asia (India, Sri Lanka, Nepal, Bangladesh, Pakistan or Bhutan).

British Library Cataloguing-in-Publication Data
A catalogue record for this book is available from the British Library

Library of Congress Cataloging-in-Publication Data
A catalog record has been requested

ISBN: 978-1-032-02216-1 (hbk)
ISBN: 978-1-003-18239-9 (ebk)

To my parents
Late Sh. R G Gupta and
Late Smt. Savitri Devi Gupta

Preface

Steel, as they say, is a part of everybody's life. From a small kid whose diapers need to be pinned with the help of a safety pin to use in very complex structures like bridges, steel is omnipresent.

To summarize, steel is perhaps one metal with most versatile usage; that's why steel is so popular and steel producing industry is perhaps the biggest out of all the metals.

The overall, steel production in India during 2010 was about 60 million tonnes; it is projected that it will touch a level of 150 million tonnes by 2020. Likewise, there will be tremendous growth and developments in rolling mills, all over the world.

Roll Pass Design is a method to design a rolled section, augment the production level of a rolling mill and to make the production with minimum possible cost. The technology of "Roll Pass Design" was generally kept as a family business in India, as well as over the world also. In England itself, it was treated as family business over the year, passing down from one generation to another. Presently, majority of rolling mills in India are still managed by illiterate foremen with their demanding terms and conditions. Due to this type of management, most of rolling mills in India are presently not cost viable, because of high operating cost, low yield and also due to huge inventory of rolls, other spares etc. They are also producing their products with inferior quality.

The main reason behind such type of work-culture is the insufficient information available to the management of rolling mills, mainly because of the non-availability of desired courses/literatures available to know about rolling technology. No proper literature is now available, which can provide guidelines/ tools to the student of rolling technology. There may be, a few literature or books available on mechanical metallurgy, they deals only with the metallographic and theoretical part of it and are suiting only to academic purpose. No handbook is available, which may focus on elements of rolling technology, how to set a mill for rolling of a particular product, mill setting for various profiles, roll pass design of a product and how to control a section etc., that's what a mill person's requires.

Who This Book is for

This book "Roll Pass Design and Rolling Practices" will be a useful handbook for shop floor personnel to the higher management of rolling mill industry; as well it will cater to the polytechnic/engineering students of metallurgical/mechanical/ process engineering. It has covered all aspects of rolling, which includes rolling of all major sections; where in roll pass design is required. This book may also be useful as reference book for students/professionals of rolling technology.

The book has a numbers of illustrative chapters. This book deliberated on the fundamental of mechanical working and, its theory in a very simpler way. In addition it has also provided coverage on rolls, rolls cooling, roll turning, roll reclamation, investigation of roll breakage, roll management and roll bearing also, these contribute to the highest in the production cost.

It also provide the description of vital elements of roll pass design in a very simple and systematic manner. It has a chapter on operational management of a rolling mill, which includes safety, inventory. Packaging of the finished products and modern operating mill practices and technologies are also discussed in details.

Above all, first time it has been illustrated to how to design an individual rolling section step-by-step with the help of formulas, graphs and various aspects based on experience. With the help of that a layman can design a round, billet or angle section, such content is not available anywhere else. In addition, this book will also explain how to do the mill setting and to control of an element of section during rolling.

Author

Acknowledgement

First of Late Shri R.G. Gupta and Late Smt. Savitri Devi Gupta, I express my sincere gratitude to almighty and my wonderful parents, without their guidance and affection since from childhood it was not possible for me to come up to such a stature of writing this book.

I am also thankful to my wife Usha, daughter Radhika, son Sarvesh, son-in-law Anurag, daughter-in-law Manie and sister Vandana. They have done a great job, making sure that not only the contents of this book are lucid and unambiguous from the beginning but also have motivated me to complete this book.

The following individuals deserve a special mention for making of this book and providing me with many useful information, comments and suggestions:

Shri A.G. Rama Rao was my first boss and mentor, who made me learn the fundamentals of roll pass design and rolling. His guidance has helped me to become well versed with subject.

Shri V.S. Verma has worked all along in the field of rolling mill. He retired from the post of Director (Production & Marketing). His suggestions has helped me in shaping this book.

I would also like to thank Madam Jyoti, Ali, Banjare, all my friends and family members, who have always helped and encouraged, while I was working on this book.

I would especially like to thanks Mr. Beenu Bhalla, Publishing Director and Mr. Nishant Saini of Manakin Press Pvt. Ltd. and his team for publishing this book and bringing my dream to reality.

Last and not least: I beg forgiveness of all those who have been with me over the course of the years and whose names I have failed to mention.

Author

Brief Contents

1. Roll Pass Design	1–58
2. Mechanical Working and Rolling Process	59–90
3. Rolls, Roll Cooling and Roll Management	91–134
4. Rolling of Blooms and Slabs	135–164
5. Rolling of Billets	165–216
6. Rolling of Rounds and TMTS Bars	217–258
7. Rolling of Flats and Squares	259–282
8. Rolling of Angles	283–320
9. Rolling of Channels	321–358
10. Rolling of Beam	359–396
11. Rolling of Rail	397–438
12. Operation, Safety and Quality Management	439–458
13. Packaging of Steel Products	459–468
14. Energy Efficient Practices and Technologies	469–494
Index	495–500

Detailed Contents

1. Roll Pass Design   1–58

1   Introduction   1
2   Roll Pass Design   1
    2.1   Object of Roll Pass Design   1
    2.2   Basic Purpose of Roll Pass Design   2
    2.3   Characteristics of Efficient Roll Pass Design   2
    2.4   Specified Conditions Under Which Roll Pass Design is Based Upon   3
    2.5   Constraints to Efficient Roll Pass Design   3
3   Characteristics of Hot Rolling   4
    3.1   Draft/Draught   4
    3.2   Elongation   5
    3.3   Spread   5
    3.4   Types of Spread   7
    3.5   Computation of Spread   9
    3.6   Contact Area, Contact Angle, Angle of Bite
        and Roll Bite Condition   15
    3.7   Methods to Improve Roll Biting Conditions   17
    3.8   Forward Slip, Neutral Plane and Neutral Angle   18
    3.9   Forward Slip in Groove Rolls   19
    3.10   Tension   19
    3.11   End Thrust   20
    3.12   Mill Spring   21
4   Pass   22
    4.1   Type of Passes   22
    4.2   Classification of Passes   23
    4.3   Elements of Passes   24
    4.4   Neutral Line of the Pass   26
5   Diameter of Rolls   27
    5.1   Average Diameter   27
    5.2   Redressing Co-efficient   29
    5.3   Roll Collars   29
    5.4   Determination of Rolling Diameter   30
    5.5   Pitch Line, Line of Rolling   31
    5.6   Top and Bottom Pressure (Over or Under Draught)   32
6   Distribution of Draught In A Pass   34
    6.1   Principle of the Distribution of Draught   34
    6.2   Maximum Reduction   35

xiv Detailed Contents

7 Roll Pass Design for Non-Uniform Deformation 36
   7.1 Analysis of Non-Uniform Deformation 36
   7.2 Computation of Draught In Non-Uniform Deformation 39
8 Characteristics of Rolling in Flanged Shapes 40
   8.1 Direct and Indirect Draught 40
   8.2 Effect of Taper 41
   8.3 Effect of Friction 43
   8.4 Speed of Rolling in Flange Profile 43
   8.5 Asymmetry of Deformation in a Flange Pass 44
   8.6 Stages of Rolling in a Flange Pass 45
   8.7 Distribution of Draught Among the Elements of Flanged Profile 46
   8.8 Features of Deformation in a Closed Web-Cutting Pass 46
9 Consideration for Design of Finishing Passes 48
   9.1 Co-efficient of Expansion 48
   9.2 Rolling with Negative Allowance 49
10 Specific Feature of Roll Pass Design of A
Continuous Mill 50
11 Computation of Power Requirement 51
   11.1 Rolling Load 51
   11.2 Power Requirement and Rolling Torque 52
   11.3 Factors Deciding The Size and Type of Main Drive Motor 56
   11.4 Factors Affecting Rolling Load 56

2. Mechanical Working and Rolling Process   59–90

1 Introduction 59
2 Hot and Cold Working 59
3 Theory of Microscopic Plasticity 60
4 Methods of Metal Working 60
5 Allotropy of Iron 62
6 Iron-Carbon Diagram 62
7 Transformation in steel 65
   7.1 Transformation in Eutectoid Steel 65
   7.2 Transformation in Hypo Eutectoid Steel 65
   7.3 Transformation in Hyper Eutectoid Steel 66
   7.4 Critical Temperature 66
8 The Effects of Various Elements on Plastic
Properties of Steel 66
9 The Effect of Rolling on Grain Structure 67
10 Defination and Classification of
Thermo-Mechanical Processing 68
11 Quenched and Self-Tempered Rebars (Qst) 69
12 Stages of Quenching and Self-Tempering Cycle 71
   12.1 First Stage 71
   12.2 Second Stage 72
   12.3 Third Stage 72
13 Main Equipments of TMT Process 73
14 Different Metal Forming Operations 74
15 Rolling 76

Detailed Contents

15.1	History of Rolling	76
15.2	What is Rolling	76
15.3	Methods of Rolling	77
16	Classification of Products	78
16.1	General Purpose Rolled Sections	79
16.2	Special Purpose Rolled Section	79
16.3	Main Categorization of Rolled Products	79
17	Classification of Rolling Mills	80
17.1	Classification as per Number of Rolls	81
17.2	Classification as per Nos. of Drives in the Mill	82
17.3	Classification as per Products Rolled	83
17.4	Classification As Per Layout	85
17.5	Continuous/Non-continious Rolling Mill	87
17.6	Classification wrt Direction of Rolling	88
18	Special Types Mill	88
18.1	Double Duo Mill	88
18.2	Sandzimar Planetary Mill	88
19	Determination Of Mill Size	89
19.1	Where Rolls are Individually Driven	89
19.2	Where Rolls are Driven by Single Motor Through Pinion Stand	90
19.3	Flat Rolling	90

3. Rolls, Roll Cooling and Roll Management 91–134

1	Introduction	91
1.1	There are Three Parts in a Roll, Namely	91
2	Classification of Rolls	92
2.1	Classification wrt Purpose	92
2.2	Classification wrt Specific Duty	93
3	Rolls Qualities	93
3.1	Effect of Alloying Elements in Iron and Steel Rolls	93
3.2	Effect of Casting Practice	94
3.4	Quality Classification of Rolls	95
4	Cast Iron Rolls	96
4.1	Grain Rolls	96
4.2	Clear or Definite Chill Rolls (CCCI)	96
4.3	Indefinite Chill Cast Iron Roll (ICCI)	97
4.4	Spheroidal Graphite Rolls (SGCI)	99
4.5	Double Poured or Composite Iron Rolls	102
5	Steel Rolls	102
5.1	Forged Steel Rolls	103
5.2	Cast Steel	103
6	Cementite Carbide Roll (Rings)	104
7	Sleeve Roll	104
8	Selection of Rolls for different category of Rolling Mills	105
8.1	Blooming and Slabbing Mills	105
8.2	Structural Mill	106
9	Proper Care of Rolls	107
9.1	Care During Design	108

xvi   Detailed Contents

9.2	Proper use During Rolling	109
9.3	Care of Roll by Mill Crew	109
9.4	Care to Avoid Over Loading	109
9.5	Other Precautions	110
10	Proper Cooling of Rolls	110
10.1	Methodology of Wear-Out	110
10.2	Factors Affecting Wear-Out of Rolls	112
10.3	What is Proper Roll Cooling	112
10.4	Conventional Roll Cooling	113
10.5	Effective Roll Cooling System	113
10.6	Roll Cooling Parameter	114
10.7	Basic Requirements of Effective Roll Cooling System	115
11	Roll Management	118
11.1	Objective of Roll Management	119
11.2	Roll Inventory	119
11.3	Proper Layout of the Shop	119
11.4	Investigation of Roll Failures	120
12	Reclamation of Rolls	123
12.1	Automatic Submerged Arc Welding	123
12.2	Factors Affecting the Building up of Roll	125
12.3	Pre-heating of Rolls	125
12.4	Post heating or Stress Relieving of Rolls	125
12.5	Process of Arcing	125
12.6	Welding Machine	125
12.7	Welding Head	126
12.8	Fluxes	126
12.9	Preparation of the Job for Weld Deposition	127
12.10	Welding of Rolls	127
12.11	Defects of Welding	128
13	Roll Bearings	129
13.1	Open Journal Bearing	130
13.2	Roller Bearing	131
13.3	Enclosed Oil Film Bearing	133

4. Rolling of Blooms and Slabs   135–164

1	Introduction	135
2	Ingot	136
2.1	Classification of Ingot	136
2.2	Selection of Dimensions and Design of Ingot	137
3	Mill Proper	138
3.1	Layout	138
3.2	Roll Stand	139
3.3	Stand Housing	139
3.4	Rolls for Primary Mill	140
3.5	Selection of Rolls for Primary Mill	141
3.6	Bearing for Primary Mill	142
3.7	Top Roll Adjustment System	142
3.8	Spindles for Primary Mill	144

Detailed Contents   xvii

3.9 Roll Speed	144
3.10 Main Drives of Motor	145
3.11 Roll Tables, Manupalaters of Mill	146
3.12 Shears for Primary Mill	146
3.13 Scale Disposal for Mill	148
4 Comparative Study of Blooming Mills	148
5 Roll Pass Design for Blooming Mill	149
5.1 Pattern of Reduction	149
5.2 Arrangement of Grooves in Blooming Mills Rolls	150
6 Types Of Passes used In Blooming Mills	151
7 Size and Shape of Pass	152
7.1 Height of Pass (H)	152
7.2 Width of Pass	152
7.3 Taper	152
7.4 Corner Radii	152
7.5 Pass Convexity	153
7.6 Roll Collars	153
8 Bottom Pressure	153
9 Turning of the Workpiece	153
10 Roll Pass Design Data Sheet of Blooming Mills	153
10.1 Rolling Schemes for Rolling Different	
Size of Blooms/Beam Blanks	155
11 Pass Design of Blooming Mill	156
11.1 Barrel or Bullhead Pass	156
11.2 First Pass	157
11.3 Second or Finishing Pass	157
11.4 Beam Blank	158
12 Defects in Bloom Rolling	158
12.1 Defects due to Rolling	158
12.2 Defects due to Steel	160
12.3 Defects due to Bad Heating Practice	161

5. Rolling of Billets   165–216

1 Introduction	165
2 Types of Mill and General Layout	165
2.1 Open Train Billet Mill	167
2.2 Cross Country Billet Mill	168
2.3 Continuous Billet Mill	168
3 Specific Features of Continuous Mill	173
3.1 Advantages of Individually Driven Stand Over Group	
Combined Driven Stand	174
4 Pass Design Details of Billet Mill	175
4.1 Open Box Pass	175
4.2 Diamond-Diamond System	175
4.3 Diamond-Square Sequence	177
4.3.1 Advantages of Diamond-square System	178
4.3.2 Disadvantages of Diamond-Square Sequence	178
4.4 Oval-Square System	178

	4.4.1 Advantages of Oval-square System	179
	4.4.2 Disadvantages of Oval-square System	179
5	Selection of a Pass Sequence	179
6	Comparative Study of Billet Mills	180
7	Layout of Continuous Billet Mill	180
8	Roll Pass Design of Billet 90 mm × 90 mm	181
	8.1 Details Regarding Input Bloom and Finished Product	182
	8.2 Details Regarding Mill	182
	8.3 Tolerance of Billet	183
	8.4 Computation of Reduction	183
	8.5 Roll Pass Design of Finishing Group	185
	8.6 Roll Pass Design of Roughing and Intermediate Group	199
	8.7 Rolling Scheme of Billet 90 × 90 mm From Bloom 325 × 325 mm with Intermediate Size of 150 × 150 mm	214
9	Defects in Billet Rolling	215
	9.1 Defects Due to Bad Rolling	215

6. Rolling of Rounds and TMTS Bars   217–258

1	Introduction	217
2	Selection of Rolling Sequence	218
	2.1 First Method	219
	2.2 Second Method	220
	2.3 Third Method	220
	2.4 Fourth Method	220
	2.5 Fifth and Sixth Method	220
	2.6 Seventh, Eighth, Ninth and Tenth System	220
3	Rolling Tolerances of Rounds as per BIS Specification	221
	3.1 On Size of Round	221
	3.2 Ovality or Out of Square	221
	3.3 Weight	221
4	Design of Passes	221
	4.1 Design of Finishing Pass	221
	4.1.1 Height of Finishing Pass	222
	4.1.2 Width of Pass	222
	4.1.3 Roll Gap	223
	4.1.4 Radius of Finished Pass	223
	4.2 Design of Pre-finishing Oval Pass	224
	4.2.1 Height of Oval	224
	4.2.2 Width/Height Ratio of Oval Pass	225
	4.2.3 Radius of oval	225
	4.2.4 Outer radius r of Oval Pass	225
	4.2.5 Area of Oval Pass	225
	4.3 Design of Strand Pass	226
	4.3.1 Strand Plug oval/edge Oval	226
	4.3.2 Strand Square	227
	4.3.3 Strand Round	227
	4.4 Design of Following Passes in Rolling Sequence	227
5	Rolling Scheme of Round–32 mm	228

Detailed Contents

6	Roll Pass Design R-32 mm	228
	6.1 Stand-12 (Finishing Pass)	228
	6.2 Pass-11 (Pre-finishing Pass)	229
	6.3 Pass-10 (Strand Pass)	229
	6.4 Pass-8 (Rd-32-36 Pass)	229
	6.5 Pass-6 (Rd-32-36 Pass)	230
	6.6 Pass-4 (Rd-32-36 Pass)	230
	6.7 Roll Design of Finishing Pass	230
	6.8 Roll Design of Pre-finishing Pass	231
	6.9 Roll Design of Strand Pass	231
7	Layout of Merchant Mill	232
8	Rolling of TMT Bars	232
	8.1 Process of TMT	232
	8.2 Comparison of Different TMT Grades	233
	8.3 The Effect of Carbon Equivalent on Weldability	234
	8.4 Rib Design	234
	8.5 Control Data for Thermo Mechanically Treated Wire Rods/Bars	235
	8.6 Roll Pass Design of TMT-32	235
9	Olling Defects and its Solution	236
	9.1 Ovality	236
	9.2 Diagonal Difference	237
	9.3 Fin	237
	9.4 Under-Filled	238
	9.5 Fish-tail	238
	9.6 Lap	238
	9.7 Worn out Mark	238
	9.8 Pipe Formation in TMT	239
	9.9 Pass Mark	239
10	Mill Setting of Rounds/TMTS	240
	10.1 Twisting	240
	10.2 Bending of Bars (Side Ways)	242
	10.3 Bend up and Down	242
	10.4 Slipping in Stands	243
	10.5 Rolling Scheme	243
	10.6 Pass Positioning and their Dimensions	244
11	Major Points during Rolling	244
	11.1 Setting of Roughing group	244
	11.2 Setting of the Intermediate Group	244
	11.3 Setting of The Finishing Group	245
12	Slit Rolling of Bars	247
	12.1 Why Slitting is Required	247
	12.2 Sections Produced by Slit Rolling	248
	12.3 The Process of Slitting	248
	12.4 Control of Loop Heights between Stands 4 and 5 and 5 and 6	251
	12.5 Method of Slitting	251
	12.6 Slitting and Slit Guide	253
	12.7 Problems Faced During Slit Rolling and its Solutions	255
13	Mill Setting problem during Slitting	256

xx Detailed Contents

13.1 Rib Height below Tolerance on Both the
Top and Bottom of The Bar — 256
13.2 Rib Height Below Tolerance on Either the
Top or Bottom of the Bar — 256
13.3 Oval With Unequal Fill Due to Misalignment of the Entry Guide — 256
13.4 Small Kinks Every 1 to 1.2 m — 256
13.5 Lap — 256
13.6 Variations in Filling Between Strands — 257

7. Rolling of Flats and Squares — 259–282

1 Introduction — 259
2 Selection of Rolling Sequence for Rolling Flats — 259
3 Roll Pass Design for Flat 50 × 12 – 16 mm — 264
  3.1 Selection of Mill, in which section is to be rolled — 264
  3.2 Customer's Requirement *wrt* Specification, Tolerance etc. — 265
  3.3 Rolling Scheme for Flat 50 × 12 & 50 × 16 — 266
  3.4 Pass Design of Flat 50 × 12 & 50 × 16 mm — 266
  3.5 Design of Pre-finishing Pass (Pass-11-Vertical Stand-Edging Pass) — 267
  3.6 Pass-7 (Vertical Stand-Intermediate Group) — 267
4 Mill Setting and Section Control — 268
5 Inspection and checking of dimensions of flat sections — 269
6 Defects and its Rectifications — 270
  6.1 Round Edges (convex) — 270
  6.2. Concave Edges — 270
  6.3 Dished — 270
  6.4 Dished and Convexed — 271
  6.5 Wedge — 271
  6.6 Opposite Diagonals Rounded — 271
  6.7 Rhomboid — 271
  6.8 Sticker Mark – Due to Sticking of Metal on the Edger — 271
  6.9 Wear out Mark on Top and Bottom Edges–
Due to Worn out Flat Pass or Edger. — 271
  6.10 Lap– due to Fin from edger. — 272
  6.11 Diagonal Difference — 272
7 Rolling of Spring Steel Flats — 272
8 Guides used in Flat Rolling — 273
  8.1 Entry Guide — 273
  8.2 Delivery Guide — 273
9 Rolling of Squares — 275
10 Selection of Rolling Sequence for Rolling Square — 275
11 Roll Pass Design of Square–50 mm — 278
12 Rolling Scheme for Square — 281
13 Defects and its Rectifications — 282
  13.1 Overfill — 282
  13.2 Under Fill — 282
  13.3 Lap — 282
  13.4 Offsquare — 282

Detailed Contents   xxi

8. Rolling of Angles   283–320

1 Introduction	283
2 Types of System used for Angle rolling	283
3 Selection of Rolling System	285
4 Salient Features of Rolling Angle	286
4.1 Selection of Mill Size	286
4.2 Design of Finishing Pass	287
4.3 Computation of the Size of Billet	287
4.4 Co-efficient of Deformation	288
5 Modern Method of Rolling Angles	288
5.1 Roll Pass Design of Equal Angle	289
5.2 Selection of Mill	289
5.3 Customer's Requirement *wrt* Specification, Tolerance etc.	291
6 Pass Design of Angle 55 × 55 × 6 mm	291
6.1 Design of Finishing Pass	291
6.2 Computation of Thickness of other Passes of Angle Rolling	292
6.3 Pre-finishing Pass (Stand -10)	293
6.4 STRAND PASS (Stand-09)	295
6.5 Intermediate Group	297
6.6 Roughing Group	302
7 Rolling Computation Sheet for Angle 55 × 55 × 6 mm	306
8 Roll Pass Design of Unequal Angles	306
8.1 Design of Finishing Pass	307
8.2 Determination of Sizes of other Passes	310
8.3 Computation of Prefinishing Pass	310
9 Mill Setting of Angle Rolling	311
9.1 Setting of Roughing and Intermediate Stands	311
9.2 Setting of Finishing Stands and Profiles for Angle Rolling	315
10 Defects and its Rectification	316

9. Rolling of Channels   321–358

1 Introduction	321
2 Methods of Pass Design	322
2.1 Conventional Method of Rolling Channel	322
2.2 Modern Method of rolling Channel	325
3 Salient Feature of Channel Pass Design	326
3.1 Draughting	326
3.2 Work Feature	327
3.3 Fillets	327
3.4 False or Counter Flange	328
3.5 Profile Radii	329
3.6 Spread in Channel Rolling	330
3.7 Pitch Line Location	333
3.8 Condition of Entry of Strip Into Pass	333
3.9 Guides Used in Channel Rolling	334
4 Roll Pass Design for Channel 100 × 50 mm	335
4.1 Selection of Mill	335
4.2 Customer's Requirement wrt Specification, Tolerance etc.	336

4.3	Selection of Billet	337
4.4	Determination of Co-efficient of Reduction for the Different Elements of Channel Design	338
5	Computation of Elements of Pass Design	339
5.1	Design of Finishing Pass	340
5.2	Pre-finishing Pass (Pass-10)	344
5.3	Strand Pass (Pass-9)	345
5.4	Pass-8	346
5.5	Pass-6	347
5.6	Pass-5	348
5.7	Pass-3	349
5.8	Pass-2	350
5.9	Stand-1	350
6	Mill Setting of Channel	351
7	Defects of Channels and its Rectification	353

10. Rolling of Beam
359–396

1	Introduction	359
2	Methods of Beam Design	359
2.1	The First Method	359
2.2	Second Method	360
2.3	Third Method	361
2.4	Fourth Method	361
3	Mill Layout	362
3.1	Conventional Rail & Structural Mill	362
3.2.	Universal Rail and Structural Mill	363
3.3	Continuous Combined Bar and Structural Mill	363
4	Special Features of Beam Design	363
5	Important considerations for Roll Pass Design of Beam	365
5.1	Selection of Bloom Size and Numbers of Passes	365
5.2	Advantage of using Beam Shaped Bloom/Billet	365
5.3	Elements of Beam Section	366
5.4	Draughting Strategy for Beam Design	367
5.5	Spread	367
5.6	Flange Design	368
5.7	Buckling of Web	369
5.8	Fillets	369
5.9	Thrust	370
6	Roll Pass Design of Beam 125 × 70 mm	371
6.1	Selection of Mill	371
6.2	Customer's Requirement wrt Specification, Tolerance etc.	373
6.3	ISMB–Details of Specification as Per BIS	373
6.4	Design of Finishing Pass	373
6.5	Selection of Billet	375
6.6	Determination of Co-efficient of Reduction for the Different Elements of Beam Design	376
7	Computation Sheet for Beam 125 × 70 MM	378
7.1	Computation Sheet for Flange	378

Detailed Contents   xxiii

	7.2 Computation for The Web	379
	7.3 Computation for Pass height, Pass Area	379
8	Pass Design of Beam 125 × 70	380
	8.1 Pass Design of Finishing Stand (12th)	380
	8.2 Design of Pre-finishing Stand (10th)	380
	8.3 Design of Strand Pass (9th)	381
	8.4 Intermediate Group (Stand-8)	382
	8.5 Intermediate Group (Stand-6)	383
	8.6 Design of Pass-5	384
	8.7 Design of Pass-3	385
	8.8 Design of Pass-2 ("V"Pass)	386
	8.9 Design of Pass -1	387
9	Rolling Scheme for Beam 125 × 70	388
10	Mill Setting of Beam	388
11	Control of Weight of Beam Section	389
12	Special Precaution is to be taken during Beam Rolling	391
13	Defects of Beam and its rectification	391

11. Rolling of Rail   397–438

1	Introduction	397
2	Development Trends	399
3	Rail Shaping Practice	400
	3.1 Conventional System	400
	3.2 Basics of Rail Design	402
4	Salient Features of Rail Pass Design	406
	4.1 Draughting	406
	4.2 Design Efficiency	407
	4.3 Roughing Pass Design	407
	4.4 Combination Rolls	408
	4.5 Proper Pass Distribution	408
	4.6 Size of Bloom	409
	4.7 Rolling Temperature	409
	4.8 General Consideration	409
	4.9 Mill Setting	410
	4.10 Diagonal Rolling	411
	4.11 Bearing Collars	413
	4.12 Roll Spring	413
	4.13 Roll Diameter	414
5	Conventional Rail and Structural MilL	414
	5.1 Rail Specification	415
	5.2 Layout of Rail and Structural Mill	415
	5.3 Details of Equipments: Mill and Furnace	415
	5.4 Details of Major Equipments in Finishing Area	416
	5.5. Welding	417
6	Rail Testing	417
7	Roll Pass Design for RAIL	418
	7.1 Rolling Scheme of Roughing line	419
	7.2 Rolling Scheme of Finishing Line	421

7.3	Pass Design of Individual Pass in Finishing Line	422
7.4	Computation Sheet for Rail Design	426
7.5	Graph for computation of Co-efficient of Reduction	426
8	Universal Roll Pass Design	428
8.1	Advantages of Shaping in Universal Pass	430
8.2	Mill Configuration with Universal Rolling Practice	430
8.3	Layout of Finishing Line	431
9	Defects and Its Rectifications	432
9.1	Steel Defects	432
9.2	Mill Defects	433
9.3	Finishing Defects	436

12. Operation, Safety and Quality Management 439–458

1	Operation Management	439
1.1	Principal Elements of the Operation Management	439
1.2	Technological Instruction on Preparation of Rolls and Rolling in Light Structural Mill	441
1.3	Precautions to be Taken During Roll Changing	443
1.4	Points/areas to be Checked after Roll Changing	444
1.5	Starting the Mill After Major Roll Changing	446
1.6	Examination of the Finished Products	447
1.7	Technical and Economical characteristics of Rolled Metal Production	448
2	Safety Management of Rolling Mill	450
2.1	Accident Prevention Regulations in the Rolled Metal Production	450
3	Quality Management in Rolling Mills	451
3.1	Quality concept of Products	451
3.2	Reliability, Availability and Maintainability of Products	452
3.3	Rolled Product Quality Control	453
3.4	Quality Control in the Production of Rolling Products	455
3.5	Requirement of TMT Rolling	457

13. Packaging of Steel Products 459–468

1	Introduction	459
2	Objective	459
3	Concepts of Packaging	460
3.1	Role of Packaging	460
3.2	Theoretical Aspects of Packaging	461
3.3	Approach to Packaging Design	463
3.4	Importance of Steel Product Packaging	464
3.5	Advantages of Packaging of Steel Products	465
4	Packaging of Long Products	465
4.1	Piling	465
4.2	Bundling	466
4.3	Binding	467

Detailed Contents

14. Energy Efficient Practices and Technologies — 469–494

1	Introduction	469
2	Role of Reheating Furnace	470
	2.1 Type of Furnace	470
	2.2 Purpose of Heating Metal for Rolling	471
	2.3 Problem Associated with Heating	471
3	Operational Energy Efficient Improvement Practices in Furnace and Mill	472
	3.1 Optimal Operation of Combustion System	472
	3.2 Optimization of Thermal Regimes	473
	3.3 Furnace Pressure Regimes	473
	3.4 Hot Charging of Material	473
	3.5 Furnace Productivity	474
	3.6 Hearth Coverage	474
	3.7 Use of Coil Box	475
	3.8 Unfired Preheat Zone	475
	3.9 Recovery of Waste Energy	475
	3.10 Stack Efficiency	476
	3.11 Thermal Cover of Roll Table in the Mill	476
	3.12 Computer/combustion Control Model	476
	3.13 Discharge Temperature	477
	3.14 Using Higher Mill Speed	477
	3.15 Using Bigger Diameter Roll	477
	3.16 Running Mill Concept	478
	3.17 Establishing Communication System Between Furnace and Mill Proper	478
	3.18 Idle Running of the Mill	478
	3.19 Automatic Stoppage of Mill Drives/Roll Tables	478
	3.20 Use of VVVF Drives with Air Blowers and Auxiliaries	478
4	Use of Energy Efficient Operating Technologies	479
5	Improvement in The Rolling of Rebar Technology	480
	5.1 Reheating	480
	5.2 Low Temperature Rolling	480
	5.3 Increased Capacity	481
	5.4 Dimensional Tolerances	486
	5.5 Endless Welding and Rolling (EWR)	486
	5.6 Improved Mill Utilization	488
	5.7 Size-free Rolling	488
	5.8 Thermo-Mechanical Treatment (TMT) Process	489
6	Recent Developments in Production of Structurals	489
	6.1 Size-Free Rolling of H-shapes	490
	6.2 Quenching and Self Tempering (QST) Process for Structural Steel	491
	6.3 Endless Welding and Rolling (EWR) for Structural Steel	493
	6.4 High Precision Rolling (HPR)	493
	6.5 In-line Head Hardening of Rails	493
	6.6 Vanadium Steels	494

Index — **495–500**

Roll Pass Design

1 INTRODUCTION

The highly increased demand of steel by all the sectors of industries, construction and agriculture, has necessitated for the development of a variety of rolled products with the change in metal rolling technology.

The ultimate goal of a roll pass designer has to ensure the production of a desired shaped product with appropriate internal structure and are with desired mechanical properties. The product should be defect free surface with good surface finish and should have to produced with lowest cost of production.

Fig. 1.1 Rolling of a Product in a Rolling Mill.

2 ROLL PASS DESIGN

2.1 Object of Roll Pass Design

Steel sections are generally rolled in several passes, whose number is determined by the ratio of initial input material and final cross section of finished product. The cross section area is getting reduced in each subsequent passes and gradually the form and size of the stock approaches to the shape of designed profile.

Plates, sheets and flats are rolled in plain-barreled roll. Here the thickness of the stock getting rolled is reduced in each passing by reducing the gap between rolls.

Steel sections and shapes are rolled in grooved rolls, *i.e.,* the annular slot cut in a single roll. The profile of this slot is referred as the "Groove or Roll Groove".

Roll pass design means the calculation and design of the rolling schedule and passes to obtain a given rolled section, as well as to design the rolls to accommodate these passes.

2.2 Basic Purpose of Roll Pass Design

- To shape the metal to the designed profile.
- The profile design indicates a system of consecutive passes to produce the profile to the desired size and shape within the stipulated tolerance limit and with the specified mechanical properties and surface finish.
- Pass design also provides the shape and size of changing cross section of rolled material at every rolling stage.
- It locates the passes on the roll by designing rolls, working and non-working collars and also specifies the roll material etc.
- Pass design provides the highest possible production capacity of the rolling mill with minimum cost of rolled production.
- Pass design also ensures the smooth entry and exit of the strip and guaranteed the strength of rolls and motor capacity.

In the case of flat rolling (for slabs, sheets, strips), the pass designer has to provide the solution of the concerned problems with the distribution of reduction. According to the number of passing available and to make the sequence of rolling of flats with different sizes, to be rolled in that particular barrel. Such optimum schedule will help to minimize the frequency of the roll changing and to get the maximum production.

2.3 Characteristics of Efficient Roll Pass Design

- To make a profile with a smooth surface and of a correct dimensions within the stipulated limits of standard.
- To ensure the minimum expenses of energy, power and roll consumption.

2 Roll Pass Design

- To give deformation in such a way to generate minimum internal stress in the finished product.
- To create a simple and convenient work culture in shop floor of rolling mill, which has to minimize the manual operation to the minimum possible extend to introduce the automation of technological process.
- To optimize number of passes, which required for rolling, to reduce the total rolling time cycle, with the minimum time spent for changing and adjusting of rolls.

2.4 Specified Conditions Under Which Roll Pass Design is Based Upon

- Characteristics of finished product *i.e.,* dimension of section, tolerances and specifications concern to mechanical properties and surface finish of rolled product.
- Characteristics of initial input material *i.e.,* dimension and weight of billet, grade of steel and the metal temperature before and in the course of rolling.
- Specification of rolling mill *i.e.,* number of stands, roll diameter, rolling speed, available power of the drive motor, other components of the mill, available mill equipments with their position and spacing with preceding and succeeding stand and facilities available for tilting of stock. It is also required to have a details of other requisite requirement of mill like furnace, finishing, shipping etc.

2.5 Constraints to Efficient Roll Pass Design

Constraints which influence the roll pass design of a section:

- Temperature of feedstock available (heating rates).
- Roll barrel space available (or number of stands).
- Diameter and strength of rolls.
- Power of mill motors.
- Run out space available (at various parts of the mill).
- Complexity/weight of finished product.
- Finishing end capacities.
- Campaign length can also affect roll design.

3 CHARACTERISTICS OF HOT ROLLING
3.1 Draft/ /Draught

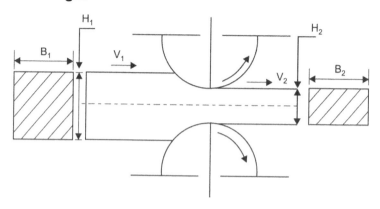

Fig. 1.2 Computation of Draught

3.1.1 Absolute Draught

Draught is expressed in linear unit and is the difference between the entry stock height/thickness to exit height/thickness *i.e.*, $\Delta H = H_1 - H_2$.

3.1.2 Relative Draught

Relative draught is expressed as a percentage *i.e.*,

$$\Delta H \% = \frac{H_1 - H_2}{H_1} \times 100 \%$$

Fundamental principle of "Roll Pass Design" says that the uniformity of reduction is to be maintained at all parts of the profile. And in case if some non-uniformity in reduction still exists, may be due to disparity in the shape of initial and finished product. It is advisable to do away with it in early forming stages itself, as steel will be more malleable at this stage, due to high working temperature and the tendency to twisting of metal at delivery end will be less pronounced at this point of time, due to higher cross-sectional area of the work piece.

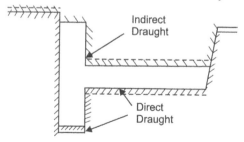

Fig. 1.3 Direct vis-à-vis Indirect Draught.

3 Characteristics of Hot Rolling

Draught can be either direct or indirect. Direct draught is a reduction in vertical direction, whereas the indirect draft exerts pressure in non vertical direction. It is applicable mostly in flange profile.

In Fig. 1.3 dotted lines show profile entering into the pass. Direct draught takes place on horizontal leg and to the bottom flange and indirect drafting reduces the thickness of top flange.

How this indirect draught does take place? Basically, it is a grinding action between collars of different rolls, which can be illustrated by placing block of clays between palms of hands and if hands are brought together with no rotation, a plain squeezing action results which is represent to direct drafting, but if at the same time hands are also being rotated in opposite direction, the clay is reduced in thickness by a action, representing indirect draft.

3.2 Elongation

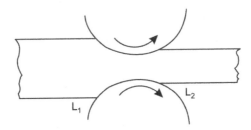

Fig. 1.4 Computation of Elongation

It is expressed either as a ratio of final length to initial length or as percentage of initial length.

$$\text{Elongation factor} = E = \frac{L_2}{L_1} \quad (E \text{ is always greater than } 1)$$

$$\text{Elongation percentage} = \% E = \frac{L_2 - L_1}{L_1} \times 100\%$$

3.3 Spread

Spread in rolling is the most important and complex deformation factor, which must be taken into account, while designing a pass. It is the fundamental principle that metal flows in the direction of least resistance, hence metal flows not only in the longitudinal direction, but also it flows in lateral direction also. This is called "Spread".

3.3.1 Absolute Spread

$$\Delta B = B_2 - B_1$$

3.3.2 Percentage Spread

$$\Delta B \% = \frac{B_2 - B_1}{B_1} \times 100 \%$$

3.3.3 Factors Effecting Spread

Spread in rolling depends upon a number of simultaneous acting factors. The effect of each particular factor is considered separately below, assuming that all other factors remain constant.

1. **Reduction:** Spread has got a linear relationship with reduction. Spread increases with increase in reduction.

2. **Width of work-piece:** The maximum spread occurs with width of work piece between 20 to 40 mm for a particular reduction. Spread starts reduce with width upto 200 mm and beyond 200 mm, it become negligible.

3. **Rolling speed:** Maximum spread occurs at rolling speed upto 2 M /Sec., it gradually reduces and become negligible beyond 10 M/Sec.

4. **Rolling temperature:** With the increase in rolling temperature, spread decreases, as co-efficient of friction reduces with increase in temperature. It is seen that the maximum spread occurs at temperature of rolling between $1050°C - 1100°C$.

5. **Friction:** Increases of co-efficient of friction gives greater spread .The uneven surface of roll prevents flow of metal in longitudinal direction *i.e.*, at elongation, it moves towards sides, causing increase in the spread.

6. **Roll Diameter:** "Large roll squeezes, but small roll one pulls" is an old saying. With the increase in length of contact area (*i.e.*, $\sqrt{R \times \Delta h}$, where R is roll radius), the sum of frictional forces acting in the longitudinal direction get increases and elongation decreases. This is to give rise increase in lateral deformation *i.e.*, spread.

7. **Composition of metal being rolled:** Spread increases with increase in Carbon and Manganese and other alloying elements in the metal being rolled.

8. **Influence of number of passes:** For the same reduction, with the increase of number of passes, spread get reduces. This is due to decrease of length of deformation zone *i.e.*, the contact area ($\sqrt{R \times \Delta h}$, where Δh is amount of reduction)

9. **Height of work piece:** Height of work piece, not only effect the amount of spread, but also the pattern of spread, as shown in Fig. 1.5.

3 Characteristics of Hot Rolling

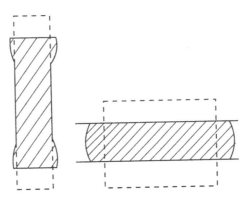

Fig. 1.5 Effect of Height of work-piece on Spread.

Rolling high pieces ($H > B$), deformation concentrates in layer near contact surface, which leads to local spreading of the metal in these area. When rolling thin piece ($B > H$), their edges usually become convex, in case of unrestricted spread.

10. **Type of pass:** The design of pass affects considerably on the amount of spread. Pressure exerted by side walls of a box pass reduces the spread considerably. In other words, elongation will get increase. Amount of spread decreases with the increase in taper of side wall. Due to high amount of taper given in diagonal pass design of rail or beam profile, either spread may not occur at all or may even have a negative value.

3.4 Types of Spread

(a) Free or Unrestricted Spread

Free or unrestricted spread is one where-in a rectangular work piece is rolled in a plain roll.

Advantage of rolling with unrestricted spread:

- Consumption of power will be less, due to absence of frictional forces against sides of pass.
- Consumption of rolls will be less due to decrease of specific pressure of metal on rolls.
- There is no danger of formation of fins.

(b) Restricted Spread

Where side walls of roll restrict the lateral movement of metal, then in such case, spread is called as the restricted spread. The deformation at different portions of the work piece is carried out with different effective diameters of the rolls which provide different reductions. Hence, the draughts for these portions will be different.

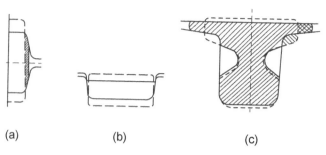

Fig. 1.6 Spread in (a) box pass, (b) in close pass, (c) Forced spread of Rail flange.

Even in comparatively simpler box passes [Fig. 1.6 (a)], the distribution of metal between elongation and spread is complicated. The pressure exerted by the side walls of the pass reduces the spread and on the contrary, elongation gets increases. In other words, spread reduces with the increase of taper of the side faces of the box pass. In diagonal passes [Fig. 1.6 (b)], spread sometimes may not occur or may even have a negative value.

Restricted spread is used in designing rail passes for rolling rails. In this case, [Fig. 1.6 (c)] the whole mass of the section subjected to a shallow draught and the base is vigorously reduced naturally. It cannot ensure lengthening of the whole piece. Therefore, the entire reduced metal can go only for the spreading the base.

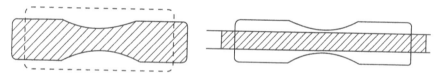

Fig. 1.7 Slitting passes for increasing width of work-piece.

The system of the so-called slitting pass (Fig. 1.7) is applied for obtaining a large width of a strip. In this case, a flat or square billet is first rolled in a slitting pass and then in a flat pass. In the later pass, the sides of the section get drastically reduced and thus obtain a larger spread. However, this method should be used cautiously since the non-uniform draught along the width of the work piece may even cause tearing.

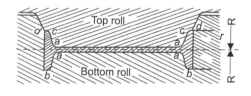

Fig. 1.8 Metal Deformation Pattern in Flange Passes.

3 Characteristics of Hot Rolling

Advantage of rolling with restricted spread:

- Production of accurate finished profile and good quality of side surface of profile.
- Metal stock in pass has better stability when reduction is not high.
- Cracks near to metal surface get welded, when these cracks are pressed against side surface.

3.5 Computation of Spread

There is number of formulas available for the computation of spread. The selection of a particular formula is based upon various factors *viz.,* type of product, mill, and of pass. It also depends upon roll diameter and stages of forming *i.e.,* whether it is at roughing, intermediate or finishing stage of rolling etc.

For the passes of a Blooming mill, the computation of spread requires only approximation and is taken as percentage of total reduction. It is generally, taken around 10-15% of total reduction. In Merchant Mill, in the earlier scale breaking passes, it is taken as 15-20% of the reduction and then 20-30% in the following box passes. While in barrel passes, where there is no restriction of spread, it is taken as 50-60%

The most widely accepted formulae for the computation of spread is "Bakhtinov" formula *i.e.,*

$$\Delta B = 1.15 \, \frac{\Delta H}{2H} \, (\sqrt{R \times \Delta h} - \frac{\Delta H}{2f})$$

ΔB is the total spread, where

$$\Delta h \;=\; \text{total Reduction,}$$

$$R \;=\; \text{radius of Roll,}$$

$$H \;=\; \text{initial height of metal stock,}$$

$$\sqrt{R \times \Delta h} \;=\; \text{Length of contact}$$

$$f \;=\; \text{friction, which can derive from the following formula } i.e.,$$

$$\;=\; K_1 \times K_2 \times K_3 \, (1.05 - 0.0005 \times t), \text{ where}$$

$$K_1 \;=\; \text{Co-efficient for condition of surface of rolls}$$

$$\;=\; 1.0 \text{ for steel rolls}$$

$$\;=\; 0.8 \text{ for cast iron rolls}$$

$$K_2 \;=\; \text{co-efficient depending upon speed of rolling, can be derived from Table 1.1.}$$

if V is the speed of rolling in m/sec.

$$V = \frac{\pi \times D_k \times n}{60}$$

Where D_k = Rolling Dia.

n = Speed of rolls *i.e.*,

$\quad = \dfrac{N}{i}$

N = Speed of motor

i = Reduction ratio of motor

Table 1.1 Computation of Value of K_2

V m/sec	1.0	2.0	3.0	4.0	5.0	6.0	7.0
K₂	1.0	1.0	0.9	0.8	0.7	0.65	0.60
V	8.0	9.0	10.0	11	12	13	14
K₂	0.575	0.55	0.525	0.500	0.480	0.460	0.450
V	15	16	17	18	19	20	
K₂	.430	.425	.420	.415	.410	.405	

K_3 = co-efficient depending upon the composition of rolled metal

$\quad$ = 1.0 for carbon steel

$\quad$ = 1.24 – 1.63 for alloy steel

"GRISHKOV" Formula holds good for the complex pass design like Wire Rod Mill. It gives very accurate results, especially when Oval-square system is used.

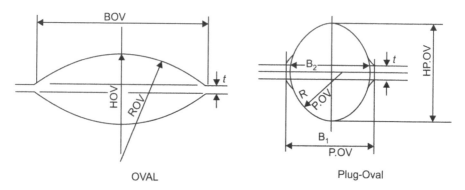

Fig. 1.9 Rolling of Plug oval in oval.

Spread $\Delta B = K \times A \times C \left(\sqrt{R \times \Delta h} - \dfrac{\Delta h}{2\mu}\right) \times 2.3 \log \dfrac{H}{h}$

Where Δh = Absolute Reduction

$\quad \mu$ = co-efficient of friction

3 Characteristics of Hot Rolling

H & h are the initial and final height of metal stock

R = Rolling Dia i.e., $D_k/2$

(a) Where $K = \dfrac{8\sqrt{3.5}}{0.5 + V}$

V is the Rolling speed in m/sec.

Value of K can be calculated from the graph, Table 1.2

(b) $A = \dfrac{\Delta h}{3H} + 0.5$

where Δh is the absolute reduction i.e., $\Delta h = H - h$

H & h are initial and find height of metal.

(c) $C = \left(\dfrac{4(1-\Delta h)}{H} \times \dfrac{B}{\sqrt{R \times \Delta h}} - 0.15 \right) \times e^{1.5} \left(0.15 - \dfrac{B}{\sqrt{R \times \Delta h}} \right) + \dfrac{\Delta h}{H}$

Value of C can be derived directly from graph a, b, c of Fig. 1.10, where the value of C can be derived from $\dfrac{B}{\sqrt{R \times \Delta h}}$ and Elongation $\dfrac{\Delta h}{H}\%$

(d) B is the width of the metal stock i.e., $= F/H$

(e) R is the Rolling Diameter, $R = (D_C + t - H)/2$, DC is the collar diameter, t is roll gap and H is height of pass.

Table 1.2 Determination of Value of K

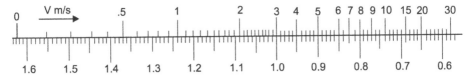

12   Chapter 1 Roll Pass Design

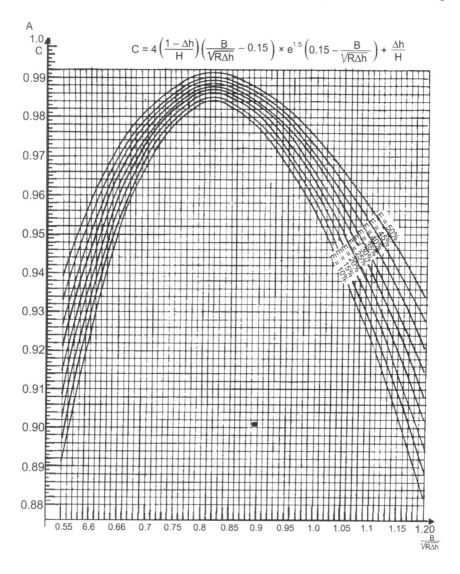

Graph (*a*)

3 Characteristics of Hot Rolling

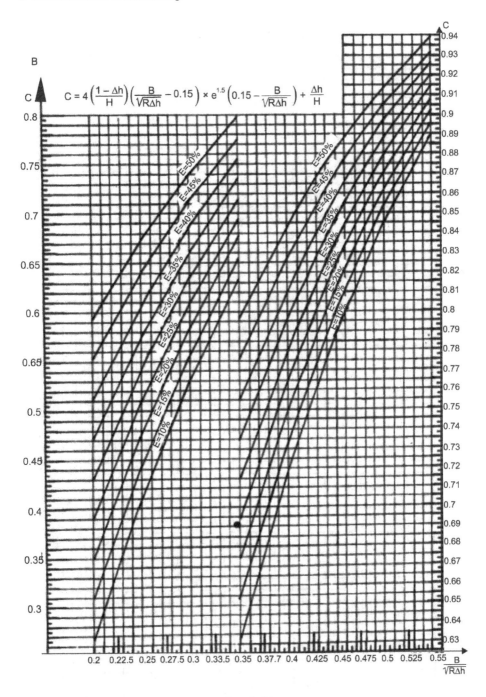

Graph-(b)

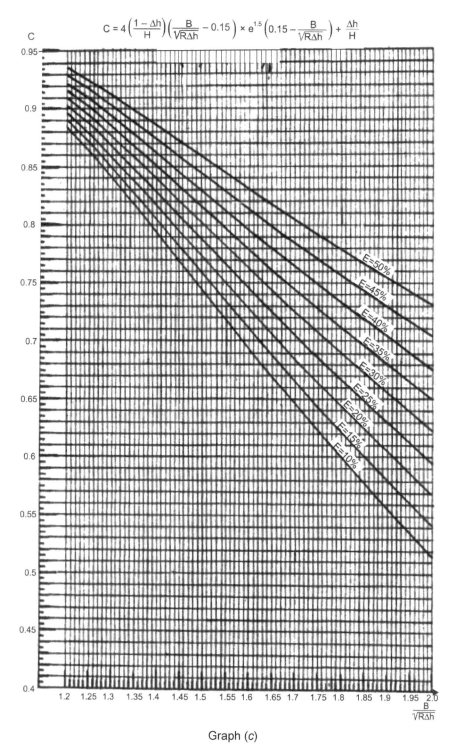

Graph (c)

Fig. 1.10 Determination of the Value of C, can be obtained from the graphs a, b & c

3.6 Contact Area, Contact Angle, Angle of Bite and Roll Bite Condition

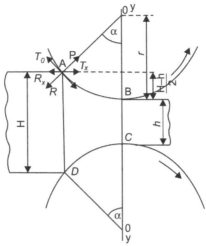

Fig. 1.11 Roll Bite Condition.

In Fig. 1.11 above, the process of rolling takes place between two plain cylindrical rolls of same diameter, which revolves at the same speed, in the opposite direction to each other.

Region *ABCD* is called "Zone of deformation". Plane of entry of in going stock; first contact the rolls at point "*A*". Angle α is made by radius *AO* and axis of roll *Y–Y*. This is called as 'Angle of Contact'. The angle of contact increases as size of in going stock increases, till a point comes, where roll will not grip or bite the stock. The terminology "Angle of Bite", often used for contact angle in general.

Arc *AB* and *DC* are the contact area. The area over which stock contacts the roll is "Area of Contact". It is equal to length of arc *i.e.*, *AB* × mean width of stock *i.e.*, *B* and *b*, where *B* and *b* are the in going and outgoing width of stock.

If length of Arc *AB* are projected to rolling plane *A′ B′*,

Then, $A' B' = A'' B'' = AO^2 - BC^2$
$$= \sqrt{R^2 - R^2 + R\Delta h - 0.25 \Delta h^2}$$
$$= \sqrt{R^2 - R^2 + R\Delta h - 0.25 \Delta h^2}$$

Length of contact $= \sqrt{R\Delta h}$, leaving $0.25 \Delta h^2$ being negligible

Area of Contact $= \sqrt{R\Delta h} \times \dfrac{B+b}{2}$

Angle of Bite $\cos \alpha = \dfrac{OB}{OA} = \left(R - \dfrac{\Delta h}{2}\right)/R$

$\cos \alpha = 1 - \Delta h / 2R$ or $1 - \Delta h/D$

It is the basic principle of mechanism, when two bodies having relative motion are placed in contact, one force acts along the common tangent in the direction of relative motion of other body. If bar moves slowly than the peripheral speed of the roll, then frictional force will tend to pull it into the roll gap. If bar moves quickly than the peripheral speed of roll, frictional force will oppose the entry of work piece into the rolls.

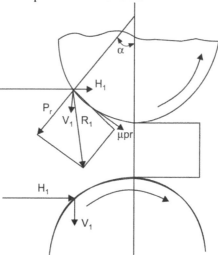

Fig. 1.12 Computation of Angle of Bite.

In Fig. 1.12 above, the radial force P_r tends to compress the stock, while tangential force μP_r (where μ is the friction between rolls and metal), acting towards the roll gap. As μ is always less than 1, then μP_r is some fraction of P_r. The resultant of these two forces μP_r and P_r is R_1 as formed by parallelogram. This force R_1, itself resolved to H_1 and V_1. Equal force H_1 and V_1, also acts on the stock from the top and bottom rolls. So, it can be said that two equal forces V_1 will give rise to a balanced compressive action on the stock, while two H_1 pulls stock into the roll gap. But, if suppose the angle of bite get increases to 40 degree, then two H_1 may act in opposite direction. At that time, it can be said that rolls will not bite the stock.

To determine the possibility of the biting of the stock by rolls, derivation of force H_1 is to be considered. Force H_1 is the algebraic sum of all horizontal components of two basic forces P_r and μP_r. The horizontal component of Radial force P_r is $P_r \sin \alpha$, which acts away from roll gap, while that of μP_r is $\mu P_r \cos \alpha$, acts towards the roll gap.

For biting the work piece

$\mu P_r \cos \alpha > P_r \sin \alpha$

$\mu > \tan \alpha$

3 Characteristics of Hot Rolling

By substituting the tangent of angle of friction β for the co-efficient of friction (f = tan β), the following conditions of biting is obtained *i.e.*, α < β *i.e.*, biting will be ensured, when angle of bite is less than the angle of friction β.

Hence, in initial period in order to get a grip, it is necessary that the angle of friction should be greater than the angle of grip or bite *i.e.*, α < β

But once, process of rolling has become stable, as grip has obtained, then relationship between co-efficient of friction and angle of bite changes. When metal completely involves itself in process of deformation, then position of resultant force, will help in normal rolling such that, even β ≤ α/2 *i.e.*, if the initial biting of the metal by roll is ensured, then subsequent steady rolling process provides the greater possibilities for increasing the metal reduction in rolling.

3.7 Methods to Improve Roll Biting Conditions

If there is compulsion on Roll Pass Designer to give a higher reduction, then ways and means are to be derived out, either to increase μ *i.e.*, the friction or to momentary reduce the α *i.e.*, the angle of bite. The temporarily decreasing the angle of bite, without reducing the draught can be achieved with ramming the stock into the roll gap by shear momentum on the approach roll tables of a mill or pushing it in by hand in hand operated mill. It can enhance the angle of bite to considerable extend to go upto 30°–34°. Sand on rolls in hot rolling may even permit angle of bite may go beyond to 30°–34°. Similarly, in a continuous mill, the preceding stand creates condition of forced biting of the work piece for the following stand.

Fig. 1.13. Knurling of Rolls.

Roughening of rolls by ragging or knurling are meant for increasing friction and may raise angle of bite upto 34°. Once bar starts entering the roll, then μ P_r cos α begins to increase and simultaneously P_r sin α will get decrease. The process of ragging or knurling should be made with utmost care, as it may lead to the appearance of defects on the surface of finished products. These processes of enhancing the angle of bite will certainly increase the cost of rolls production, besides increasing the power consumption during rolling, due to enhanced friction.

On the other hand, quite often notches on rolls help make easier to roll soft steels, but it may lead to appearance of defects on the surface of finished products, especially when hard steel is also be rolled from the same roll set. Thus, it is advisable to change the rolls, before the rolling of hard grades.

The prime aim of Roll pass designer is to distribute reduction in such a way that there shouldn't be any need of ragging or knurling on rolls at all. Inspite of these disadvantages, it is quite common to use ragging or knurling on rolls, especially in mill like Blooming or Billet Mill, Rail or structural mill for increasing the angle of bite. Where in such mills, output get severely affected with increase in numbers of passings. In addition, it is quite common to use the process of knurling/ragging to increase the angle of bite, if there is a limitation for increasing the rolling temperature of input billet.

At the same time, it is also desired, to organize a constant vigil over the condition of passes and of the quality of production.

It is to be noted that the temperature appears to have only a small effect on angle of bite, while roll speed has a marked effect. For example, in a particular case, it was found $26°$ at 50 rpm, while at 100 rpm; it was $24°$, thus giving a marked difference of $2°$.

3.8 Forward Slip, Neutral Plane and Neutral Angle

Bar issuing from the rolls are at higher speed than the peripheral speed of the rolls. If V_1 is in-going bar speed and V_2 is the roll peripheral speed,

$$\text{Then, forward slip will be } \frac{V_2 - V_1}{V_2} \times 100\%.$$

A vertical plane within the zone of deformation, where in the speed of work-piece is equal to peripheral speed of roll, is called "Neutral plane" and the angle corresponding to this plane is known as "Neutral angle". The position of neutral point is determined by the amount of forward slip. It is dependent upon the co-efficient of friction and given reduction.

"Ekelund" formula, calculated the position of neutral point by equating horizontal forces and an angle will be

$$a_n = \frac{\sqrt{H-h}}{2\,D} - \frac{1}{\mu} \times \frac{H-h}{2\,D}$$

3 Characteristics of Hot Rolling

Where, a_n is no slip angle in radian.

H and h are initial and final height of the work-piece, and $H - h$ is Draft. D is the rolling diameter and μ is the co-efficient of friction.

It may be concluded that the position of neutral point depends upon the co-efficient of friction and draught for any given diameter.

a_n increases with increase in co-efficient of friction and decreases with increase in draught. Forward slip increases with higher diameter roll and with the smaller thickness of work-piece. It also decreases with increase in lateral spread.

3.9 Forward Slip in Groove Rolls

Forward slip will be more at the deepest point of the pass, with compare to a point of a pass, which is nearer to the roll gap or break. At this point, the difference of the peripheral speed of roll and that of stock is of minimum extend. At some point of roll, where, there will be no slip at all, roll radius at this particular point is referred as "Effective roll radius" and the corresponding diameter is known as "Effective working diameter" of the pass. Furthermore, the Roll designer has to compute the effective working diameter, in a continuous mill so that the stand speed will be at balance.

The effective working diameter may be used for balancing a continuous train with groove passes by keeping the rolling constant same for the entire stands of the continuous group.

3.10 Tension

Tension, on either front or back reduces both rolling load and required torque. Front tension moves neutral point towards the plane of entry and increases the forward slip; likewise back tension makes neutral point to shift to plane of exit and in turn it decreases the forward slip. In both the cases, the tension is doing work on the stock, so less work is required from rolls.

If there is no change is observed in rolling load or in power consumption in a stand of a continuous mill, then it can be assumed that there is no tension. Tension can be eliminated either by changing the speed or reduction between stands. Eliminating tension between stands in such mill leads to improve the quality of finished product. It is easier to eliminate tension in a continuous mill. With having individually driven stand, where speed adjustment in individual stand is made to eliminate the pushing or pulling between stands, that is difficult in case of a combined drive of a continuous mill *i.e.*, having one drive for more than one stand in a group of stands. Power consumption by individual stand can be used as a measure to control the tension in a continuous mill.

Tension stretches the bar like an elastic band and may place the main body (middle) of the bar under the minimum size specification. The only part of the billet never subjected to tension or compression is very tail end. It never reside in two stands at the same time, so it can be considered as the "True Size".

3.11 End Thrust

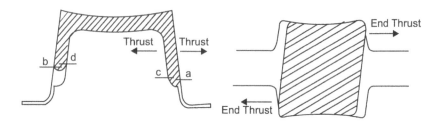

Fig. 1.14 End Thrust.

When hot steel is reduced between two rolls, due to the nature of reduction or temperature of rolling, it gives rise to forces, to attempts to push one roll axially in one direction and other roll to other direction. This axial force is called "End thrust" and it will result in imbalance side thrust in that particular pass, as shown in Fig. 1.14 above results in varying sizes of flanges, the flange *bd* become thicker and shorter, while flange *ca* become thinner and longer, of a profile.

If the amount of end thrust is minor, then it will be take care by thrust bearing of roll chock. However, bearing collars are required to be made in section profiles to take care of end thrust in section rolling:

There are two types of bearing collars.

Working bearing collars as shown in Fig. 1.15. These collars mates and prevent side shifting of rolls.

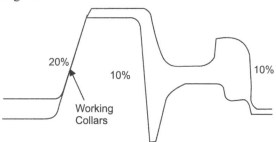

Fig. 1.15 Working Bearing Collar or Runner.

In roll design of section mill, especially for rail design, this aspect can be take care by designing working collars in rolls to avoids end thrust, and preventing shifting of rolls.

3.0 Characteristics of Hot Rolling

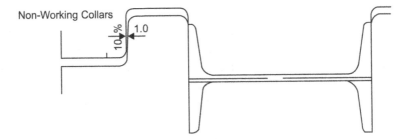

Fig. 1.16 Non-working Bearing Collar or Runner.

Non-working collars, which have a gap, generally of 0.5-1.5 mm. Gap is given to act as safety device, to avoid shifting of rolls during rolling.

There are two drawbacks of having bearing collars in roll design:
- Bearing collars are wasteful from point of view of roll usage, otherwise available space would have been used for the working pass.
- Bearing collars wear rapidly, because of grinding action of one surface of roll against an opposite turning surface of the mating roll. Only for this reason, collars are made inclined. This inclination reduces working area between two rolls, as well as the restoration of working collars are possible with minimum loss of roll diameter.
- To avoid wear-out of bearing collars, lubricants are sometimes used on the collars. Sometimes, surface of bearing collars are swapped with rolls. The colloidal suspension of graphite in oil also find some application for reducing wearing of roll collars.

3.12 Mill Spring

This is the most important element to decide the roll gap in roll pass design. During the process of rolling, the deformation creates high pressure on rolls by metal. As a resultant of these pressures, a gap between rolls gets formed during the working of metal between rolls, it is called as "Mill Spring". The magnitude of mill spring depends upon the elastic limits of different metals, parts of the stands, the bending of rolls, the compression of bearing, bearing supports, screw down device, nuts and of bending and expansion of stand frame, the design of stand and magnitude of rolling force applied on the stand.

Type of stand influences mill spring to a greater extends. Close top housings are preferred to open-top housing for the finishing stand of a section profiles, wire rods, flat products, since rolling of these profiles demand high degree of rigidity *i.e.,* less spring for the finishing stand. Housing-less stands are now-a-days used to overcome the problem of mill spring in rolling. Due to high rigidity of housing-less stand, wire rods are now produced with the close tolerances even upto ± 0.10 mm.

For any given mill, an approximate relationship can be established between rolling force and mill spring. The ratio of rolling force to mill spring is called as "Mill Module". This consideration is especially important for selecting mill for flat products. A highly rigid mill ruled out, any variation in gauge thickness. Modern plate mill is generally designed with mill modules of 1000 T/mm. Mill spring is proportional to rolling load, which is limited by the strength of rolls.

Mill spring varies from decimal of a millimeter in a sheet and wire drawing unit to 5–10 mm at a Blooming and Billet mill stand and other primary reducing mill.

To overcome the problem of mill spring and to roll the section with close tolerances, now-a-days housing less stands are replacing the conventional close top or open top housings.

In conventional housing, working on textolite bearings, the amount of mill spring is taken upto 1% of the diameter of rolls at finishing stand and upto 2% of the diameter of rolls at roughing stand. In case of rolls installed on roller bearings, gap may be taken upto 0.4%.

Rolling close tolerance bar and rounds in a modern rolling mill is possible only because of the use of housing less stands, in which mill spring is almost zero.

4 PASS

4.1 Type of Passes

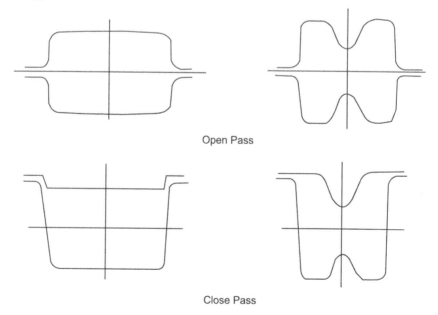

Fig. 1.17 Open and Close Type of Pass.

4.2 Classification of Passes

Passes are classified into four main categories:

4.2.1 Breakdown or Reducing Pass

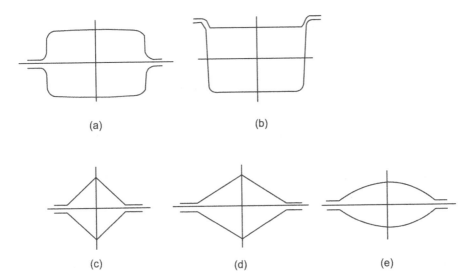

Types of breakdown Passes:
a and b—box, c—square, d—diamond and e—oval.

Fig. 1.18 Breakdown Passes.

Passes, which designed only for reducing the cross section of stock, are called "Breakdown or reducing pass". The most widely employed breakdown passes are box, square, diamond and oval passes.

4.2.2 Roughing Passes

The subsequent passes after the break-down passes in rolling are called roughing passes. These passes change the form, so that stock gradually approaches to the shape of final profile. There are different types of roughing passes, which depend only upon the variety of shapes are to be rolled.

4.2.3 Leader or Pre-finishing Passes

The next pass to the final pass is called the leader or pre-finishing pass.

4.2.4 Finishing Pass

The final or finishing pass imparts the final size and form the rolled stock.

4.3 Elements of Passes

4.3.1 Pass, Joints or Parting, Open or Close Pass

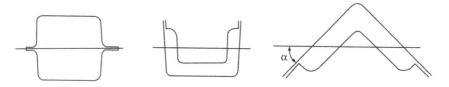

Fig. 1.19 Different Elements of Pass.

Roll with grooves arranged one above another, form the "Pass". Thus. if a pass lies partly in one and partly in another roll, the point where boundary of pass changes from one roll to another is called the "Joints or Parting". If this is formed by lines parallel to the roll axis, then pass is called as "Open Pass", but if it is formed by lines approximately perpendicular to the roll axis, then the pass is called as "Closed Pass", as shown in the Fig. 1.19 above.

In case, when the roll parting is inclined, when the inclination α is more than 60° to the roll axis, then it is called as open pass, but it is called as closed pass, if it is less than the 60°.

4.3.2 Roll Gap

To determine roll gap, following factors are to be considered:
- Type of pass *i.e.,* whether, it is used at finishing, pre-finishing or at roughing group of stands.
- Type of products to be rolled from particular pass. Higher value is choosen for semi-finished products.
- Minimum and maximum size of products are require to be rolled from that particular pass or rolled from the same pass.
- Tolerances for roll turning.

  The roll turning tolerances will be taken from lesser value, if roll is turned in a CNC lathe, than in conventional lathes.
- Mill Spring (Explained earlier).

As mill spring is proportional to the rolling load, roll gap has become function of rolling load. For example for the cogging and break-down stands of a structural mill, it is taken around 1 to 1.5% of and for finishing mill, it varies from 0.05 to 1% of the roll diameter.

4.3.3 Outlet or Collar Taper of Pass

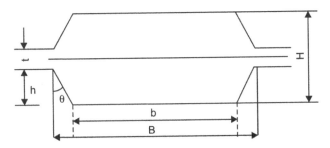

Fig. 1.20 Computation of Collar Taper.

The side walls of a pass are never be perpendicular to the roll axis. As a rule they should be slightly inclined to the roll axis. Tangent of the angle of slope of the sides of the pass is called "Outlet" or "Collar taper", as shown in Fig. 1.20.

$$\tan \theta = \frac{B-b}{2h}$$

The absolute value of this inclination is expressed in mm. It may also be expressed in degree θ.

The advantage of a proper outlet will not only allow the smooth, correct and convenient entry of work piece, but also it facilitates the straight and smooth exit of stock from the pass. Passes with vertical walls may create more difficulties in entry, as even with a slightest error in guiding, bar will run into the vertical wall of the pass. In such passes, the bar may get jammed due to spread in the pass or even get wind up on the roll. Entry of a bar will be smooth with properly designed collar taper and it will not involve any difficulty while rolling. Higher taper box pass helps in the breaking of the scale in initial passing and that's why sometimes it also called as scale breaking pass.

The most important advantage of the side taper is the restoration of the pass, which can be accomplished with the minimum loss of diameter. It is not possible to restore the lateral dimensions of a pass with vertical walls. Higher the outlet or collar taper, the less diameter shall be require to be turned to redress the pass.

The following inclination values are widely used:
- For breakdown box passes – 10 to 20 percent.
- For closed bar passes – 5 to 10 per cent.
- For roughening passes in rolling rails, beams, and channels – 5 to 10 percent.
- For finishing mill – 1 to 1.5%.

4.3.4 Split of passes, Rounding of corners

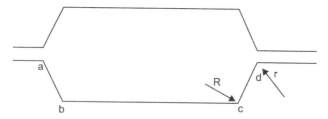

Fig. 1.21 Split, Rounding of corners of Pass.

Split of pass is the place, where contour of one groove of one roll merges on the other roll. The split is a dangerous zone, when there is an extra metal. It tends to fill up and if it become more extra, it get oozes out of this place. The direction of split is changed in every consecutive pass to avoid the formation of fins. Rounding of corners is also of great importance.

Following problems can be solved by giving a proper rounding, as explained below:

(*a*) Curves at the crossing of vertical and horizontal sides of pass (*a, b*); prevent drastic cooling of the edge.

(*b*) The curves in the pass makes possible for metal to expand smoothly (for passes for rolling square, rectangle, and rhombus).

(*c*) Due to proper rounding, outer angle of the pass prevents the cut on the surface of the strip, particularly, when side reduction of shaped profile is not known.

It is studied that there is a direct relationship between the fillet radius and stress concentration. A 10 mm fillet increases the stress about 1.5 times at the corner, while 2 mm fillet radius increases it to about 6 times.

4.3.5 Depth of Pass

The maximum depth on roll in the pass or collar should not exceed:
- For finishing pass – 0.167 – 0.20.
- For cogging or roughing pass – 0.25 – 0.30.

4.4 Neutral Line of the Pass

Neutral line of a pass is such a line, towards which the moments of force applied to the profile from where the top and bottom roll are equal. In other words, neutral line is a horizontal line, which coincides with the central line between the rolls. It provides the straight exit of metal from the pass, without any tendency to get moved towards top or bottom direction of the roll.

5 Diameter of Rolls

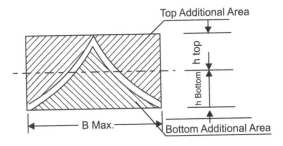

Fig. 1.22 Computation of Neutral Line.

The neutral line coincides with the axis of symmetry of the pass in a simple design passes *viz.*, squares, rectangles, rounds, ovals, rhombus etc. To determine the neutral line of a close pass, there are several formulas to determine the accurate neutral line. The most convenient method to compute neutral line is explained below:

$$h_t = \frac{F_{tg} + 0.5 \times F_k}{B\,\text{max}}$$

$$h_b = \frac{F_{bg} + 0.5 \times F_k}{B\,\text{max}}$$

$h = h_t + h_b$

h_t = Distance of neutral line from the top

h_b = Distance of neutral line from the bottom

F_{tg} = Top additional area

F_{bg} = Bottom additional area

F_k = Pass area

B max. = Maximum width of pass.

5 DIAMETER OF ROLLS

5.1 Average Diameter

Average diameter of roll is a factor, which decides the mill size. The mill which has only one stand and if each roll of the stand is driven individually by separate motors, then the average diameter is called as mill size. The maximum and minimum diameter of rolls are to be selected, based upon the maximum and minimum opening of stand and of the inclination of spindles. The selection of diameter should be such that there should be always provision of further opening of stand in the event of cobbles and other emergencies.

Table 1.3 The selection of max. / min. diameter vis-à-vis stand size

Type of stand	Dia. of pinion mm	Stand Opening Max. mm	Stand Opening Min. mm	Dia of Roll Max. mm	Dia of Roll Min. mm	Max. Take Off mm
Horizontal	500	600	480	535	480	55
Vertical	450	520	380	420	380	40
Horizontal	400	500	380	420	380	40
Horizontal	350	470	340	370	340	30
Vertical	350	460	340	370	340	30

Maximum diameter of roll is sometime called as" Paper size "and any diameter fall after redressing is called as diameter "Below paper size".

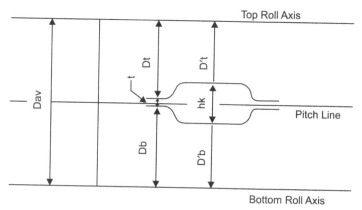

Fig. 1.23 Computation of D_{av}.

D_{av} of roll is computed with following formula

5.1.1 D_{av} at the Collar

$$D_{av} = \frac{(D_t + D_b + 2 \times t)}{2}$$

D_t = top diameter of roll

D_b = bottom diameter of roll

t = roll gap or split

5.1.2 D_{av} at the Pass

$$D_{av} = \frac{D'_t + D'_b + 2 \times h_k}{2}$$

h_k = pass depth or height

5.1.3 When top and Bottom Diameter are Equal

$$D_{av} = \frac{2 \times D + 2 \times t}{2}$$

$$= D + t$$

or, $\quad D = D_{av} - t$

5.2 Redressing Co-efficient

The fresh roll get deployed in a mill at a particular diameter *i.e.*, D_{max}. During the course of service, the roll may be redressed at a number of times, then gradually its diameter get reduces from D_{max} and when diameter comes at a minimum *i.e.*, at D_{min}, it is taken out of the service and is called as "scrap roll".

The "Redressing co-efficient" of rolls may be determined from the following;

$$K = \frac{D_{max.} - D_{min.}}{2}$$

For primary mill, the K is taken as 0.08 – 0.12, but for sheet and plate mill, it is assumed as 0.05 – 0.08.

The diameter may be determined from the above equations as under:

$$D_{max.} = D_{av.}(1 + k/2)$$
$$D_{min.} = D_{av.}(1 - k/2)$$

5.3 Roll Collars

Each pass is separated from its neighbours by collars. The collars between the outmost pass and the face of the roll barrel is called "End Collars" and rests are called as "Inner Collars". The end face of the roll between roll neck and periphery of the outer collar is called the shoulder of the roll.

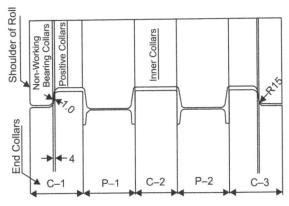

Fig. 1.24 Position of pass *vis-à-vis* collars.

The number of collars in a roll set would always be (*n* + 1), where *n* is the number of passes in the roll set. As shown in Fig. 1.24 there are two passes (P_2) of Beam 125 in the roll, and then numbers of collars will be (2 + 1) *i.e.,* three.

If the collar protrudes beyond the boundary of the pass, then it is called as positive collar or a collar. If, it does not, then it is called as a collar hole or a negative collar.

The width of outer collars is determined, which is based upon the stock is be easily manipulated into the outer passes, without interfering it with the housing. The sufficient width of outer collars is also required not only to place the guide box, but also for the sufficient availability for manual adjustment of guides.

The width of inner collars is determined by strength of rolls and of the depth of pass.

Width of collars may be kept as half of the total depth of groove in steel rolls, while in case of cast iron, width should be equal to the depth of groove.

5.4 Determination of Rolling Diameter

The diameter of rolls, where metal contacts the roll is called as "Rolling Diameter".

Very often, the metal is rolled in such a manner when its contact with roll takes place at a section of roll, which has different diameter and surface speed. As shown in Fig. 1.25 below:

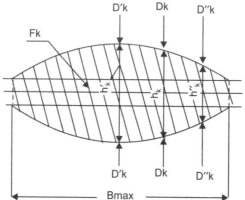

Fig. 1.25 Determination of Rolling Diameter.

$$V' = \frac{\pi \times N \times D'k}{60}$$

$$V'' = \frac{\pi \times N \times D''k}{60}$$

5 Diameter of Rolls

The work-piece comes out of the roll, not with a velocity V' or V'', but with some other speed, which varies between V' and V''. The diameter, which corresponds to this specified speed is called as "Rolling Diameter". The computation of the rolling diameter is more important in case of continuous mill.

The computation of average rolling diameter is based upon average reduced height of the pass. It is obtained by dividing the area of pass by its maximum width.

$$D_k = D_{av.} - h_k$$

$$D_k = D_{av} - \frac{F_k}{B_{max.}}$$

Where h_k is the reduced height of pass.

F_K is the area of the metal.

$B_{max.}$ is the maximum width of the pass, and D_{av} is the average diameter of the roll.

5.5 Pitch Line, Line of Rolling

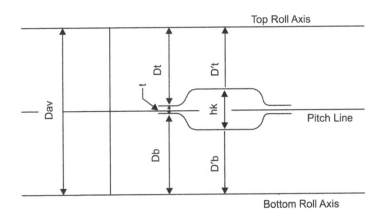

Fig. 1.26 Pitch line, line of rolling.

The line passing in between the axis of rolls divide the distance between pinion centers into two parts is called "Pitch line of rolling". In Fig. 1.26 above, line midway diameter ($D_{av.}$) between the axis of two rolls is called the Pitch line of rolling.

If pass is to be cut into the rolls, symmetrical to the horizontal axis, there should be neither over draught or top pressure nor under draught or bottom pressure, then it will be necessary to coincides the pitch line to the axis of symmetry of rolls *i.e.*, neutral line. Then this line is called as "Line of rolling".

5.6 Top and Bottom Pressure (Over or Under Draught)

Straight delivery of bar from the rolls is of extreme importance, both for obtaining higher output and for safety measures. There is always a tendency of the front end of bar to go up, especially during rolling of light flange sections. If metal finds a gap between the roll and nose of stripper guide, due to breakage or displacement of the stripper guides from its position. It causes wrapping up of rolls, which ultimately may lead to breakage of roll.

Likewise, heavier section like bloom, slab or billet has got a tendency to go down.

Going up or down of stock may increase mill down time due to above explained reasons. In addition, it may also cause breakage of spindles, roll table etc. It also leads to serious accidents.

The tendency of work-piece to go up or down can be manipulated by adjusting either the peripheral speed of top or bottom roll, it can also be done by either by making difference in diameter or speed of top or bottom rolls, as explained in formula below:

$$V = \frac{\pi \times D \times N}{60}$$

It is evident from the above formula that the peripheral speed of individual roll can be changed either by changing the diameter or by changing the speed. The scope for the change of speed of individual roll is limited, as top and bottom roll in a stand is generally driven by single motor and making difference in speed of top and bottom roll is impossible. Exception is only the primary mill like blooming or slabbing mill, where individual motor is provided separately for top and bottom roll and with this both rolls are driven independently.

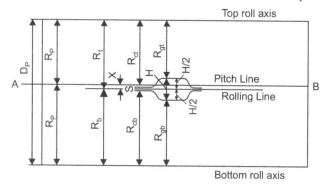

Fig. 1.27 Over draught or Top Pressure.

5.0 Diameter Of Rolls

With this, diameter is the only factor available, with which difference in peripheral speed can be made.

$$D_t - D_b = \Delta D$$

$$R_t - R_b = \frac{\Delta D}{2}$$

Then according to Fig. 1.27, above

$$R_t = R_p + x$$
$$R_b = R_p - x$$
$$R_t - R_b = 2x$$
$$2x = \frac{\Delta D}{2}$$
$$x = \frac{\Delta D}{4}$$

Consequently, the rolling line for an over draught of ΔD passes at a distance of $\Delta D/4$ is below pitch line and for under draught, same will passes at the same distance on above pitch line. The difference in diameters of top and bottom rolls is called" Pressure or Draught".

If the diameter of top roll is 5 mm more than the bottom roll, there is said to "Top pressure or Over draught" of 5 mm. In this case, bar will turn downwards due to higher speed of top roll.

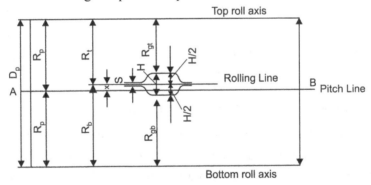

Fig. 1.28 Under draught or Bottom Pressure.

Similarly, in case of "Bottom pressure or under-draught," the diameter of bottom roll will be more than the top roll. The bar will turn upwards due to higher speed of bottom roll.

The value of top and bottom pressure is selected with due regard to the speed of rolling. If a mill operates with higher speed, then difference in speed can be arrived with the lesser difference in roll diameter.

Most section mills are rolled with top pressure or over draught. The following over draught values, expressed in percentage of roll diameter, are recommended:

(*a*) For Break down passes–not over 2 or 3 percentage.

(*b*) For other types of open roughening and breakdown passes–not over 1 percentage.

(*c*) For finishing pass of any form–top pressure should be minimized to maximum possible extend.

6 DISTRIBUTION OF DRAUGHT IN A PASS

In rolling process, the draft cannot be remained constant throughout the rolling process. As a rule, the draught gets gradually reduced from the first pass to the subsequent passes in accordance with the increase in resistance to deformation of the metal. In hot rolling, this increase in resistance to deformation is mainly associated with the cooling of the stock and its reduction in thickness. The intensity of cooling of stock is also proportional to the reduction in stock thickness (due to increase in surface area at constant volume) and inversely proportional to the degree of deformation. The less the amount of deformation the more rapidly the stock is cooled, as deformation itself adds a certain amount of heat to the stock. It is impossible to apply a mathematical approach to the problem of distribution of draft between the passes because of the extremely complex character of variation of plasticity and resistance of deformation under the combined action of various conditions *i.e.,* stressed state, temperature, deformation rate, co-efficient of external friction, form of stock etc.

6.1 Principle of the Distribution of Draught

In initial passes, draft is given in consideration of the angle of bite. It is always advisable to give higher reduction in initial passings, when metal is hot and the resistance to deformation is less. Further on during the course of reduction, when "Bite", no longer limit the draught, and the draught is given on the basis of the strength of the rolls and available power of the drive motor.

In last passing, when wear out of the pass and surface finish of the product is of prime importance, then draught is reduced to the minimum possible extend.

Figure 1.29 shows the distribution of draught in a rolling sequence.

6 Distribution Of Draught In A Pass

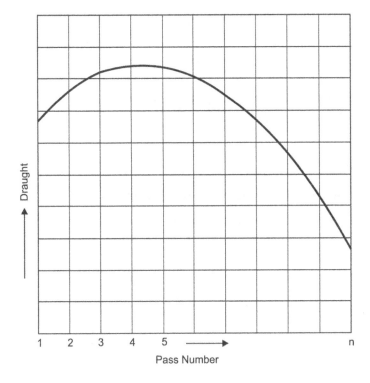

Fig. 1.29 Distribution of elongation among Passes.

This method is proved to be the most useful, as it enables the capacity of a given mill to be utilized to a greatest possible extent. Thus, it is necessary to assign the maximum draught to each pass *i.e.*, from the point of view of a more complete utilization of the plastic properties of the metal, bite limit, roll strength and available power of rolling.

6.2 Maximum Reduction

Bakhtinov's derives a formula to computing the maximum reduction in a pass;

$$\cos \alpha = 1 - \frac{\Delta h}{D}$$

$$\Delta h = D(1 - \cos \alpha)$$

As angle of bite depends upon co-efficient of friction, it is convenient to make use of the direct relation of reduction to the co-efficient of friction

$$\Delta h_{max} = D\left(1 - \frac{1}{\sqrt{1+f^2}}\right)$$

The relationship is derived on the basis of conditions that;

tan α = f and direct relationship can be made between Δh and the co-efficient of friction and diameter of roll.

7 ROLL PASS DESIGN FOR NON-UNIFORM DEFORMATION

Mostly all sectional products are subjected to non-uniform deformation during the rolling process.

(a) Non-uniform deformation due to Geometrical factors

Non-uniform deformation in a section profile is mainly due to its shape of the pass, form of the input and output stock, and difference in effective working diameter etc.

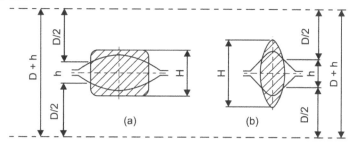

Fig. 1.30 Examples of Non-uniform Deformation.

The simplest case for rolling with non-uniform draught is the flattening of a round bar between plain-barreled rolls. If a square bar is rolled in an oval pass, non-uniform deformation is affected by the shape of the pass [Fig. 1.30 (a)]. A more complex case of non-uniform deformation is the rolling of an oval bar in a square pass [Fig. 1.30 (b)].

(b) Non-uniform Deformation due to Physical Characteristics

Non-uniform draught is due to the variation in physical characteristics, due to non-homogeneous chemical composition and structure of the metal, and may be due to the non-uniform heating. The variation in properties along the cross section of bar is sometimes due to influence of friction, work hardening, residual stresses etc.

7.1 Analysis of Non-uniform Deformation

Non uniform deformation in rolling is of a complex nature and has to be analyzed separately alongwith the width, height of bar and also along the length of deformation zone:

7 Roll Pass Design for Non-uniform Deformation

(a) Along the Width of Bar

Fig. 1.31 (a) shows the draught diagram is plotted to evaluate the non-uniform deformation alongwith the width of the bar. In plotting this diagram, the cross section of the bar is superimposed on the pass and is divided into a number of parts of equal width. The H/h ratios, corresponding to the parts, are plotted as ordinate along the bar width. The draught curve depicts the degree of non-uniform width of bar.

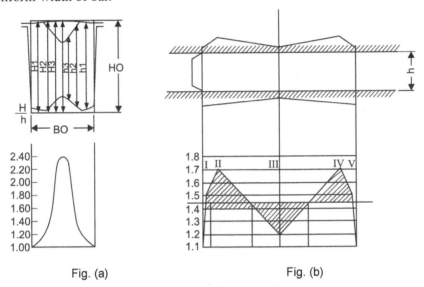

Fig. (a) Fig. (b)

Fig. 1.31 Non-uniform deformation along the width of Bar.

Fig. 1.31 (a) shows the principle of plotting draught diagram for a web-cutting pass.

Fig. 1.31 (b), shows a rolling of a bar between plain barreled rolls with the varying draught along the width. The maximum H/h ratio is 1.70 and the minimum is 1.18. The average elongation is 1.43. The region of increased draught (II and IV) and that of low draft (III) are clearly seen on the diagram.

It can be concluded that regions of low draught are subjected to tensile stresses, while those of high draught are to compressive stresses. Non-uniform deformation alongwith the width of pass may be the cause of induced spread and also for the formation of wrinkles (buckles), results from surplus elongation of metal in the region of high draught in comparison of mean elongation of the bar, especially in the rolling of sheets and strips. In low draught region, stretching may be observed, to reduce the size of the cross section of the bar. Tears may occur in this region, when metal with low plasticity is rolled.

Pulling-down is often observed in the rolling of flanged shaped (rails, channel, beam etc.), leads to the reduction of height of flange in closed passes.

(b) Non-uniform Deformation Along the Height of Bar

There is much common similarity between height non-uniformity and lateral non-uniformity. If lateral non uniformity is caused by geometrical factors, non-uniformity along height is caused due to variation in mechanical properties of metal. This is turn may be the result of non uniformly heated metal, surface work hardening, contact friction, segregation etc. Since, there are zones of different strength along the height of bar which are to be subjected to different draught. The example of non-uniform draught along height is bulging on the sides of a bar being rolled between a plain barrelled rolls.

In actual practice, the bar is subject to non-uniform deformation simultaneously along its width and height.

(c) Non-uniform Deformation along the Length of Zone of Deformation

When a bar is rolled between grooved rolls, its form gets gradually altered, as it passes through the zone of deformation. This result is in corresponding variation in the character of the deformation and in the stresses in the bar. This type of non uniformity is more pronounced in case of flanged profile like rolling of beam as shown below in Fig. 1.32.

During the initial stages of rolling in a beam pass, the open flange is worked by indirect draught, while close flange gets worked by direct draught and while the web portion is not subjected to draught at all.

On the contrary at the final stages of rolling, web should be worked extensively, while flange is to be given lesser draught. Deformation of such nature generates a complex state of stresses in the bar.

In initial stages, flanges are subjected to compression and web to the tension, while at the end of process, flanges will be under tension and web undergoes compression. Thus, deformation of this nature produces a complex state of stresses in the beam pass.

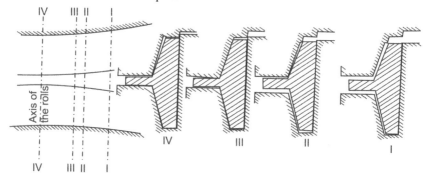

← Direction of Rolling

Fig. 1.32 Stages of reduction along the length of the deformation in a Beam pass.

7 Roll Pass Design for Non-uniform Deformation

7.2 Computation of Draught In Non-Uniform Deformation

There is no popular formula available for the computation of draught in non-uniform deformation; which is required for the purpose of computation of spread and for the rolling load. It is proper to convert it into the simple case of rolling a rectangular bar between plain–barreled rolls.

Two methods are generally used to convert the complex shape into the simple rectangular shape are:

The ratio-corresponding bar method

Two profiles are considered to be ratio-corresponding if their area and ratio of the similar axis (sides) in their cross-section are equals.

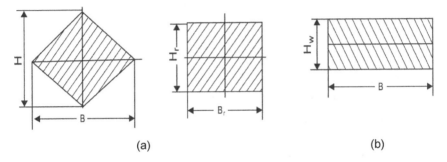

a– Ratio Corresponding Method   b–Width Corresponding Method

Fig. 1.33 Methods of Conversion of Width.

F and F_r are the cross section areas of given and ratio corresponding bar.

B and B_r are the width of given and ratio corresponding bar.

H and H_r are the height of given and ratio corresponding bar.

Then, in accordance with the definition:

$F = F_r$ and $B/H = B_r/H_r$

(b) Width-corresponding bar method

This method has proven to a simpler and more convenient in use. In this method, firstly convert the given bar and to the pass into the rectangle of the same cross sectional area, while maintaining the bar width as it enters into the pass as a basic dimension. The height of the width-corresponding rectangular is determined as:

$$H_w = F / B$$

This method is used not only to compute spread, but to compute roll load as well.

8 CHARACTERISTICS OF ROLLING IN FLANGED SHAPES

8.1 Direct and Indirect Draught

It is not only the vertical, but also the horizontal components of forces exerted by the rolls act on the metal being rolled in a pass as shown in Fig. 1.34 below.

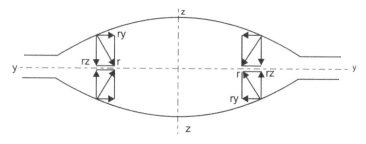

Fig. 1.34 Direct and Indirect draught

In cases, where horizontal components are directed towards the centre of bar, they restrict the spread and increase co-efficient of elongation. In addition, they also create more favourable conditions for developing plastic properties of metal being rolled, but at the same time, increase the resistance to deformation and roll wear.

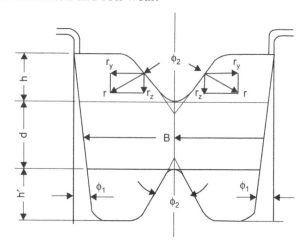

Fig. 1.35 Rolling of Flanges in a Web Cutting Pass.

In such case, where the horizontal component of rolling forces is larger in magnitude than the vertical components, then metal is said to be worked by indirect draft. An example of indirect draft (side working) is the rolling of flanges in a web cutting pass (Fig. 1.35, above).

8 Characteristics of Rolling In Flanged Shapes

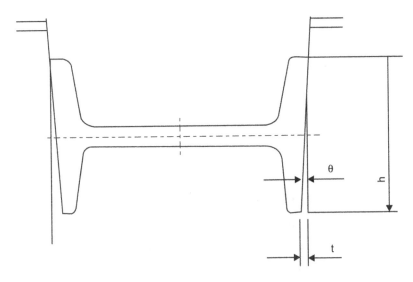

Fig. 1.36 Dead and Live Flange.

In the Fig. 1.36 above which shows a flange profile, the bottom flange lies completely in one roll and there is no grinding action. This part of pass is called the "Dead" as incoming stock will not flow easily into dead flange. Hence, here the incoming stock has to be longer and narrower. It means "Dead" flange is filled in by direct draft on height and leaving a scope of spread in width.

The top flange, as it lies in between collars of both rolls will be "Live" for grinding action, while reducing the metal by indirect draft. The in-going stock has to be flatter but shorter to avoid overfilling.

The ratio of horizontal and vertical components of the rolling forces varies with the angle of inclination of the web cutting element of the pass. If the cutting wedge is inclined less than 45° to the horizontal (horizontal components are less than the vertical components), then the metal is said to be worked by direct draught and at an angle above 45°, (horizontal components exceeds vertical components), then it is said to be worked by indirect or side work.

Indirect draught is often seen in the rolling of flanged shaped profiles *i.e.*, Channels, beams, rails etc.

8.2 Effect of Taper

Flange shaped profiles create difficulties for roll pass designer to obtain the desired results. Fig. 1.37 below shows a closed beam pass which comprises the web cutting elements *i.e.*, live and dead flange holes.

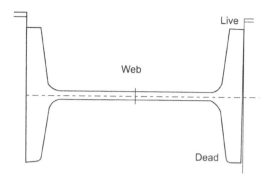

Fig. 1.37 Design of a closed Beam Pass.

A live flange is formed by two rolls, the inner face belong to one roll, while the outer side face belong to the other. Rolling is possible in such pass only if the inner and outer side of pass is tapered.

Importance of a wall taper in a pass become evident, if a cross-section is examined in a plane parallel to the roll axis (Fig. 1.38, below). This cross section shows a slot flaring out both at the entry and exit sides. The curves forming this slot are hyperbolas. This form of the flange holes enables the flange to be reduced from thickness t_1 to thickness t_2. The hyperbolas form is due to the taper of the pass walls without which no draughting of the flange takes place and rolling of metal would become impossible.

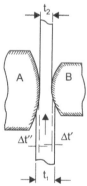

Fig. 1.38 Form of a flange hole.

A-side wall of the pass and B-side of web cutting element

Taper of the pass is a very important factor, without which no draughting of the flange takes place and formation of shape would become impossible. After wear of the beam pass, only the size of flange may be restored. The width of the pass is not restored and it increases in the course of flange wear from the minimum to the maximum size.

8 Characteristics of Rolling In Flanged Shapes

The inclination of the side wall in breakdown pass is usually taken as 5 to 8% (3° to 4°), in leader passes as 1.5 to 2% (1° to 1.5°) and upto 1.5% (1°) in the finishing pass. Taper on the internal faces of flange hole is taken from 12 to 16% (7° to 9°) and from 40 to 70% (22° to 35°) in the web cutting pass.

The taper angle of the inner faces of flange has a great effect on the resistance of dead hole for deformation. As angle reduces, the resistance of dead hole is intensively increased in comparison to the live hole. For example, in leader and finishing passes, the angle is equal to 10° and the resistance of the dead holes is 4 to 5 times as large as that of live holes.

Under these conditions, it is necessary to take measure to facilitate the filling of dead hole in order to avoid excessive height reduction of the flange. For this purpose, no side draught is given in the leader and finishing passes, to facilitate entry of bar into the dead hole without any restriction. As a designer point of view, the flange must enter the dead hole from 1/2 to 2/3 of their depth without restriction and for live hole, from 1/3 to ½ of their depth. Less indirect or side draught is given to the dead than the live hole.

8.3 Effect of Friction

External friction has a large influence on the ratio of the resistance of the dead and live holes. When the friction will be higher than resistance will be also higher in the dead hole, as compared to the live hole. As the resistance becomes zero, then it becomes equal to the both. Therefore, all measures which are reducing the co-efficient of friction, such as selecting the proper roll materials and matching it to high quality finish will also promote the deformation in the dead holes, especially in the leader and finishing pass.

8.4 Speed of Rolling in Flange Profile

Ascertaining the speed conditions in rolling flanged profiles, in particular to the beam profile, is of prime importance in roll pass design.

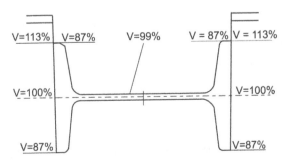

Fig. 1.39 Speed Variation in Beam Pass.

As shown in Fig. 1.39 above, due to higher diameter at live side, the speed in the live holes are higher than that of web and the web speed exceeds that the dead hole. The stock is delivered from the roll at the mean speed, say equal to the web.

In live hole, the flange enters between two moving surface at different speed and their mean speed exceeds that of section. The thickness of flange gets reduced by the working surface of rolls.

Pulling action is observed in the dead hole, where there is no movement of the relative surface of two rolls, this is accompanied with reduction or pulling down of the height of the flange. If the web is long heavier than there will be less error in assigning the speed of the web as the speed of beam profile. In web cutting and break-down pass, where web is not developed properly, the delivery speed of the section cannot be assigned to the web speed.

8.5 Asymmetry of Deformation in a Flange Pass

With the result of unequal forces acting in the live and dead holes, the asymmetry of deformation is observed in flanged profile rolling. This asymmetry is especially pronounced in web cutting pass in rolling and the initial billet of rectangular cross-section, may be causing the unequal height to obtained for live and dead flanges.

The asymmetry of profile obtained in the same pass, varies with the following conditions:

Maximum asymmetry is found in a pass, where there is no spread. On the contrary, when comparatively narrow bars are rolled, asymmetry reduces or disappears and flanges come out with almost equal height.

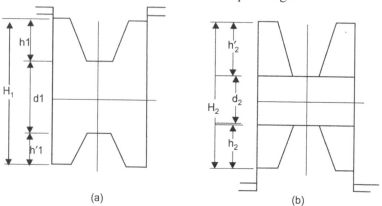

Fig. 1.40 Diagram for Pass Design for Beam Rolling.

The height of input billet also influences asymmetry of initial web cutting pass. Higher is the height of billet, higher will be the asymmetry.

8 Characteristics of Rolling In Flanged Shapes

This asymmetry is to be taken into account in computing and designing the web cutting and breakdown beam passes.

In web cutting pass, flange is formed due to the penetration of the web cutting element into the rectangular billet. The height of the live flange will found to be higher than the dead flange [Fig. 1.40 (a)].

In designing web cutting pass, the ratio of the height of live h_1 to dead h'_1 are usually taken as

$$h_1/h'_1 = 1.5 \text{ approx.}$$

In succeeding breakdown passes [Fig. 1.40(b)], though total height of the stock decreases (from H_1 to H_2), but at the same time flange height will be increased ($\Delta h_2 = h_2 - h'_1$) and ($\Delta h'_2 = h'_2 - h_1$).This increase will be at the cost of reduction of web thickness. If d_1 and d_2 are the web thickness of preceding and succeeding pass, then total increase in height will be

$\Delta h_2 + \Delta h'_2 = d_1 - d_2$, if no pulling down in the height of profile is assumed and $\Delta h_2 + \Delta h'_2 = (d_1 - d_2) - (H_1 - H_2)$, if pulling down is assumed in height.

Increase in height of live and dead flanges are due to web reduction. It will continue as long as web draft exceeds the pulling – down (drawing) action on the height of the profile. As soon as web draft becomes less than this drawing effect, further beam rolling will be accompanied by a decrease in height in dead holes. Height of live hole may also get increases slightly by 1-2 mm, that's mainly due to side draft.

8.6 Stages of Rolling in a Flange Pass

Characteristics features of the rolling process for a typical flanged shape *i.e.,* Beam section is that all elements are not worked simultaneously. These features may be made clear by dividing the working process in the zone of deformation into four stages of rolling, *i.e.,*

- First, the live flanges are side worked; the web and dead flange are not subjected to draught.
- Upon further travel of the profile into the zone of deformation, in addition to the side draught of the live flange, the height draught of dead flange begins.
- In the third stage, the flanges are worked in the same ways as in the preceding stages, but at the end of this stage, the web comes in contact of rolls.
- In last stage all work is concentrated on to the web only, here flanges are subjected to little work only.

Due to intensive web reduction causing pulling down of flanges, it is desired that massive reduction of the web should be undertaken at the earlier stage of rolling, when flanges are thicker and temperature is very high. As the bar approaches the finishing end, the reduction of web should be decreased to minimum.

The difference in time between working of the flange and web, it also depends upon taper of angle of the inner face of the flange. The larger the taper angle, the time will be the less between the working of web and flange and less will be the non-uniformity of deformation. On the contrary, for small taper angles, the time between working of web and flange will increase, as will non-uniformity of deformation increase . In this case, web should be draughted only slightly, otherwise flange will be pulled down.

8.7 Distribution of Draught Among the Elements of Flanged Profile

The distribution of draft among various element of a flanged profile is the most essential element in pass design.

Total beam profile is divided into three elements *i.e.,* web, live flange and dead flange. The flanges are again sub-divided into tip and root of flange. In addition ratio of total web area to flange area should also be taken into consideration. Thin web flange cools faster than the thick web flange. In case of heavy beam, where, web is capable of pulling the flanges because web area exceeds that of four flanges. In such cases, beam should be rolled with draught for web should be kept less than that of flange, to avoid pulling down of flanges.

On the contrary, in case of light and medium beam, where web draft should be always more than the flanges, both in initial and in the following intermediate passes. Only in the leader and finishing passes, web reduction should be less than that of flanges. This ensures the proper filling of the finishing pass. There is no considerable pulling down is observed in such design of thin web beam.

It is common to all drought schedule is that web is reduced less than flange such distribution of draught among the elements of a flanged shape ensure proper filling of the passes.

8.8 Features of Deformation in a Closed Web-Cutting Pass

Closed web-cutting passes are designed to obtain the rough beam blank from the rectangular billet. It is characterized mainly by the form of the wedge shaped collar for cutting the web.

In contradiction to the ordinary beam pass, this web-cutting collar is made with a rather sharp edge, so that there is almost no web between the flanges. The angle of this wedge shaped collar is taken from 45° to 70°, depending on the beam width. The top and bottom collars usually differ, the sharper one will be in the closed part of the pass.

8 Characteristics of Rolling In Flanged Shapes

The apex of the collar is rounded by a circular arc or a parabola. These sharp collars easily cut into the rectangular billet and form triangular grooves, above and below, to serve as, the beginning in forming the web of a beam profile. The flanges, due to their massive form are subjected to the comparatively little pulling down.

If the web is cut by blunt collars then considerable drawing of the flanges will takes place.

The more difficulty will be encountered in obtaining long flanges, when the large angle of the web element cutting will be in the pass. Therefore, for web-cutting elements, the optimum angles lie in a range from 45° to 70°. Larger angles are rarely applied. The deformation process in a web-cutting pass comprises the following stages (Fig. 1.41).

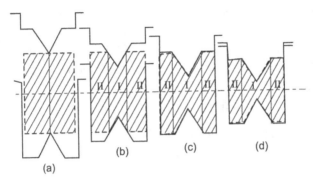

(a) First Stage   (b) Second Stage   (c) Third Stage   (d) End of Third Stage

Fig. 1.41 Stage of Rolling in a Web Cutting Pass.

In the first stage, the element cuts into the rectangular billet if space has been provided for spread of the bar. At this moment only the central region of the cross-section (region I) is subjected to draught; the side regions II are not subject to draft. Under the action of the wedge-shaped elements the billet is expanded to fill up the pass along its width.

The second stage begins when further cutting of the web occurs under conditions in which there is no more space for spread in the pass. Here side draught begins for the regions II.

The third stage begins when height draught of the side regions II is added to the side draught. Most frequently, the web-cutting passes are designed with height to spare on the flanges and, consequently, no height draft is imposed on the flanges in this first pass.

In the first state of deformation, induced spread is easily developed and the bar is expanded to the sides so that the bar may be subject to no elongation. The flanges formed at this moment by the cutting element are not subject to pulling down and completely maintain their height.

48   Chapter 1   Roll Pass Design

The possibility of spread is absent in the second stage and draught of the web and flanges causes elongation of the bar. Since the central region I of the cross-section is more heavily draughted than the side regions II, there may be induced pulling down of the less draughted regions.

9 CONSIDERATION FOR DESIGN OF FINISHING PASSES

The dimensions of finishing pass differ from the finished profile due to following reasons:

9.1 Co-efficient of Expansion

Rolling in pass usually takes place between the temperature range from 850 °C–1000 °C, in case of hot rolling of the, rounds and structural profiles, while it is more than 1100 °C, in case of continuous wire rod mill. At a mean co-efficient of linear expansion of iron, $\alpha = 0.000012$. The relationship between the size of hot profiles, *i.e.*, h_z, b_z, l_z to the dimension of cold profile *i.e.*, h_x, b_x, l_x will be

$$H_z = h_x \left(1 + 0.000012 \times t\,°C\right)$$
$$b_z = b_x \left(1 + 0.000012 \times t\,°C\right)$$
$$l_z = l_x \left(1 + 0.000012 \times t\,°C\right)$$
$$h_z = b_z = l_z = 1 + 0.000012 \times t\,°C$$

Table 1.4   The relationship of cold section and its hot section at various temperature of hot rolling.

Temp. of rolling in °C	Cold section	Hot section
800	1.0000	1.0096
850	1.0000	1.0102
900	1.0000	1.0108
950	1.0000	1.0114
1000	1.0000	1.0120
1050	1.0000	1.0126
1100	1.0000	1.0132
1150	1.0000	1.0138
1200	1.0000	1.0144
1250	1.0000	1.0150

Thus, it is necessary to know the finishing temperature of the rolling to determine the exact dimension of profile, which will come out after cooling. It is therefore, a designer generally consider hot section higher by 1.011 to 1.015 times more than the cold section.

9.2 Rolling with Negative Allowance

Modern trend in rolling is to roll product in negative field of tolerance with the following objectives:

Economic necessity arises, from point of view of rolling in negative field of tolerance. It helps to make it possible to increase the length of finished bar *i.e.,* metal economy for consumers.

Technical necessity arises from the further processing requirement at machine parts or manufacture units. For example, wire drawing units demand for wire rod with negative tolerance, since over dimensional wire rods causes quicker wear out of their wire drawing dies.

It is therefore, necessary to design a new finishing pass, taking into consideration of negative allowances from the point of view of life of rolls. Since, fast wear out of width of their passes lead to final rejection of product. Profiles such as beams, channels and rails require more attention.

As rolling cannot be made with exact dimensions, there must be some tolerance has to be given on each part of product. According to all standards, some allowance has been provided on the positive and negative side of each part of rolled profile and on its standard weight also. These are called as tolerances as per standards. Generally, majority of standards give a minimum tolerance of a section to an average about 1.5% on both positive and negative side *i.e.,* equal to 1.015 times of the profile.

It is customary practice to design a pass with the minimum possible tolerance on the negative side:

- From point of view of life of rolls. Especially, profile such as beam, channel and rails require more attention, as fast wear out of width of pass leads to final rejection of metal. It will increase the efficiency of the mill because of fewer change overs of rolls.
- From point of view of customers, as it will ultimately reduce the cost of project. Finished length of stock will come more with the same weight of product; in turn project will require less weight of material.

As mentioned earlier, majority of standards gives minimum tolerance of a section to about 1.5% *i.e.,* equal to 1.015 times of the profile. At the same time, to obtain hot dimension, designer has to multiply the cold dimension with the factor of co-efficient of expansion, which comes to 1.011–1.015 at the finishing temperature of rolling of long products. It is based upon the above facts, which conclude that initial dimension of finishing pass can be designed with the cold dimension of profile.

For section profiles like beam, channel, angle etc., following procedure has to be followed:

According to the finishing temperature of the finished product, hot dimensions of profiles can be obtained by multiplying the cold dimension by 1.011–1.015.

Out of the obtained dimensions, subtracts the minus tolerance, leaving some allowances for adjustment of rolls.

The obtained dimensions are again corrected in such a way, taking into possible consideration of uneven shrinkage and uneven wear out of certain elements of profile etc.

10 SPECIFIC FEATURE OF ROLL PASS DESIGN OF A CONTINUOUS MILL

In case of continuous rolling, the work of piece is rolled simultaneously in all or several stands of the mill, as shown in figure below. The condition of the continuous rolling process is to have equality of rolling constant in all stand of the continuous group.

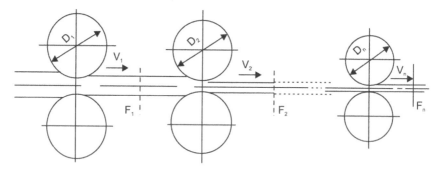

Fig. 1.42 Rolling in a continuous Mill.

If rolling constant is C,

Then $C_1 = C_2 = C_3 = \ldots\ldots\ldots = C_{n-1} = C_n$

Where, $C_1 = F_1 \times D_1 \times n_1$, is the rolling constant of stand-1

$C_2 = F_2 \times D_2 \times n_2$, is the rolling constant of stand-2 and

$C_n = F_n \times D_n \times n_n$, is the rolling constant of nth stand

Where F, D and n are area, rolling diameter and roll speed of respective stands.

$$n = N / i,$$

N and i are the motor speed and reducer ratio of the respective stands

11 COMPUTATION OF POWER REQUIREMENT

11.1 Rolling Load

Computation of Roll Pass Design also involves calculation of loads.

$$P = P_m \times F$$

Where, $\quad P =$ Total rolling load

$\quad P_m =$ Mean unit pressure of metal on rolls

$\quad F =$ Projected contact area

$\quad = (B + b)/2 \times L,$

Where, B & b are the initial and final width of stock and L is length of arc of contact, which is computed by the formula *i.e.,*

$$L = \sqrt{R \times \Delta H},$$

Where, Reduction, $\Delta H = H - h$, H and h are the initial and final height of metal

And, $\quad R = D_k/2$

$\quad D_k =$ Rolling Diameter, for box pass, it will be

$$= \frac{D_t + D_b + 2 \times t}{2}$$

When diameter of top and bottom roll are different,

then $\quad L = \dfrac{\sqrt{2 \times R_1 \times R_2}}{\sqrt{R_1 + R_2}} \times \Delta H$

Where R_1 & R_2 are the radius of top and bottom roll respectively.

$$P_m = (\text{Қ} + \text{ŋ}\,\mu)\,(1 + m)\ \text{kg}/\text{mm}^2, \text{ where}$$

Қ =Unit resistance to static compression (yield point in static compression at rolling temperature) in kg/mm²

ŋ = Viscosity of metal being rolled kg- sec/mm²

μ = Mean rate of deformation 1/sec.

$m =$ Co-efficient accounting for increase in resistance to deformation of metal due to friction between stock and the rolls.

On the basis of experience and based upon experimental data, it is proposed the following formulas for the determining the values of these co-efficients:

$$\text{Қ} = (14 - 0.01 \times t\,°C)\,(1.4 + \%\ C + \%\ Mn + \%\ 0.3\ Cr)$$

$t\,°C =$ Rolling temperature

$$\text{ŋ} = 0.01\,(14 - 0.01 \times t\,°C) \times Cv,$$

Where, Cv is a co-efficient, whose value depends upon peripheral speed of rolls.

Table 1.5

V M/Sec.	Upto 6 m/sec.	6 to 10 m/sec.	10 to 15 m/sec.	15 to 20 m/sec.
Cv	1.0	0.8	0.65	0.60

$$\mu = \frac{2 \times V \times \sqrt{\Delta H/R}}{H + h}$$

$$V = \frac{\pi \times D_k \times n}{60}$$

$n = $ Roll rpm

$\quad = N/i$, where i is the reducer ratio of pinion stand

$$m = \frac{1.6 \times f \times \sqrt{R \times \Delta H} - 1.2 \times \Delta H}{H + h}$$

$f = (1.05 - 0.0005 \times t\,°C)$

11.2 Power Requirement and Rolling Torque

11.2.1 For non-reversible Constant Speed Motor

The required power N_m will be

$$N_m = \frac{N_r + N_f}{\eta_d}$$

N_m = Required power of the drive motor on a non reversible motor

N_r = the power required for rolling process (effective power required to deform the metal with rolls)

N_f = Frictional power is power loss due to frictional loss in roll bearings

N_d = Efficiency of roll drive *i.e.*, spindle, coupling box etc.

The power required for rolling (N_r) is found from the rolling torque (M_r) by the:

$$N_r = \frac{M_r \times N_r}{0.716} h_p$$

M_r = Rolling torque required to drive rolls (without bearing loss) intons-mts

N_r = Rolling rpm corresponding to given speed of rolling

The rolling torque (M_r) is determined in various ways for various cases of rolling. But for simple case of rolling process when the two diameters are equal and rolls revolve with same speed.

11 Computation of Power Requirement

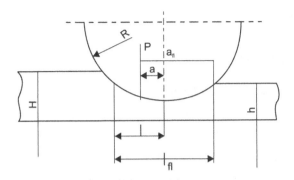

Fig. 1.43 Determining Rolling Torque.

$$M_r = M_1 + M_2$$

where M_1 and M_2 are rolling torque required for the top and bottom rolls respectively. If both roll diameter are equal, then, $M_1 = M_2$ or

$$M = 2 \times a \times P$$

P is the rolling load in tons.

a is the lever arm of the resultant total load applied to the arc of contact

$$a = \alpha \times r \times \psi$$

α is angle of bite in radian, r is roll radius.

ψ is torque arm co-efficient, which is equal to the ratio between the length of the torque arm and that of contact area. On the basis of the experience based on the experimental data, the value of ψ is taken as: For hot rolling, $\psi = (0.45$ to $0.50) \times L$

For cold rolling, $\psi = (0.35$ to $0.45) \times L$

When rolls undergo a considerable deformation (flattening), the torque arm gets increased, as shown in Fig. 1.43

α is the angle of bite in radian

$$N_f = \frac{M_f \times n_r}{0.716} h_p$$

Where, $M_f = f \times d_{neck} \times P$

f is frictional co-efficient for Roll neck bearing and dneck is diameter of roll neck.

$f = 0.003$ to 0.005 for anti-friction roller bearing

$= 0.005$ to 0.01 for textolite bearing

$$N_d = n_1 \times n_2 \times n_3$$

Where, n_1 = co-efficient for frictional losses in spindles

$= 0.96 - 0.97$ for wobbler type spindles

$= 0.95 - 0.97$ for universal type spindles

54   Chapter 1   Roll Pass Design

n_2 = co-efficient for frictional losses in reducing gears and main motors coupling *i.e.,* 0.93 – 0.96

n_3 = co-efficient for frictional losses in pinion stand *i.e.,* 0.93–0.95

11.2.2 For Variable Speed Motors

Blooming mill, Plate mill or Reversible three high mills, which are running at a variable speed, it is necessary to take dynamic torque into consideration.

Motor torque, *i.e.,* $M_m = M_s \pm M_d$

Where, M_s = the motor torque required to overcome static load (constant in the rolling process).

M_d = Dynamic torque on the motor shaft required to accelerate or decelerate the rotating component of main drive in the course of one pass.

Here, plus sign corresponds to acceleration, while minus one to deceleration of variable speed motor. Static torque is required to overcome the static load (constant during the rolling process), determined by formula. When roll undergo a considerable deformation (flattening), the torque arm is increased, as per Fig. 1.43

$$M_s = \frac{0.716 \times N_r + N_f}{N_d \times n_m}$$

Where, n_m = the motor speed corresponds to constant speed of rolling and equal to

$$= n_r \times i,$$

where *i* is the gear ratio of reducer in drive train

$$M_d = \frac{j \times \pi \times \varepsilon}{30}$$

ε = Angular acceleration (or deceleration) of the motor in rolling (ε = 40–60 rpm/sec)

j = Moment of inertia of rotating parts referred to motor shaft

$$= \frac{G \, D^2 m.\text{sh}}{4 \, g}$$

G = Mass of rotating parts/unit

D = Diameter of gyration

g = Acceleration due to gravity, *i.e.,* 9.81 m/sec²

ε = Angular acceleration, for electric motor of rolling mill, the value varies from 40–60 rpm/sec.

11 Computation of Power Requirement

$G\,D^2m$-shaft = Moment of gyration of all the rotating parts of drive train, (including the motor), referred to the motor shaft. This moment is calculated by summing up the moment of gyration of each rotating parts of the drive train with due to the regard of gear ratio "i" from motor to each rotating parts.

$$G\,D^2m.\text{ shaft} = \frac{\Sigma\,G_n\,D_n{}^2}{i_n{}^2}$$

When stock enters to roll, during at that movement, the speed of roll decreases, because motor slows down under the heavy load. The flywheel supply, kinetic energy stored during acceleration (E_1), thus supply of the drive to get extra energy in rolling and smoothening out the peak load at the time of entering the work piece to the rolls.

There is definite relationship exists between moment of gyration of all parts including flywheel to the required kinetic energy.

$$G\,D^2m\text{-shaft} = 540 \times \frac{E_1}{N_m{}^2} \text{ tons-m}^2$$

N_m = motor speed,

For flywheels also, it is equal to motor speed, since flywheel is mounted on input shaft of reducer gear.

It is assumed from the practical data that,

$$E_1 = (15 \text{ to } 20) \times N \text{ rate hp-sec.}$$

Where, N rate is the rated horse power of the motor.

During idle run of the rolling mill, acceleration will be positive and it become negative when mill gets slows down (as metal enters the rolls). The above formula indicates that in the idle run of the mill and during movement of metal from one pass to another, the flywheel will increase the load of the drive. On the other hand, at the time of biting of metal by the roll, the motor slows down under the load, the flywheel will reduce the motor torque, as it imports the kinetic energy previously stored. To enable the flywheel to give the stored energy, the motor mounted on the mill must be of a type that permits a reduction in speed upon increase in load. Speed generally drops by 5–10 %, under load.

11.3 Factors Deciding the Size and Type of Main Drive Motor
Table 1.5

Sl. No.	Power Required For	Percentage
1.	Rolling	59.00
2.	Friction of Pinion and Mill	05.90
3.	Loss in Reversing Motor	10.84
4.	Loss in Generators	11.44
5.	Loss in Flywheel	01.32
6.	Loss in Slip Regulators	02.69
7.	Loss in Induction Motor	06.40
8.	Loss in Exciters, Blowers etc.	01.66
9.	Loss in Electrical connections	00.75
	Total	100.00

11.4 Factors Affecting Rolling Load

1. **Width of bar:** Rolling load is directly proportional to the width of bar being rolled. Load requirement for rolling flat products is much more than rolling of long products. Hence, power requirement of motor is the main criteria for selecting the width of flat products. It is seen that except for narrow strips, the rolling load is always nearly proportional to the strip width for any given reduction.

2. **Roll size:** Rolling load increases with the increase of the roll diameter. At 50% reduction, the rolling load, corresponds the largest roll will be 1.5 times than the small roll. As with larger roll the arc of contact for the given reduction will be more and consequently more work has to be done to overcome the surface friction. Fig. 1.44 shows the effect of roll size on the friction hill.

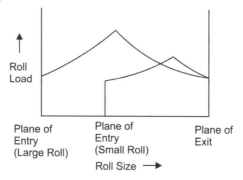

Fig. 1.44 Effect of Roll Size.

3. **Work hardening:** The affect of work hardening is to increase the value of yield stress of metal, as bar processes through rolls, as shown in Fig. 1.45.

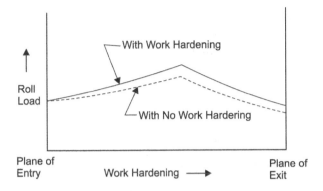

Fig. 1.45 Effect of Work-Hardening.

4. **Chemical composition of metal being rolled:** The magnitude of metal stress depends upon the chemical composition of metal being rolled. The power requirement and rolling load will be related to the tensile strength of rolling material and thus increases with tensile strength. In other words, load increases with the increase of carbon and alloying elements in steel.

5. **Temperature of bar being rolled:** With increasing temperature, the yield stress decreases. In other word, the rolling load increases with drop in rolling temperature.

6. **Roll speed:** Speed has a very little effect on either rolling load or mean specific roll pressure. Actually, at higher speed, slightly less power per ton is required than at lower speed.

7. **The co-efficient of friction:** The horizontal stress increases with increase in frictional stresses, as shown in Fig. 1.46. Rolling load varies substantially with increase in co-efficient of friction.

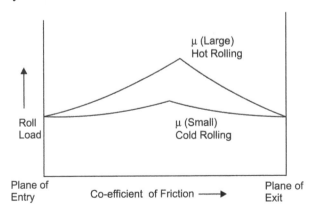

Fig. 1.46 Effect of co-efficient of Friction.

8. **Reduction in pass:** For a given roll size and initial bar thickness, the length of the arc of contact varies with the percentage of reduction. The effect will be similar to the change in roll size. Roll load increases with reduction in pass.

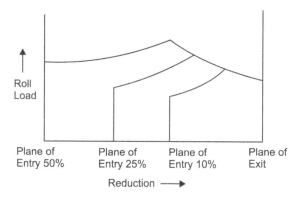

Fig. 1.47 Effect of Reduction.

9. **Bar thickness:** The horizontal stress increases as bar become thinner. In case of thin strip, the horizontal stress will become several times, therefore rolling load increases, as the initial thickness of bar decreases for a given draft and roll size.

10. **Tension:** The effect of front and back tension on the friction hill curve is shown in Fig. 1.48 (a & b). In each case, it is seen that rolling load is reduced and the line of resultant is moved either forward or backward, depending upon whether back or forward is applied.

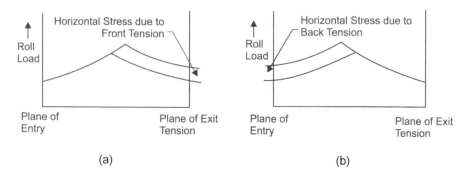

Fig. 1.48 Effect of Front (a) and Back Tension (b).

2

Mechanical Working and Rolling Process

1 INTRODUCTION

1.1 The aim of mechanical working operation is not only to get the required shape, but to enhance the mechanical properties of material. Hundreds of mechanical forming methods have been evolved for particular metal working applications.

1.2 Mechanical working involves plastic deformation of either cold or heated metal by the external action of specific tools, such as mill rolls, forging dies, flat hammer dies or extruding dies.

1.3 The mechanical deformation in cold state leads to strain or work hardening. Strain hardening is, not only increase the tensile strength and hardness of the metal, but is also responsible for a measurable reduction in some of its characteristics mechanical properties like, elongation and impact strength. If the forming of metal is still required to be continued, then increase in strength of metal necessities to apply higher strain over the metal for further plastic deformation. After heavy strain hardening of metal, further deformation in cold state may be become impossible due to development of discontinuities of metal *i.e.*, crack, tears and other defects. Deformation then must be immediately stopped. Metal will then ready for further plastic deformation, when it further softened by heating to a definite temperature.

2 HOT AND COLD WORKING

Forming process is commonly classified into hot working and cold working.

2.1 Cold working is the plastic deformation of metal at a temperature and rate such that Strain hardening is produced. When the temperature at which deformation takes place is high enough to produce softening of metal, simultaneously with the strain hardening, which effect on the working process gets nullified. Plastic deformation under this condition is called hot working. So, hot working can be defined as "Plastic deformation at a temperature and rate such that no strain hardening is produced."

2.2 Softening principally is due to the phenomenon known as "Recrystallization", in which old strained crystals get altered into new strain-free crystals.

2.3 The minimum crystallization temperature for the given working conditions creates the dividing line between hot working and cold working. The recrystallization temperature is not constant for a particular metal, but it depends upon the time, temperature, the amount of previous deformation and other variables also.

2.4 There is no specific or defined temperature of deformation, wherein a distinction can be made between hot or cold working. Hot working operation is performed at a high temperature to get a speedy rate of recrystallization in most of the cases. However, for lead and tin, working at room temperature called as hot working, similarly working Tungsten at 750°C, which is the hot working range for steel, but it is a cold working range for tungsten.

3 THEORY OF MICROSCOPIC PLASTICITY

A few important observations, based upon stress and strain is placed below:

3.1 During plastic deformation, volume of metal continues to remain the same. In other words, the sum of principal strain rate, stretches and compression is zero. This is due to plastic flow effects because of mechanism of slip, which does not require any volume change.

3.2 When maximum shear stress in some direction or plane attains the critical value, then yielding takes place at that particular point.

3.3 The direction of the greatest shear strain rate coincides with direction of greatest shear stress.

3.4 The amount of plastic working is generally enhanced by using compressive rather than tensile methods.

4 METHODS OF METAL WORKING

Steel is an alloy generally made of iron *i.e.*, Fe (approx. by 99% wt) and carbon *i.e.*, C (<1.0% wt), sometimes it also contain small amounts of various other alloying elements.

4 Methods of Metal Working

- Iron atoms are much bigger than carbon atoms, which fill the gaps (interstitial).

Fig. 2.1 Carbon atoms in gaps.

- Iron atoms try to pack as closely to one another as possible.
- The most efficient way to achieve this is through face-centered cubic packing.

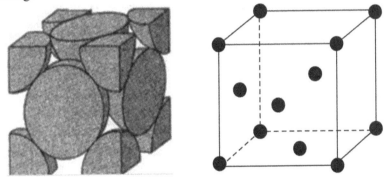

Fig. 2.2 Face centered cubic packing.

- Pure Fe is in its most stable form at high temperatures in a FCC arrangement.
- FCC has 26% free volume.

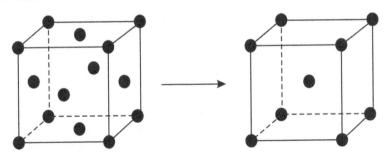

Fig. 2.3 Phase transformation from Austenite (FCC) to Ferrite (BCC).

5 ALLOTROPY OF IRON

The effects of carbon addition are closely associated with the allotropic changes of iron. At temperature above 1540°C, iron is in liquid form. Liquid iron starts solidifying at 1540 °C. The temperature remains constant till solidification is completed. When solidification is completed, the iron is in delta (δ) form, with a BCC structure. When the temperature drops down to 1395°C, an allotropic modification of iron takes place from delta (δ) iron to gamma (γ) iron with FCC structure. No change takes place with the further drop in temperature from 1395°C to 910°C. However, at 910°C the gamma (γ) iron changes into alpha (α) iron with BCC structure. The alpha iron at 910°C is non-magnetic and remains so till temperature of 768°C is reached. Beta (β) iron is the non-magnetic version of alpha (α) iron. At 768°C, it is termed as curie temperature which transform the non-magnetic beta iron into magnetic alpha iron.

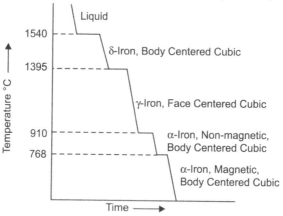

Fig. 2.4 Allotropy of Iron

6 IRON-CARBON DIAGRAM

The iron-carbon phase diagram is the most important subject in the study of ferrous metallurgy. It provides the basis for understanding the properties and heat treatment of steel and the effect of alloying elements in alloy steel.

The part of iron-carbon phase diagram plotted for the concentrations (weight percent) upto 6.67% carbon is of commercial significance. This part between pure iron and iron carbide (Fe_3C) is called as iron-iron carbide diagram. This is not a true equilibrium diagram because the compound iron carbide does not form a true equilibrium phase. Under proper conditions, iron carbide decomposes to form a more stable carbon (graphite). But this decomposition is almost never observed in ordinary steels. So iron carbide is considered stable and therefore treated as an equilibrium phase. The iron-iron carbide diagram is divided into two parts. The alloys contains solid phases

6.0 Iron-Carbon Diagram

with less than 2% carbon is known as steels and those which contains more than 2% carbon are known as cast irons.

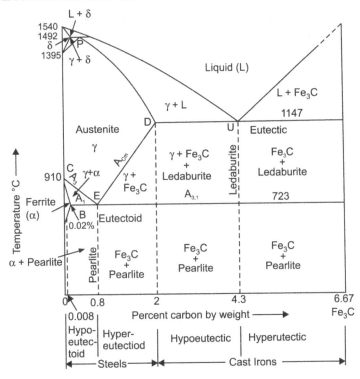

Fig. 2.5 The iron-carbon phase diagram.

- **(a) α-Ferrite (α) :** Ferrite is an interstitial solid solution of carbon in BCC alpha (α) iron. It is almost pure iron and the name comes from the Latin word "ferrum" (meaning iron). Carbon solubility in α-Ferrite at 0°C is 0.008% and it increases with temperature, reaching a maximum of 0.02% at 723°C. It is relatively a soft and ductile phase.
- **(b) Austenite (γ):** Austenite (γ) is an interstitial solid solution of carbon in FCC gamma (γ) iron. The solid solubility of carbon in austenite is a maximum of 2.08% at 1147°C and decreases to 0.8% at 723°C. It is a soft, ductile and malleable phase.
- **(c) δ-Ferrite:** The interstitial solid solution of carbon in high temperature BCC δ iron is called δ-ferrite, the maximum solubility of carbon in δ ferrite is 0.1% at 1492°C.
- **(d) Cementite (Fe$_3$C):** The intermetallic compound iron carbide (Fe$_3$C) is called cementite. Cementite is from the Latin word 'Caementum' (meaning stone chips). This compound has a fixed carbon content of 6.67% C. It is extremely hard and brittle compound.

There are three important phase transformations that occur in a Fe-Fe$_3$C phase diagram.

(*i*) **Peritectic Transformation:** The peritectic transformation region is expanded and is shown in Fig. 2.6 below:

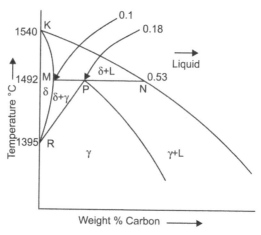

Fig. 2.6. Peritectic Transformation Region In Fe-Fe$_3$C diagram.

At the peritectic point P, all liquid combines with δ-ferrite to form γ austenite. This transformation takes place at 1492°C and is written as

$$\text{Liquid} + \delta\text{-ferrite} \xrightarrow[\text{Cooling}]{1492°C} \gamma \text{ austenite}$$
$$(0.53\%\text{C}) \quad (0.1\%\text{C}) \qquad\qquad (0.18\%\text{C})$$

(*ii*) **Eutectoid Transformation:** At the eutectoid reaction point E, at iron-carbon diagram, solid austenite produces very fine eutectoid mixture of α-ferrite and cementite. The eutectoid mixture have the alternate layers of ferrite and cementite. So it is termed as "Pearlite" because of its resemblance to mother-of-pearl, when it is It viewed at low magnification under an optical microscope. This eutectoid reaction which occurs at 732°C is written as:

$$\gamma\text{ - austenite} \xrightarrow[\text{Cooling}]{723°C} \underbrace{\text{Ferrite} + \text{cementite}}_{\text{Pearlite}}$$
$$(0.8\%\text{ C}) \qquad\qquad (0.02\%\text{ C}) + (6.67\%\text{ C})$$

Steel which contains 0.8%C is called as eutectoid steel. Steels, with containing less than 0.8% C are known as hypo eutectoid steels and the phases present after transformation are ferrite and pearlite. Steels containing more than 0.8%C are known as hyper-eutectoid steels and the phases present after transformation are cementite and pearlite.

7 Transformation in steel   65

(*iii*) **Eutectic Transformation:** At the eutectic reaction point U, liquid transforms and gives a eutectic mixture of austenite and cementite. This eutectic mixture is called as ledaburite.

This reaction which occurs at 1147°C is written as :

$$\text{Liquid} \xrightarrow[\text{Cooling}]{1147°C} \text{austenite} + \text{Fe}_3\text{C}$$

$$(4.3\% \text{ C}) \qquad \underbrace{(2\% \text{ C}) + (6.67\% \text{ C})}_{\text{Ledaburite}}$$

Cast iron that contains 4.3% C is known as eutectic cast iron. Cast irons that contain less than 4.3% C are termed as hypoeutectic cast irons and those which contain more than 4.3% C are termed as hypereutectic cast irons.

7 TRANSFORMATION IN STEEL

Plain carbon steels are generally defined as alloys of iron and carbon. Plain carbon steel is treated as iron-carbon binary alloys, but in practice the carbon steels contain small amounts of other elements also.

7.1 Transformation in Eutectoid Steel

Let us consider a steel containing 0.8%C. Temperature is to be kept above 750°C and for such a period of time to become a homogenous austenite. Then the steel is cooled slowly. It does not begin to transform until the eutectoid point E is reached. The transformation begins and ends at the same temperature. *i.e.,* at 723°C, where all austenite get transform into 100% pearlite. During slow cooling, there will be simultaneously transformation of gamma to alpha iron and also the precipitation of cementite because of due to the low solubility of carbon in alpha iron. The cementite is in lamellar form interspaced with ferrite to form pearlite.

7.2 Transformation in Hypo Eutectoid Steel

Let us consider a steel containing 0.4% C. It is heated to about 900°C for a sufficient time to become a homogeneous austenite. Upon slow cooling, there is no change until the line CE is crossed. This line is known as the upper critical temperature line (A_3) on the hypo eutectoid side. Just below this line, the structure begins to change from FCC to BCC. Consequently, small crystals of proeutectoid ferrite nucleate. As the continues cooling, the ferrite grows in size at the austenite grain boundaries, and since ferrite is almost pure iron, the left over remaining austenite will be enriched with carbon. When the line BE is reached, the remaining austenite contains 0.8% C. This line is known as the lower critical temperature line (A_1) on the hypo eutectoid side. At this point (723°C), the left over austenite transforms into pearlite by the eutectoid reaction. So, the final structure will be of ferrite and pearlite. Any steel that contains less than 0.8%C, the transformation from austenite to a mixture of ferrite and pearlite, which is made in a similar way. But the relative amounts of ferrite and pearlite will be depending upon the carbon content of the steel.

7.3 Transformation in Hyper Eutectoid Steel

Let us consider a steel containing 1.2% C. It is heated to about 950°C and held for a sufficient time to become austenite. If it is slowly cooled, there is no change until the line DE is crossed. This line is known as the upper critical temperature line (A_{cm}) on the hyper eutectoid side. Just below this line, the transformation starts. Since austenite has excess amount of carbon, cementite nucleates first at the austenite grain boundaries and it is called proeutectoid cementite. Upon further cooling, the cementite grows and this causes the austenite to become progressively less rich in carbon. When the line EF is reached the remaining austenite will contain only about 0.8% C. This is the lower critical temperature line ($A_{3.1}$) on the hypereutectoid side. At this temperature (723°C), the remaining austenite transforms into pearlite by the eutectoid reaction. So the final structure will have cementite and pearlite.

7.4 Critical Temperature

The critical temperature lines are designed as A_1, A_3, $A_{3.1}$, A_{cm}. These are the indication of transformation under equilibrium conditions. In practice, during heating the critical lines are raised and during cooling, they are lowered. The critical lines observed on heating are designed as A_c (c from French word 'chauffage' - heating) and on cooling are designed as A_r (r from French word 'refroidissement' cooling).

The gap between these lines depends on the rate of heating and cooling. When the rate of heating and cooling is slower, the gap reduces. For extremely slow cooling or heating rate, critical temperature will remain same as shown by equilibrium diagram.

8.0 THE EFFECTS OF VARIOUS ELEMENTS ON PLASTIC PROPERTIES OF STEEL

Table 2.1

Sl. No.	Element	Its Effects
1.	Manganese	Susceptible to overheating and requires strict adherence to specified temperature condition. Mn has a favourable effect on plastic properties, as unlike ferrous sulphide, Manganese sulphides does not occur at grain boundaries.

9 The Effect of Rolling on Grain Structure

Sl. No.	Element	Its Effects
2.	Silicon	At lower values, Si is in solid solution and it has no effect on the ductility. Ductility of metal get reduces, when Si in steel is > 3%.
3.	Sulphur	It is a deleterious impurity, being low melting points compounds turns liquid at high temperature and by flowing between the grain boundaries, destroy all cohesion of the grains, thus causing, called "Hot Shortness". Hot shortness is somewhat neutralize by the presence of Mn. The increase of S content in steel favours, the formation of ferrous sulphides, which is an eutectic, with a melting point of 950°C. The destruction of cohesion of grains results in the appearance of the fissures and cracks during plastic deformation.
4.	Oxygen	Oxygen also behaves like S causes Hot Shortness, because of the oxides formation, which has low melting temperature and gets concentrate near or at grain boundaries.
5.	Hydrogen	Hydrogen is also deleterious, since it causes internal fracture by flake formation.
6.	Phosphorus	The increase in P content in steel, results in "Cold – Shortness" during plastic deformations. Phosphorus get dissolved in the solid solution of iron, it does not deteriorate the ductility of steel.
7.	Copper	Pure Copper possesses high ductility, its presence in steel deteriorate the ductility of the steel. It also make the steel corrosion resistant.
8.	Chromium	It is conducive to the formation of an excess ferrite phase, which reduces ductility and it also has use for grain size control, which helps into improve yield strength.
9.	Molybdenum	It may form sulphides at the grain boundaries and also used to control grain size. The workability of steel deteriorates with increase in Mo content.
10.	Tungsten	Tungsten also favours the formation of carbides, which hampers the grain growth during recrystallisation. The resistance to deformation also get increases.
11.	Vanadium	It is added in steel to reduce the grain size.
12.	Boron	Its addition improves the rollability of high alloy steel.
13.	Others	Tin, Arsenic, Antimony and Lead weakens the cohesion of grains.

9 THE EFFECT OF ROLLING ON GRAIN STRUCTURE

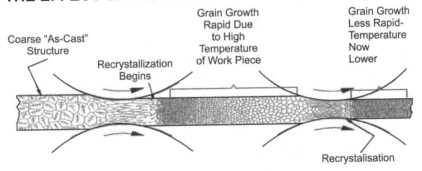

Grains broken up→New grains nucleate→Grains grow→Grains broken up

Fig. 2.7 Effect of rolling on grain structure.

10 DEFINATION AND CLASSIFICATION OF THERMO-MECHANICAL PROCESSING

As a rule, all heat treatment processes can be carried out with the metal still at rolling temperature only a few processes require heat treatment is to be given in several steps. Relevant examples are the quenching and tempering process, in which tempering takes place after hardening, for example hardening of high-speed steel is a process in which residual austenite get changed and transformed step by step into martensite, through repeated tempering of workpiece.

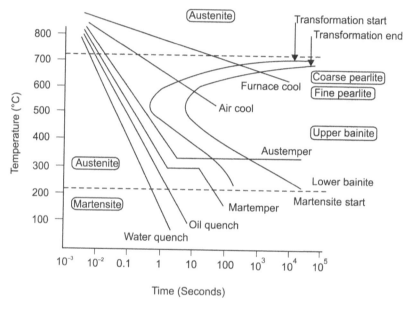

Fig. 2.8 Thermo-Mechanical Processing.

Accelerated cooling after rolling causes heat to allow to dissipate, either totally or in part to have subsequent processing steps of heat treatment such as, patenting, hardening, normalizing, austenite stabilizing or solution annealing. Thermo-mechanical processing of metal, followed by retard cooling is mainly suitable to substitute conventional heat treatments, to enhance the mechanical properties of the material to the desired level. These are soft annealing, spheroidising (to form globular cementite) to attain shearing ability or easier cold formability: in addition, its aim is to avoid the martenstic microstructure, particularly in the case of higher-alloyed steels.

The term 'thermo-mechanical processing' generally defines as any production process in which formed process and heat treatment are combined in a controlled or in a specific way.

11 QUENCHED AND SELF-TEMPERED REBARS (QST)

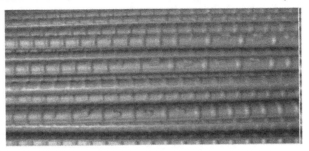

Fig. 2.9 32 MM TMT Bar.

Fig. 2.10 Popular use of TMT Bar.

TMT stands for "Thermo-Mechanically Treated" steel reinforcement bars. TMT has become popular as, TMT process not only gives, the minimum yield strength, atleast 65% higher than the plain mild steel, but the most important point is that with the increase in yield strength, there will be no major loss of ductility, thereby TMT bars will have good weldability.

Conventionally, high weldabale concrete reinforcement bars are produced in multi-pass hot rolling mills from continuously cast or rolled billets into bars of preferred size ranges from 8 to 40 mm.

The customer demand has now for improved microstructure and mechanical properties. Though the yield strength of the bars in rolled conditions can be obtained also by modifying the steel chemistry by addition of micro-alloyed expensive precipitation forming elements (*i.e.*, vanadium, niobium, titanium), but these addition will further increase the cost of production by the addition of expensive alloying elements.

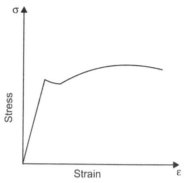

Fig. 2.11 Stress-Strain Curve For Hot Rolled Bar.

The same properties can also be achieved through a relatively simple setup of a series of water cooling tubes, which quench the surface of the bar as it leaves the finishing stand; and creating a hard martensitic structure at the surface. The heat remaining at the core of the bar, causes anneals and tempers the product, as it cools on the cooling bed, leaving a tempered martensite periphery and fine grained ferrite-pearlite at the core. The whole process can be made computer controlled, based upon finishing speed, temperature etc. With the help of desired cooling effect, the end product with the yield strength >500 mpa can be achieved, compare to normal steel with yield strength of 200-250 mpa. The soft, ductile core of the bar forms around 65-75% of the whole cross-section

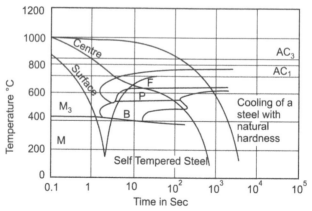

Fig. 2.12 TTT Diagram for Self Tempered Steel.

Increasing carbon and manganese content is quite effective. The addition of carbon is considered relatively inexpensive, but unfortunately an increase in carbon content results in a significant reduction in weldability.

12 Stages of Quenching and Self-Tempering Cycle

The use of the "quenching and semi-tempering" technique produce high strength concrete reinforcing bars with good ductility and weldability with the lesser addition of alloying additions. It is possible by quenching and self tempering process to increase yield strength by more than 200 mpa. As a result, it is possible to reduce the carbon contents of steel to produce high-yield stress concrete reinforcing bars with good weldability. The quenching and self tempering process consists essentially of a special heat treatment from the heat of rolling. The basic principle is systematically illustrated in Fig. 2.12.

When the bar leaves the last finishing stand of the rolling mill, it passes through the cooling line.

12 STAGES OF QUENCHING AND SELF-TEMPERING CYCLE

The special heat treating cycle of quenching and self tempering has three stages.

12.1 First Stage

This consists of a fast cooling operation applied to the bar after it enters the cooling line. The efficiency of this cooling is very high so that it quenches to a certain depth below the surface to the martensite transformation range.

Fig. 2.13 "Thermex" installation.

At the end of this operation, the bar has an austenite core surrounded by a layer composed of a mixture of austenite and martensite.

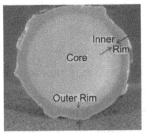

Fig. 2.14. Phase formation in TMT Rolled Bar.

(a) Outer Rim (b) Inner Rim (c) Core

Fig. 2.15 Microstructure of ATMT Bar

It is also important to note that in some cases, the layer below the martensitic ring can be transformed partially and even completely to bainite structure depend only upon the operating parameters of process. The martensite content decreases with increasing distance from the surface.

12.2 Second Stage

The bar leaves the area of drastic cooling and is exposed to air. The heat transfer co-efficient of the environment is very small and the temperature gradient within the cross section of the bar is very large. The core reheats the quenched surface layer by conduction. As a result, the martensite formed during the first stage is subjected to self tempering to ensures adequate ductility while maintaining a high yield strength level.

The second stage could be considered upto the stage of temperature equalization is dependent on the bar diameter and the cooling conditions which applied during the first stage.

12.3 Third Stage

This stage occurs as the bar lies on the cooling bed. During this stage the austenite core transforms to ferrite and pearlite with micro-structure depending on the steel chemical composition, bar diameter, rebar entry temperature to cooling-line, cooling efficiency and duration of cooling.

The main physical phenomena involved in the above mentioned three stages of the quenching and self tempering process are :

- Heat transfer from the rebar surface to the surroundings.
- Reheating of the surface from core by conduction.
- Metallurgical phenomena such as transformation of austenite to martensite, bainite, ferrite and pearlite.

Other advantages of this process lie in the fact that no expensive addition of micro-alloys is necessary on the one hand and the requirements are fulfilled even with reduced carbon contents on the other hand.

13 MAIN EQUIPMENTS OF TMT PROCESS

(*a*) TMT quenching system.

Fig. 2.16 TMT Quenching System.

(*a*) TMT quenching system.

(*b*) Pinch rolls, flying shears and cooling bed.

(*c*) Dedicated water system (Water Boxes & Equalization Troughs).

(*d*) Universal testing machine cum bend testing machine.

The success of high-speed operation and the ability to produce a superior finished product depends upon the design of the post finishing block equipments. Alignment of the water boxes and troughs are critical to continuous high-speed operation and for the elimination of cobbles. Wear resistant, split bore cast stainless steel nozzles and troughs are used throughout in the water boxes and equalization trough zones, to ensure the alignment is to be maintained. The nozzles and troughs are located on stainless steel wear pads mounted on water-cooled headers to negate misalignment due to thermal expansion during rolling.

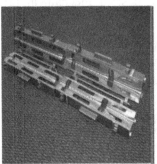 

Fig. 2.17 Cast Stainless Steel Nozzles & Dual bore troughs.

The latest designed high-efficiency split bore water box nozzles provide quick and easy access for inspection and for cobble removal, contributing to maintain of a high operating efficiency. The latest design provides reduced flows for the same cooling length as original design nozzles providing reduced water consumption or increased cooling capacity over the same cooling length (Fig. 2.17).

All the water boxes in each outlet utilize a similar design, differing only in the size of the stainless steel nozzles used to accommodate the respective product size ranges.

Due to the size of the products being rolled through the straight bar outlet to brake slide and the resulting weight of the water box nozzles, the water boxes have been made traversing with two paths of water box nozzles and one by-pass roller table to cover the complete product size range. These water boxes have been split into 3 m sections to allow the total cooling length to be quickly adjusted for optimum cooling for both TMT products and plain bar products. (Fig. 2.18).

Fig. 2.18 Water Box & Equalization Troughs.

The water boxes are installed at various points to provide controlled temperature rolling in order to get the desired mechanical properties to different grades of steel.

The result is improved mechanical properties; including finer and improved grain size suitable for getting high yield strength with optimal sectional weight. This provides consistency of mechanical properties throughout the bar and from bar to bar.

14 DIFFERENT METAL FORMING OPERATIONS

1. Direct Compression Method.
2. Indirect Compression Method.
3. Tension Type Process.
4. Bending Process.
5. Shearing Process.

In direct compression method, the application of force is made to the surface of work piece and flow of the metal will be at right angles to the direction of compression, *e.g.*, rolling, forging.

Shearing processes involve application of shearing forces of the sufficient magnitude to rupture the metal in plane of shear.

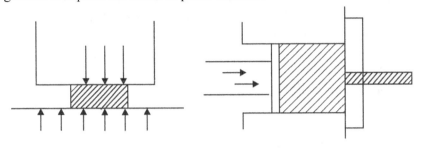

Fig. 2.19 Forging & Extrusion Processes.

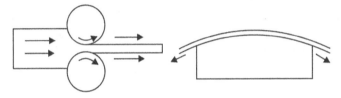

Fig. 2.20 Rolling & Stretch Forming.

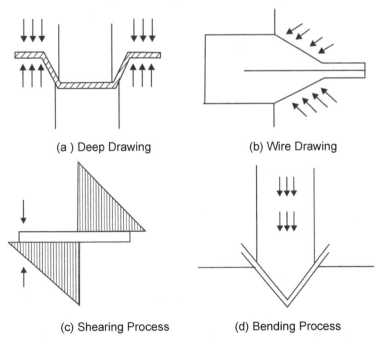

(a) Deep Drawing   (b) Wire Drawing

(c) Shearing Process   (d) Bending Process

Fig. 2.21 Other Popular Mechanical Working Process.

Forming process may be further categorized into:

(*a*) **Primary Mechanical Process** is one, in which we reduce ingot or billet to simple shapes such as sheet, plate, bar, etc. They are also called "**Processing Operations**".

(*b*) **Secondary Mechanical Process** is forming methods, which produce a part to a final shape, hence are called secondary mechanical process. This method is also called as "**Fabrication**".

15 ROLLING

15.1 History of Rolling

(a) Sketch of rolling mill by Leonardo de Vinci

(b) An ancient rolling mill to roll sheet lead

Fig. 2.22 Ancient Rolling Mills

Leonardo de Vinci invented a hand operated mill (later worked by a water wheel), using two flat rolls for rolling of precious metal in the year 1475.

However, credit for first wrought iron rolling, using two grooved rolls goes to Henry Cart, who had created a rolling mill in the year 1790 for rolling of 1½″ square wrought iron billets made under a forge press into 3/8″ rounds. Cart's mill consisted of only one single two high stand and was powered by water wheel. Angle was first rolled in the year 1819. Rails in 1820 and small tees in the following year. Zores of France had developed beam design in the year 1849. In 1867, a French rolling mill rolled beam section in a universal type of mill.

15.2 What is Rolling

In Rolling process, the metal is plastically deformed by means of rolls. Here due to squeezing action of rolls, metal is subjected to high compressive stresses and to surface shear stress, due to friction between rolls and metal. Frictional force also help in drawing the metal into the rolls.

15.0 Rolling

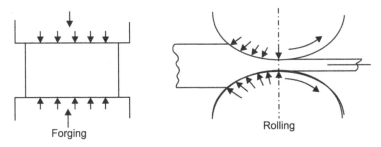

Fig. 2.23 Rolling and Forging Process.

15.3 Methods of Rolling

Depending on the direction of motion of rolls with respect to the worked metal and the direction of deformation, the following three methods of rolling are distinguished.

15.3.1 Longitudinal Rolling

It is the most commonly used method. In longitudinal method of rolling, the metal is rolled between two rolls, turning in opposite direction (Fig. 2.24).

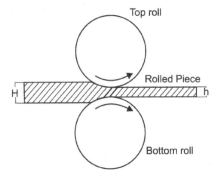

Fig. 2.24 Longitudinal Rolling.

15.3.2 Cross Rolling

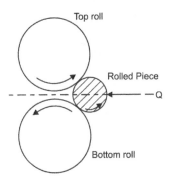

Fig. 2.25 Cross Rolling.

In cross rolling, the rolls turn in same direction (Fig. 2.25). The billet is held in place by means of special device (Force Q) and turns around its own longitudinal axis.

15.3.3 Helical Rolling

In helical rolling, the rolls also turn in the same direction, but, in contrast to cross rolling, they are arranged at an angle with respect to each other. The rolls set the round billet being worked in rotary and translatory motion. This method is used for piercing shells (roughly shaped pips) in pipe manufacturer (Fig. 2.26).

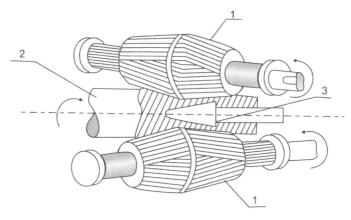

1-Roll, 2-Billet, 3-Mandrel

Fig. 2.26 Helical Rolling.

16 CLASSIFICATION OF PRODUCTS

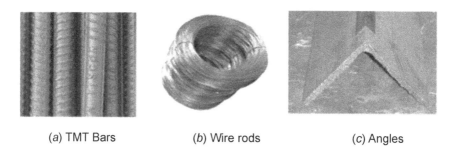

(a) TMT Bars   (b) Wire rods   (c) Angles

Fig. 2.27 Most Popular Rolled Products.

16 Classification of Products

16.1 General Purpose Rolled Sections

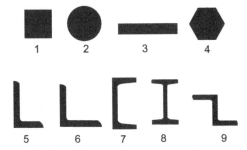

1. Square, 2. Round, 3. Flat, 4. Hexagonal, 5. Unequal angle
6. Equal angle, 7. Channel, 8. Beam, 9. Zee Section

Fig. 2.28. Different General Rolled Products.

16.2 Special Purpose Rolled Section

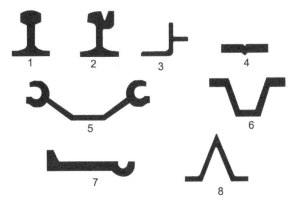

1. Rail-road rail, 2. Street car rail, 3. Zee shape, 4. Grooved spring leaf section,
5. Grooved pile, 6. Shape for mine props, 7. Shape for automobile tyre rims,
8. Shape for tractor grouser

Fig. 2.29 Rolling of Special Purpose Rolled Section.

16.3 Main Categorization of Rolled Products

All rolled products can be classified into the four main categories:
- Steel Shape or section.
- Plates and sheet section.
- Pipes and tubes.
- Special type of rolled product sections, which are widely used in general engineering purpose includes rounds, squares, flats, strips, wires, angles, channels, beams and others.

- Special purpose sections are rails and shapes employed in automobiles, agricultural machineries, rail road, car building and other industries.
- Sheets, Strips with good surface finish and improved mechanical properties are being produced by cold working. It maintain close control over dimensions of products.
- The terminologies used to describe rolled products, with respect to dimensions are fairly loose or overlapping. The product obtained after first breakdown of ingot is called as "Bloom". Generally, Bloom is square section and of dimension greater than 150 × 150 mm. Further reduction by hot rolling of Blooms results in the production of Billets. The minimum cross-section of Billet is generally taken as 35 × 35 mm and upto 150 × 150 mm. Slabs refers to a hot rolled section with the width that at least twice the thickness and cross-section area is always more than 10000 mm^2. Bloom, Billet and Slabs are known as semi-finished products, because, these sections are subsequently formed into other mill products.

The difference between Plate and Sheet is determined by the thickness of product. In general, plates are having thickness more than 4 mm, while thickness upto and including 4 mm are called as sheets and strips. Sheet refers to a product with a width greater than 300 mm, while strip refers to a rolled product with a width upto 300 mm.

17 CLASSIFICATION OF ROLLING MILLS

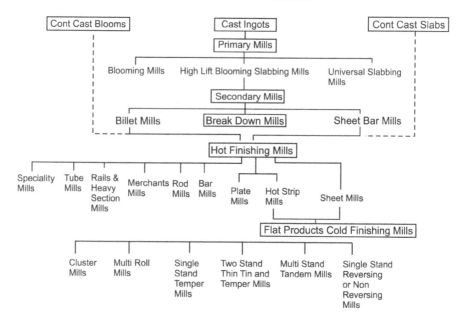

Fig. 2.30 Classification of Rolling mill.

17 Classification of Rolling Mills

Rolling mill are classified according to following criteria:

Table 2.3

Sl. No	Criteria of Classification	Example
1	Nos. of roll in mill stand	2 high, 3 high, 4 high mill
2	Nos. of drives in mill stand	1 Drive, 2 drive, 3 drive mill
3	Type of product /function	Blooming mill, Billet mill, Rail mill, Plate mill
4	Layout of mill	Open train, continuous mill, Tandem mill, cross country mill
5	Direction of rotation of roll	Reversible, non reversible
6	Special purpose mill	Cluster mill, sandzimir mill

17.1 Classification as per Number of Rolls

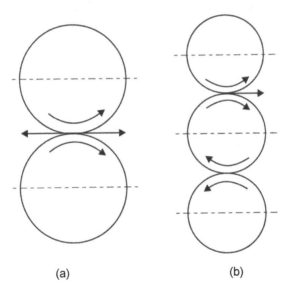

(a)   (b)

Fig. 2.31 Roll Arrangement in 2-High and 3-High mill stand.

Fig. 2.31 (*a*), shows two high rolling mill, equipped with two rolls. As character of rotation of the mill rolls, one distinguishes the two high reversing mills (rotation of rolls is reversible) and two high non-reversing mills (rotation of roll is non-reversible).

Fig. 2.31 (*b*), shows three high rolling mills, having three rolls in the working stands and rotation of their rolls is non-reversible.

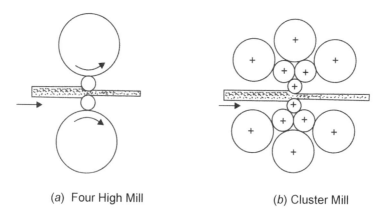

(a) Four High Mill (b) Cluster Mill

Fig. 2.32 4-High and Cluster mill.

Four high mills have four rolls, Fig. 2.33 (*a*), arranged in a vertical plane. Two smaller in diameters are working rolls, while larger one are called Back-up rolls. The most important requirement for rolling flat products is to maintain dimensional accuracy throughout at width. Back-up rolls support working roll in operation and minimize the elastic deflection. Smaller work roll reduces the power requirement.

A cluster mill is a multi roll mill to use work roll of a smaller diameter [Fig. 2.32 (*b*)]. The minimum diameter of roll is limited by the permissible flexural stresses. Therefore, in all design of multi-rolls mills, the small diameter work rolls are supported by large diameter rolls (back-up rolls).

17.2 Classification as per Nos. of Drives in the Mill

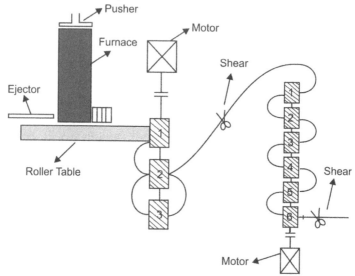

Fig. 2.33 Two-Drives Rolling Mill.

17 Classification of Rolling Mills

Here, classification of rolling mill is based upon the No. of drives in the mill *i.e.,* one drive, two drive, three drive etc. An example of 2 drive mill is shown in Fig. 2.33.

17.3 Classification as per Products Rolled

There are different types of mills according to products to be rolled or function are placed at Table 3.4 below:

Table 2.4

Mill	Size of Roll mm	Barrel Length mm	Function
A. Primary & Intermediate			
1. Blooming	800-1450	–	Reduction from ingot to blooms
2. Slabbing	1100	–	Reduction of ingots to slabs
3. Billet Mill	450-750	–	Reduction of blooms to semis billet of 50×50 to 150×150
4. Rounds	700-800	–	Rolling of blooms to semis rounds of size of 70 to 350 mm
B. Section			
1. Rail & Section	750-900	–	Rolling rails, beams & heavy sections
2. Heavy Section	500-750	–	Rolling of heavy sections, steel squares, rounds
3. Medium Section	350-500	–	Rolling of medium sections, rounds, squares
4. Light Sections	250-350	–	Rolling of light sections,rounds,squares
5. Strip (Skelp)	300-400	–	Rolling Strip 65-500 mm Wide and 1.5 to 10 mm Thick
6. Rod	250-350	–	Rolling Rod From 5 To 10 mm dia.
C. Flats			
1. Plate	–	2000-5000	Rolling plates of size 5 to 50 mm thick
2. Sheet	–	700-2500	Rolling sheets of size from 600-2300 mm
3. Hot Wide Sheet	–	700-1300	Rolling of thin sheets of 0.2 TO 4 mm thick & 500-1200 mm wide
4. Tube Rolling	–	–	Rolling seamless tube upto 650 mm dia. & more
5. Cold Rolling			
6. Sheet	–	700-2800	Rolling sheets of 600 to 2500 mm wide
7. Narrow Strip	–	150-700	Rolling of narrow strip from 0.2 to 4 mm thick & 20 to 600 mm wide, supplied in coils.
8. Foil	–	200-700	Rolling foils from 0.008 to 0.012 mm thick.

(Contd...)

D. For Special Kinds of Rolling			
1. Wheel	–	–	Rolling wheels of railway wagons & other purposes.
2. Hoop & Tyre	–	–	Rolling hoops & tyres for wheels
3. Ball	–	–	Rolling balls for ball mill.
4. For Rolling Variable Sections	–	–	Rolling various kinds of sections.
5. For Gear Wheels	–	–	Rolling teeth of gear.

Fig. 2.34 PLATE MILL.

Classification of mills particularly the division of large group of section mills into heavy, medium and light although is widely used but it is not clearly defined. In rolling of steel three types of intermediate mills are:

1. Continuous billet mill for rolling billet section from 50 × 50 to 150 × 150 mm.
2. Continuous sheet bar mill for rolling sheet bars and also other sections.
3. Tube round mills for rolling rounds from 70 mm to 350 mm diameters, which are required in production of seamless tubes.

Main types of section mills are universal mill for rolling wide flange beams, rails, medium section, light section, rod and strip or skelp (for rolling strip of width not greater than 400-500 mm). The strip mills are sometimes also called as skelp mill.

Light section mills are used for rolling finished sections in which the following are most commonly rolled: rounds and squares from 8 to 40 mm, angles from 20 × 20 to 50 × 50 mm and other sections like beam and channel of smaller sizes.

Cold rolling mills are divided into following groups:

1. Sheet mills for rolling short lengths.
2. Wide strip mills for coil rolling.

17 Classification of Rolling Mills

3. Narrow strip mills for rolling strip from 0.02 to 4.0 mm thick and foil (0.002–0.012 mm) from aluminium, steel molybdenum, tungsten and other metals.
4. Wire flattening mills for rolling very narrow strips from wires.
5. Mills for cold rolling tubes.

Fig. 2.35 Block Mill for Wire Rod.

17.4 Classification as per Layout

17.4.1 Open Train Mill

Most simple form of mill is a open train mill, wherein, stands are positioned in a line side by side as shown in a Fig. 2.36. It is driven through same set of pinions and with a single motor. One most disadvantage of this mill, is that rolls of all stands are run at a same speed, therefore, it is impossible to roll in this mill, a bar at a speed, that increases as length of bar increase.

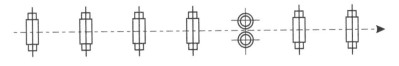

(a) Tandern Train (Plan)

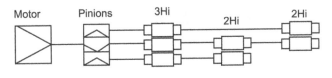

(b) Open Train (Elevation)

Fig. 2.36 Classification as per Layout.

17.4.2 Cross Country Mill

When stands are placed one after another in different rolling line parallel to each other, with the use of repeater or chain transfer to facilitate rolling with higher speed and to facilitate in smaller space, then it is called as a cross country mill as shown below:

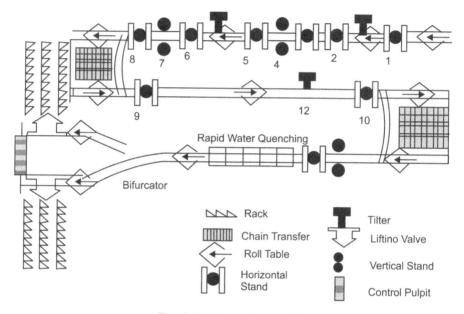

Fig. 2.37 Cross Country Mill.

17.3.3 Universal Rolling Mill

A modern mill which has a set of combination of horizontal and vertical stand is called as a universal rolling mill. Generally, a modern wide flange beam mill as shown in Fig. 2.38 below is of such type.

Fig. 2.38 View of A Universal Rolling Mill.

17 Classification of Rolling Mills

17.4.4 Combination Mill

This type of mill may have tandem, open, looping, and cross-country trains and are used in combination with many layouts, *e.g.*, an alloy bar and rod mill could commence with a two-stand, 3-high, open roughing train. It is to be followed by a cross-country train (used as a looping train on smaller bars) terminating with a tandem finishing train, as illustrated in Fig. 2.39.

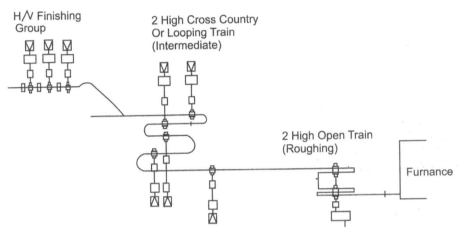

Fig. 2.39 Alloy Bar and Rod Combination Mill Layout

Multi-strand trains are those designed to roll more than one length of stock simultaneously. In these mills, two, three or even four pass line is set up across the roll barrel to roll simultaneously. The practice is adopted in bar and rod mills

17.5 Continuous/Non-continuous Rolling Mill

A Continuous rolling is in the broadest sense implies for a process, in which the stock get engaged in atleast two stands simultaneously. A "Continuous Mill" usually implies for the mill, in which, the stock need not reverse at any stage and makes only one pass per stand. The term "Fully Continuous" is used to signify that once entered stock cannot emerge freely until it get finished in finishing stand, due to the closer spacing of the stands. This is in contrast to mills such as wide strip mills with spaced roughing trains where the stock may emerge freely after each roughing stand. The "Semi-continuous mill" is characterized for a mill, where reversing roughing stand precede a tandem finishing train but the same description cannot be given to a mill, which consists of a continuous roughing train followed by a cross-country finishing train.

17.6 Classification wrt Direction of Rolling

17.6.1 Reversible Mill

Where the direction of rolling gets reversed in each passing's like 2 high reversible Blooming Mill and Slabbing mill.

17.6.2 Non-reversible Mill

Where direction of rolling is not reversed. Examples is continuous Wire mill and cross country Merchant Mill.

18 SPECIAL TYPES MILL

18.1 Double Duo Mill

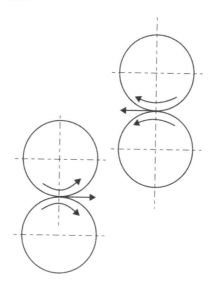

Fig. 2.40 A Double-Duo Mill.

A Double-Duo mill consists of two pair of two high roll set in the same housing, but at the different height. Although, there are 4 rolls are in same housing, 2 rolls are placed in one plane, while other 2 are placed in different plane. In other words, it serves as double 2 high stands *i.e.,* the bar after passing through the bottom pair of rolls; it turns upward or repeated to go through top pairs of rolls. Its advantage over 3 high is the flexibility of roll pass design avoiding common middle rolls of 3 high stands.

18.2 Sandzimar Planetary Mill

Sandzimar Planetary Mill is used for the production of narrow to medium width strips by hot rolling (Fig. 2.41). It consists of 2 large back up rolls, surrounded by a number of planetary work rolls. The planetary work rolls are

19 Determination of Mill Size

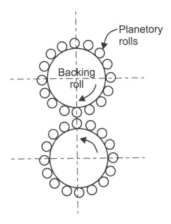

Fig. 2.41 Sandzimar Planetary Mill

driven around backup rolls, driven by the motor. The cumulative reduction by planetary make it possible to have as high as 90% of total reduction in a single pass. Feed rollers are located at the entry side of the mill, to force the stock into the mill. Proper arrangements are made at the delivery side to avoid ripple on strips.

19 DETERMINATION OF MILL SIZE

The size of the mill is determined in the following ways:

19.1 Where Rolls are Individually Driven

The mill size is determined by the mean or average roll diameter of the stand. This is applicable, where in the top and the bottom roll are driven by individual motor and in between there is no pinion stand. Blooming and slabbing mill, which are driven by individual motor for top and bottom drive, falls under this

Mean or Average dia., $D_{av} = \dfrac{(D_{max} + D_{min} + 2 \times t)}{2}$

Where D_{max} and D_{min} are max. and min. roll diameter and "t", is the roll gap. An 1150 mm Blooming mill, which has a max. and min. diameters as 1180 mm and 1080 mm respectively and roll gap is 20 mm, then Average diameter will equal to

$$D_{av} = \dfrac{1180 + 1080 + 2 \times 20}{2}$$

19.2 Where Rolls are Driven by Single Motor Through Pinion Stand

In case of Merchant mill or a wire rod mill, wherein top and bottom rolls are driven by a single motor through pinion stand, the pitch diameter of pinion stand is called as "Mill Size".

A 350 mm Merchant Mill, the 350 mm is the pitch diameter of pinion stand.

19.3 Flat Rolling

In case of Plate and Sheet mill, where the length of barrel is more significant than the actual diameter. The mill size is determined by the available roll barrel between necks. For example, a 3200 mm plate mill has roll barrel length of 3200 mm.

3
Rolls, Roll Cooling and Roll Management

1 INTRODUCTION

Rolls are the most expensive tools of the rolling mill trade and are used to execute the duty of deforming steel.

Fig. 3.1 Turning of a Roll.

1.1 There are Three Parts in a Roll, Namely

(*a*) **The Body:** The part on which rolling is made.

(*b*) **The Neck:** Which support the body and withstand the rolling pressure.

(*c*) **The Wobbler:** From where driving force is applied through loose fitting spindles and boxes, which together form a sort of coupling.

A plain surface rolls are used (in pairs) for rolling sheets, flats and plates, while for bar and shaped profiles, grooved or shaped rolls are used.

2 CLASSIFICATION OF ROLLS

Generally, rolls are classified according to the purpose or to the specific duty they have to perform, as explained below:

2.1 Classification wrt Purpose

Classification according to the purpose, for which they are used and categorized into the following groups:

(*a*) **Rolls for Longitudinal Rolling:** These rolls are cylindrical in shape. Plain and groove rolls are used for flat and section mill rolling respectively.

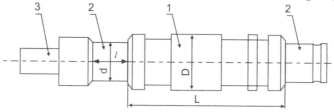

Fig. 3.2 Rolls for Longitudinal Rolling.

Rolls for longitudinal rolling comprises several elements (Fig. 3.2), 1-barrel, its diameter-D and length L; 2-neck, neck diameter-d and length-l, on both sides of barrel, mounted in the bearing and 3–tang, having a fork shape and serving to link the roll to the spindle, while other end of roll is either cylindrical or notched as a wobbler for rigging the hoisting rope, when handling with a crane or for driving the roll during redressing in a roll lathe.

(*b*) **Rolls for Cross / Spiral Rolling:** These rolls are used in the production of tubes, hollow profiles and for machine parts *viz*. barrels, disc, cone and special shapes.

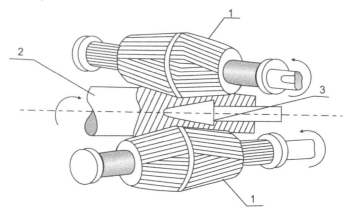

1. rolls, 2. billet, 3. mandrel

Fig. 3.3 Rolls for Tube Rolling.

2.2 Classification wrt Specific Duty

The major qualities required in rolls for rolling mill are:

(*a*) Strength/toughness: Ability to withstand shock load, including thermal shock.

(*b*) Hard Wearing: Good surface quality to ensure long life in term of tonnage rolling.

On the basis of hardness, rolls can be categorized into four types as given below:

Table 3.1

Types of Rolls	Hardness		Use	Material
	Brinnel	°Shore		
Soft	150–250	25–35°	- Primary Mill - Roughing stands of Heavy section mill	- Cast steel - Forged Carbon Steel - Grey Cast Iron
Semi Hard Rolls	250–400	35–60°	- Roughing stand of Rail & Structural Mill - Section Mill - Flat Mill - Finishing Stands of Section Mill	- Semi Chilled Iron - Cast steel - Forged steel
Hard Rolls	400–600	60–85°	- Finishing stands of strip, sheet, rail and section mill. - Back up rolls of for 4-high stand	- Chilled iron - Alloy steel
Very Hard Rolls	600–800	85–100°	Cold Rolling Mill	- Forged Alloy steel with Cr

3 ROLLS QUALITIES

Steel and iron rolls are two main categories of rolls. Their use in any particular purpose depends upon the specific duty which they have to perform. Accordingly, mechanical property of a roll is the most important for that particular application *i.e.*, toughness, resistance to thermal cracking, shock loading or wear out. The classification of rolls depends upon the carbon percentage in the roll.

3.1 Effect of Alloying Elements in Iron and Steel Rolls

The carbon content of iron or steel has an important bearing on these properties as shown in table 3.2. In general, greater the carbon content, the harder will be the roll and conversely with lower carbon content, the roll shall be more tougher and softer.

Table 3.2

Iron Rolls	Effect of Elements	Steel Rolls
Increases hardness, wear resistance brittleness and ductility of chill	Carbon	Increases hardness, brittleness and wear resistance.
Increase content of graphite, adds cleanness	Silicon	Cleanses steel, if added in proportion of 0.2–0.35, adds to hardness, deoxidizer and promotes sound casting.
Increases hardness and brittleness	Phosphorous	Increases hardness and brittleness. Decreases ductility.
Increases hardness and brittleness depths of chill.	Sulphur	Increases hardness and brittleness. Decreases ductility, must be used with discretions.
Reduce chill in lower range increase chill in higher range, increase hardness in combination with nickel and chromium, increase brittleness.	Manganese	Increases hardness and brittleness cleanser for oxides and sulphides, increases tensile strength and wear resistance.
Increase tensile strength and wear resistance, decrease depth of chill.	Nickel	Increase strength and resistance to fire cracking in combination with others.
Increase strength makes fine-grained structure; increase strength and resistance to fire cracking.	Molybdenum	Increase strength and hardness.
Hardener used in combination with Ni or Mo or both increase depth of chill.	Chromium	Hardener used in combination with Ni or Mo or both.
Similar to nickel	Copper	Similar to nickel.
Hardener used with discretion.	Boron	Increase hardness.

3.2 Effect of Casting Practice

The classification of rolls is also based upon the casting practice adopted for the manufacture of the roll *i.e.*, the form and percentage in which the free carbon is present in the structure of the roll as shown in table 3.3 below:

Table 3.3

Rolls	Steel Rolls		Steel Base Roll	Cast-Iron Rolls
Micro structure	Ferrite & Pearlite	Pearlite & Cementite	Cementite spheriodical & pearlite (*i.e.* heat treated)	Cementite or graphite + pearlite sorbite or martensite
C%	0.80		1.25	> 2.5

3.3 The distinctive line between cast iron and steel base rolls is about 2.5%. The variation in hardness and tensile strength varies directly with respect to the carbon percentage as indicated in Table 3.4 below:

Table 3.4

Types of Rolls										
C %	0.60	0.80	1.0	1.20	1.40	1.60	1.8	2.0	2.2	2.4
HARDNESS Sh ºc	34	40	41	37	42	45	46	47	47	48
Tensile strength N/mm²	74	76	77	79	74	69	66	55	50	35

3.4 Quality Classification of Rolls

Table 3.5

Rolls	Quality	Hardness ° Shore	Characteristics	Typical Application
Cast Steel	Alloy Steel (0.5–0.8%)	25–40	Good Machinability	Cogging, slabbing and Blooming Mill.
	Steel Base (0.8–1.4% C)	40–50	High tensile strength, Elongation	Roughing stand of Medium, heavy & light Section mill.
	Adamite (1.4–2.2 % C)	50–80	Resistance against fire cracking and breakages. High Bending Strength	Finishing stand of heavy flange Profile
Forged Steel	Carbon Steel (0.35–0.70%C)	25–35	High Tensile Strength	Cogging and Blooming mill.
	Alloy Steel (0.35–0.70%C)	50–60 80–90	Resistance against Fire Cracking, Breakages and spalling.	Slabbing mill, large Backup rolls of plate mill, Backup and Work roll of cold rolling mill.
	Mo. chill	60–70	Resistance against Fire Cracking, Breakages and spalling.	Large Backup rolls of plate mill, Backup and work roll of cold rolling mill.
Cast Iron	Grain Type (1.5–2.5%C)	40–50 50–60	Uniform hardness Penetration and wear resistance with tough core and neck.	Roughing stand of heavy and medium structural mill Finisher of billet mill. Intermediate of light Structural mill.
	Indefinite Chill (3.0–3.5%C)	50–60 60–70 70–80 80–90	Equal and gradual hardness penetration with high and uniform wear resistance	Intermediate and Finisher of medium And light structural Mill. Finisher of Wire Rod, Hot strip, and Merchant mill.
	Spheroidal Graphite (3.0–3.5% C)	40–50 50–60 60–70 70–80	Combination of wear resistance and high tensile strength allowing rolls to use as substitute to the steel as well as to cast iron rolls.	Roughing, intermediate and finishing rolls of heavy medium and light structural mills. Use as a finisher of Billet, Plate and sheet mill.

96   Chapter 3   Rolls, Roll Cooling and Roll Management

4 CAST IRON ROLLS

4.1 Grain Rolls

1. Grain roll has low carbon content than any other type of cast iron rolls. It is called gray cast iron, as it contains the flakes of free graphite. The structure of roll will be uniform through-out the roll and it has the property of high resistant to fire cracking. To some extend, it is self-lubricating due to the presence of free graphite in the structure. The presence of free graphite gives resistance to the scale adhesion and of the wear of thrust collars of rolls. That's why, it is always advantageous to use this type of cast iron roll in such usages, where thrust collars are used to resist end thrust during rolling.

2. **Chemical Composition:** The chemical composition of grain rolls is given below:

Table 3.6   (In %)

C	Mn	Si	Cr	Mo
1.50–2.50	0.70–1.20	0.70–1.10	0.90–1.10	0.25–0.80

3. **Mechanical Properties:** The mechanical properties of Grain rolls are given below :

Table 3.7

Tensile Strength N/mm^2	Impact Strength j/cm^2	Hardness Sh °C
590–785	1.0–3.0	40–55

To improve the quality of rolls, various alloying elements are added in Grain Rolls, so that the hardness, toughness and bending strength can be improved in such an extent. These rolls can be used for roughing and intermediate stands. When alloy additions are added in the Grain rolls, then it is called as "Alloy Grain Steel Rolls".

4. **Applications:** Grain rolls are used in:

- Medium and Heavy Section Mill.
- Intermediate and Finishing Stands of Rail and Structural Mill and Merchant Mill.
- Intermediate Stand of Bar Mill.

4.2 Clear or Definite Chill Rolls (CCCl)

4.2.1 Manufacture of clear chill rolls utilize the unique property of iron and carbon system, wherein by controlling the chemical composition and of cooling rate. The desired eutectic carbide can be obtained on the surface layer

4 Cast Iron Rolls

(upto certain desired depth).This definite chill layer will not contain any free graphite in surface layer, because of that no fall in hardness is expected in this chilled zone or layer. The chilled zone is followed by a mottled zone and then roll will have a gray core. Definite chill roll will have a hard working layer, alongwith tough and soft core to withstand the rolling load. During the casting process, neck is made in the sand and bodies are in solid chill, to get complete grey neck, without having any free cementite and bulk of carbon will present as free graphite.

Due to this structure, it has:

- Excellent wear resistance with alloying, with adequate resistance to fracture, fire cracking and spalling.
- Surface layer of white iron will be produced by rapid cooling of it through chill in the mould, which prevents the formation of free graphite. Much slower cooling in other portions of sand moulds produces gray iron structure.
- Low cost.

4.2.2 Chemical Composition

The chemical composition of Chilled Cast Iron rolls is given below:

Table 3.8 (%)

C	Si	Mn	P	S	Cr	Ni	Mo
2.90–3.60	0.20–1.00	0.20–0.60	0.20–0.40	0.055–0.40	0.20–0.40	1.80–2.20	0.30–0.40

4.2.3 Mechanical Properties

The mechanical properties of definite chill rolls are given below:

Table 3.9

Tensile Strength N/mm^2	Bending Strength N/mm^2	Impact Strength j/cm^2	Hardness Sh °C
180–250	300–400	2.0–3.0	60–75

4.2.4 Applications

Definite chill rolls are used in:

- Finishing rolls for sheets, Plate and Strip mill.
- Intermediate and Finishing rolls for small section, Bar and rod mill.

4.3 Indefinite Chill Cast Iron Roll (ICCI)

The major difference between the chilled and the indefinite chilled cast iron roll is the presence of free microscopic graphite in the structure of indefinite chilled rolls. The matrix varies from fine pearlite to bainite depending on the percentage of the alloying elements. The desired hardness is obtained by changing the carbide/graphite balance, which can be controlled by the chemistry. Indefinite chill cast iron roll does not have a clearly define depth of chill. The transition from hard surface to the soft core of roll will be gradual and there will be no mottled or transition zone, as it in definite chill cast iron rolls. With the alloying of Ni, Cu, Mo and Si in the ICCI rolls, the working surface, will no longer become complete white, but it will have a small amount of finely divided graphite, which get increase the amount and size with depth of chill with corresponding decrease in primary carbides. The core and neck of this type of roll will be of fully gray.

ICCI rolls are stronger and tougher than the clear chilled rolls. The graphite in the surface structure reduces the spalling, improves the shock-resistance and fire cracking. Gradual change of hardness in ICCI rolls makes it possible for the use in for the deep grooved rolls of structural mill. The rate of decrease of hardness from the surface to the center of roll is carefully chosen by making variation in the chemical composition and also in heat treatment. ICCI rolls are cheaper than the SG iron rolls, but, its resistance to abrasion is not very high. It is advised to avoid the violent fluctuation of temperature during rolling, when these rolls are used.

4.3.1 Chemical Composition

The chemical composition of ICCI rolls is given below:

Table 3.10 (in %)

C	Si	Mn	P	S	Cr	Ni	Mo
3.0–3.50	0.90–1.40	0.50–1.20	0.30 max.	0.10 max.	0.90–1.40	0.70–2.00	0.25–0.40

4.3.2 Mechanical Properties

The mechanical properties of ICCI rolls are given below:

Table 3.11

Tensile Strength N/mm^2	Bending Strength N/mm^2	Impact strength J/cm^2	Hardness Sh°C
200–280	350–450	2.0–4.0	55–75

4.3.3 Applications

Indefinite chill cast iron rolls are used in:

- Roughing mill rolls of Billet and Merchant Mill.
- Intermediate and finishing rolls of Wire rod, Merchant Mill and Light Structural mill.
- Edger rolls for Universal Beam Mill.

4.4 Spheroidal Graphite Rolls (SGCI)

SGCI Rolls are considered as one of the most acceptable and versatile roll material today, as its properties cover both of cast iron and of the cast steel. A combination of these properties can make it to use these rolls in roughing, intermediate and finishing group of stand. Depending upon its use, a wide range of SG iron rolls is available with pearlitic, bainitic to martensitic structure and is also available in chilled or sand cast form. The hardness of SG rolls varies from 42° to 80° shores. Here, graphite takes either the shape of spheroids or nodules, to eliminate the notch effect of flake graphite and improve mechanical properties of rolls. SGCI rolls are generally chilled cast and to get the working surface which having the cementite and spheroids of free graphite. The proportion of eutectic carbide decreases from barrel surface to core with increase in graphite nodules.

The SGCI has much greater strength and toughness than ICCI rolls, to make it to withstand the heavier draught. Presence of graphite in rolls extends self-lubrication on it and this is important when, it is subjected to heat and friction. It also increases resistance to fire cracking. Rolled surface are much better, when compared to other quality of rolls. It also provides damping effect, which helps in the prevention of spalling due to mechanical stresses and fire cracks. As graphite interrupts the continuity of matrix, silicon is added to regulate free graphite. Mn is added to neutralize the sulphur and to dioxide the metal. High Mn content act as carbide stabilizer, but too high Mn content may resist the graphitization

4.4.1 SGCI Rolls (Pearlitic)

The structure of surface is of eutectic cementite and graphite nodules in pearlitic matrix. Pearlitic structure has good thermal fatigue strength at elevated temperature.

SG Cast Iron Rolls are good substitute for many conventional steel and Cast iron rolls due to combination of superior strength, good wear resistance to spalling and of resistance to fire cracking.

4.4.1.1 Chemical Composition (%)

The chemical composition of SGCI (Pearlitic) rolls is given in Table 3.12 below:

Table 3.12 (in %)

C	Ni	Mo	Mg
3.00–3.50	1.50–2.50	0.20–0.50	0.04–0.07

100 Chapter 3 Rolls, Roll Cooling and Roll Management

4.4.1.2 Mechanical Properties

The mechanical properties of SGCI (Pearlitic) rolls are placed below:

Table 3.13

Tensile Strength N/mm²	Impact strength J/cm²	Hardness Sh°C
400–600	2.0–4.0	45–65

4.4.1.3 Applications

SGCI (Pearlitic) rolls are used in:

- Roughing and Intermediate rolls for Blooming, Slabbing and Wire Rod Mill.
- Roughing, Intermediate and Finishing rolls of heavy structural mills.
- Finishing rolls of Sheet Mill.

4.4.2 SGCI Rolls (Acicular)

The structure of SGCI (Acicular) rolls consists of graphite nodules in the matrix of bainite. It has high alloy content to get a greater strength. For a given degree of wear resistance, the acicular iron roll is tougher and has more resistance to impact strength than the pearlitic type of SGCI rolls.

4.4.2.1 Chemical Composition (%)

The chemical composition of SGCI (Acicular) rolls is given below in the Table 3.14.

Table 3.14 (in %)

C	Ni	Mo	Mg
3.20–3.50	1.80–3.00	0.40–1.00	0.04–0.07

4.4.2.2 Mechanical Properties

The mechanical properties of SGCI (Acicular) rolls are placed below :

Table 3.15

Tensile strength N/mm²	Impact Strength J/cm²	Hardness Sh°C
500–600	3.0–5.0	65–85

4.4.2.3 Applications

SGCI (acicular) rolls are generally used in for the Intermediate and finishing rolls of heavy structural mills.

4.4.3 Salient Features of SG Rolls

- Imperfect and/or inadequate spherodization may give rise to quasi flake or flake graphite in the structure. Which will thus impair the mechanical properties of SGCI *i.e.*, tensile strength, elongation, hardness and impact strength of rolls.
- SGCI rolls are extremely sensitive to thermal shock.

4.4.4 SPECIAL SG CI ROLLS (Combination of Pearlitic and Acicular)

It is a combination of both Pearlitic and Acicular SG. Spheroidal Graphites are surrounded by Ferrite (Bull's eye). It gives the enhanced resistance to fire cracking and thermal failure. Special heat treatment is given to achieve the desired mechanical properties.

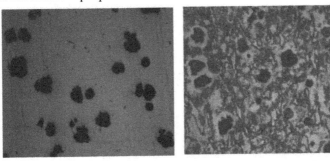

Unetched   Etched X 50

Fig. 3.4 Microstructure of Special SGCI Rolls.

Chemical Composition: The chemical composition of special SGCI rolls is given below in the Table 3.16.

Table 3.16

C	Si	Mn	S	P	Ni	Cr	Mo
2.80–3.50	1.40–2.30	0.50–0.65	< 0.025	< 0.10	2.00–3.50	< 0.30	0.50–1.00

Mechanical Properties: The mechanical properties of special SGCI rolls are placed below:

Table 3.17

Hardness	Tensile Strength
45–55 °shore	450–600 N/mm^2

Special SGCI rolls are generally used in the roughing and intermediate stand of Bar and Rod mill. It is used in all positions in medium and heavy structural mill.

4.5 Double Poured or Composite Iron Rolls

Endeavour to obtain rolls with higher wear resistance and toughness led to the development of double pour rolls. Such rolls are generally composite rolls of having hard shell and tough core. It is possible to obtain bainitic or martensitic structure for high wear resistance. Rolls will have hard-alloyed iron on surface and soft gray iron or SG iron core, depending upon its applications. Shell matrix with proper distribution of carbides and graphite imparts high wear resistance and good surface finish. Softer core gives good mechanical properties and resistance to impact loading.

4.5.1 Chemical Composition

The chemical composition of composite iron rolls is given below :

Table 3.18: Shell   (in %)

C	Ni	Mo	Cr
3.15–3.45	3.80–4.50	0.30–0.50	1.40–1.90

Core

C	Ni	Mn	Mg
3.00–3.50	1.50–2.50	0.45–0.75	0.04–0.07

4.5.2 Mechanical Properties

Table 3.19

Barrel Hardness °Shore	Tensile strength N/mm²	Impact Strength J/cm²	Neck Hardness Sh°C
65-85	200-500	2.0 MAX.	35-45

4.5.3 Applications

- Hot strip mill, Plate mill.
- First stand of tandem cold mill.
- Finishing stand of Wire rod mill.

5 STEEL ROLLS

Steel rolls may be either cast or forged. They are much stronger and tougher than the iron rolls. Mill can have higher draught with steel rolls, as it also gives a better biting of stock due to higher friction of its rough surface. Resistance to fire cracking is an added advantage of using steel rolls.

5 Steel Rolls

5.1 Forged Steel Rolls

Forged steel roll are used, where rolls are subjected to sudden shock loading and where higher degree of surface finish of finished product is desired. Forged steel rolls are classified into carbon steel forged rolls and alloy steel forged rolls. The carbon content in forged roll varies over a range from 0.35-0.80 %.

Table 3.20

Type of Forged Steel	Symbol	Hardness ° Shore	Characteristics	Typical Applications
Carbon Steel	FCS–30	25–35	High Tensile Strength	Cogging, Blooming mill.
Alloy Steel	FCS–50	50–60	Resistance against fire Cracking.	Slabbing mill, large Backup rolls of plate mill.
	FCS–80	80–90	Resistance against, Breakage, Spalling.	Backup rolls and Work roll of cold Rolling mill.

5.2 Cast Steel

Cast steel rolls are classified into alloy cast steel, steel Base and of adamite quality, which generally depends upon the percentage of carbon and alloying elements. Addition of alloying elements increases resistance to fire cracking and wear resistance of cast steel rolls.

Table 3.21

Type of Cast steel Rolls(%C)	Symbol	Hardness ° Shore	Characteristics	Typical Applications
Alloy Steel (0.5–.8%)	CSB – 30	25–40	- Good Machinability	Cogging, Slabbing, and Blooming Mill.
Steel base (0.8–1.4%)	CSB – 40	40–50	- High tensile strength - Elongation	Roughing stand of Medium heavy & Section mill.
Adamite (1.4–2.2)	CSB – 50	50–80	- Resistance against fire Cracking and breakages. - High Bending Strength	Finishing stand of heavy flange Profile

Table 3.22   (in %)

C	Mn	Si	Cr	Mo	W	V	Co
1.5–3.5	0.3–1.5	0.3-3.0	< 5	<9	< 20	3.0–15.0	<10

Hardness –75–90 °Shore

5.2.1 *Applications*

HSS rolls are generally used in:

- Roughing, intermediate and finishing train of hot strip mill.
- Finishing stand of universal mill for Rail and Structural rolling.
- Intermediate stand of Bar and Rod Mill.

6 CEMENTITE CARBIDE ROLL (RINGS)

6.1 An effort to sort out the problem of fast wear-out of rolls of "No twist block" stands of high speed bar and rod mill. The use of HSS rolls was proved to an intermediate step and ultimately the final answer comes from the use of "Cemented carbide roll or rings". The application of these rolls in hot rolling bar and rod mills are permitted the rapid growth in the output rates of these mills. These rolls give a life 10 to 15 times higher, compare to the cast iron rolls.

6.2 It is formed by bonding of Tungsten carbide (WC), which act as a hard face and cobalt, which act as a binder in proportion of 70–90% to 10-30% respectively. It has got the basic cementite/carbide structure. It may contain varying proportion of alloying elements to form different phases of cementite carbides *i.e.*, the basic WC form the á phase, Cr and Ni forms the β phase and γ phase contains Tic (Titanium carbide),Tantalum carbide Tac (Tantalum carbide) and Niobium Carbide Nbc (Niobium Carbide). These additions will help in improving the mechanical properties of the roll.

6.3 Cemented carbide roll withstands heavy deformation, high impact load, high compressive and tensile strength, high corrosion resistance and high wear resistance. In other words, it has got the unique combination of strength, hardness and toughness.

6.4 Applications

Cementite carbide rolls are used for:

- Bar and Rod mill intermediate and finishing train.
- Merchant mill–rounds, rebars, shapes.
- Pinch rolls–guide rolls.
- Roller entry guide.

7 SLEEVE ROLL

Sleeve rolls are used when it is desired to have high wear resistance and requirement for resistance to roll breakage, in a single roll. These requirements are conflicting to each other. It is not possible to have both properties in a single roll.

With composite sleeve rolls, it is possible to get wear resistance and breakage resistance both in a single roll. Simultaneously, it is by choosing high hardness material for sleeve and high toughness material for the core. When sleeve get worn out, then only sleeve will be replaced and the core (Arbor) will be again reused. With this, the roll cost will get reduced.

These rolls are mainly produced by three ways of manufacturing

- Centrifugal Casting.
- Forging.
- Sintered Alloying.

8 SELECTION OF ROLLS FOR DIFFERENT CATEGORY OF ROLLING MILLS

8.1 Blooming and Slabbing Mills

Main problem faced, while for the selection of primary mill roll is the breakage of rolls. Rolls of these mills are subjected to the very high "Roll separating force" and of large alternating bending stresses, as there is no provision of backup roll in primary mills. Due to above reasons, forged steel rolls are only used in the primary mill.

While reducing ingot in to the slab in Slabbing mill, the same portion of roll is subjected again and again to the deformation, causing the development of fire cracks in the rolls. On the contrary, in case of Blooming mill, the heating zone keeps on changing as deformation keeps on shifting position on barrel length during the course of rolling.

Addition of Molybdenum in forged rolls will certainly increase resistance to fire cracking. It is to note that a better roll cooling system of roll in primary mill helps in reducing the fire cracking on rolls, which is always subjected to severe thermal stresses.

It is always advisable in primary mill, to give a higher reduction at start of rolling of ingot, when metal is more malleable and resistance to the deformation is less. In later stages of rolling, heat added by the resistance to deformation makes rapid heat up of roll locally and in turn decreases the strength of roll.

To give more draught at early stages of rolling and to improve angle of bite, knurling or ragging on roll surface of rolls is generally recommended for Blooming and Slabbing mill.

Knurling reduces the possibility and propagation of circumferential cracks, compare to ragging of roll surface. Knurling is also helpful in effective

106 Chapter 3 Rolls, Roll Cooling and Roll Management

descaling of slabs for subsequent operation. In addition, it also decrease slipping of work pieces in rolls, because of increase in the co-efficient of friction between roll and slabs.

8.2 Structural Mill

In the structural rolling, the aim of roll pass designer is to give reduction as high as possible in earlier forming passes, as non-uniformity in reduction existed at this stage due to the difference in the shape of input material and of the desired finished Product. In early stage of reduction, higher reduction is recommended, when steel is more malleable due to higher temperature. The tendency towards twisting is less pronounced because of greater cross sectional area of the input material. Reduction, at this stage is only limited by the strength of rolls and of the angle of bite. Because, of these reasons only, the rolls of roughing group of structural mill should be either of forged or cast steel (with hardness ranges from 32°-40° shore).

In intermediate group of stands, generally roll becomes weak, as rough shape of flange profile has already taken place and flanges of profile get protrude inside the roll, criteria at this juncture for roll pass designer are to recommend such a roll, which will have a combination of toughness and wear resistance. Spheroidal cast iron rolls are generally preferred for these reasons. Sometimes, alloy cast steel or steel base roll are also used in the intermediate group of stands, as the cost of SGCI roll are high.

Finishing group of stands, where the wear resistance is the only most important criteria for the selection of rolls, there indefinite chill cast iron roll of hardness over 55° shores are generally preferred because of low cost.

Table 3.23 Permissible Limit of Diameters for Scrapping Rolls in a Structural Mill

(a) Roughing Group (Stand 1-5)

	Stand No	I	II	III	IV	V
Rounds	D_{Max}	535	535	535	462	535
	D_{Min}	490	490	490	392	490
	Rounds B.P.	45	45	45	70	45
	Material	CS	CS	CS	CS/CI	CS/CI
Angles	D_{Max}	535	535	535	462	535
	D_{Min}	490	490	535	462	535
	Angles B.P.	45	45	45	70	45
	Material	CS	CS	CS	CS/CI	CS
Beams	D_{Max}	540	540	540	–	340
	D_{Min}	490	490	490		490
	Beams B.P.	50	30	50		50
	Material	CS	CS	CS		CS

Contd...

9 Proper Care of Rolls

Channels	D_{Max}	535	540	540	–	540
	D_{Min}	490	490	490		490
	Channels B.P.	45	50	50		50
	Material	CS	CS	CS		CS

(b) Intermediate and Finishing Group (Std 6-12)

	Stand No	VI	VII	VIII	IX	X	XI	XII
Rounds	D_{Max}	425	460	425	425	425	325	325
	D_{Min}	395	390	390	390	390	350	350
	Rounds B.P.	30	70	35	35	35	25	25
	Material	CS	CI	CI/CS	CS/CG	CI	CI	CI
Angles	D_{Max}	425	460	425	425	425	–	390
	D_{Min}	395	390	390	390	390		350
	Angles B.P.	30	70	35	35	35		40
	Material	SG	CI	SG	CI	CI		CI
Beams	D_{Max}	440	–	440	440	440	–	390
	D_{Min}	400		400	400	400		360
	Beams B.P.	40		40	40	40		30
	Material	FS/SB		FS/SB	SG	CI		CI
Channels	D_{Max}	440	–	440	440	440	–	385
	D_{Min}	400		400	400	400		360
	Channels B.P.	40		40	40	40		25
	Material	FS/SB		SB	SG	CI		CI
Max. Pass width Allowed					+2.0	+2.0		

NOTE: 1. FS – Forged Steel, 2. SB – Steel Base, 3. SG CI –Spheroidal Graphite Alloy Cast Iron, 4. CS – Cast steel, 5. B.P.-Below Paper size *i.e.*, how much redressing can be taken up.

9 PROPER CARE OF ROLLS

Roll is the most expensive items and has the major contribution towards the cost of production in a rolling mill. It is very important that roll should be use very carefully in rolling to reduce wear and breakages.

Roll breakage, not only means of the full loss of the expensive rolls, but may also damage to guides, rest bars, chocks and to the driving system of stand also. In addition there will be a huge loss of production due to stoppage of mills.

To avoid above mentioned losses, rolls should be properly designed by the roll pass designer and should used carefully. Roll cooling system should also be made very effective during structural rolling. Following precautions are to be taken:

9.1 Care During Design

1. *Neck fillet Radius*: Neck radius should be very generous to avoid the stress concentration at any particular point of the roll.
2. *Pass fillet radius*: In case of angle rolling, finishing pass should be given a fillet radius 1 to 1.5 mm to avoid the breakage of roll due to sharp corners, otherwise it will become a place of stress concentration.
3. In rolling beam and channel, the sharp side of closed flange pass should be given a radius of 1 to 1.5 mm to avoid breakage at that juncture.
4. In beam rolling, if roll diameter of one roll at dead flange become very small due to flange height in close pass, then consideration may be given for having a semi-closed type of pass, in place of closed pass, as shown in figure 3.5 below:

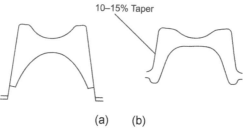

Fig. 3.5 Closed (a) and semi closed pass(b).

5. The roll pass design can play a major role in avoiding excessive loading of rolls. The excessive loading may be due to slower rolling rate, which makes the bar cold due to more rolling cycle time. If rolling temperature of the rolling at that particular stand gets falls from the normal 1200°C to 1000°C, then it is seen that rolling load for the same draft, will become double.
6. Design change of the wobblers of rolls can also help into reduce the breakage of wobblers.
7. If roll pass design does not help to resolve the problem of the overloading of rolls and the problem of roll breakages still persists, then roll material of that particular stand should be changed with the roll material of higher tensile strength and higher elongation. Though, by this exercise, we may sacrifice a bit of wearing properties of rolls, but it will certainly help to reduce the risk of breakages.
8. The quality of rolls of the intermediate group of the structural mill is very important and as far as possible, the same quality of rolls should be provided in the intermediate group. If inferior quality roll is used in the preceding stand, then due to fast wear out of preceding roll, it may give

9 Proper Care of Rolls

overload in the succeeding stand. This also is the reasons for combining pass changing of the two consecutive stands. Pass condition of the each stand should be checked in the beginning of the each shift.

9.2 Proper use During Rolling

1. Damage of rolls can be avoided by using rolls with utmost care and by avoiding mishandling during transportation, especially for cast iron rolls. It is the fact that due to continuous rolling, roll gets weaker due to a process, called fatigue *i.e.*, by every revolution of roll during rolling the stresses get reversed and continual reversal of stresses lead to fatigue and fire cracks. In other words, it formed due to alternate heating and cooling cycles. If by chance, cooling water stops during rolling, then roll should be allowed to rotate for sometimes and speed of roll should be dropped down slowly, only after that water should be opened otherwise roll will get cracked.

2. Proper care should also be taken for roll diameter also, once it crosses the specified limit of minimum diameter, it should not be used.

9.3 Care of Roll by Mill Crew

1. The driving coupling should be properly maintained and should be matched with the contour of roll wobbler. If coupling does not drive the roll on its all faces, then the coupling or roll wobblers will get damaged.

2. The roll bearing and chocks should be properly maintained and if they are not properly assembled and there is a play, then it will jump and shift with rolling of every bar, causing shock load on rolls.

3. The shape of bearings should be matched with shape of the roll neck, especially with taper neck roller bearing. The inner races of roller bearings and diameter of journals should be properly matched. In case of fabric bearing, water circulation should be optimized; otherwise bearing gets burnt and may damage to roll neck.

4. The alignment between driving coupling and roll neck should be made accurate to avoid the problem of breakages of rolls and wobblers.

9.4 Care to Avoid Over Loading

The over loading may be due to:

(*a*) Entry of double bar, a foreign body, split end bar.

(*b*) Entry of bars at places other than pass, like collars.

(*c*) Entry of cold bar to roll especially because of wrong judgement, particularly during the night time.

(*d*) Roll collaring: This may not only damage or break rolls but often break other mill equipment also.

(*e*) Incorrect turning of pass.

(*f*) Abnormal setting of pass.

(*g*) Insufficient roll neck cooling.

(*h*) Guide displacement.

9.5 Other Precautions

Which are required to be taken during rolling to avoid the damage of rolls are appended below:

(*a*) Preheat roll by idle running of mill, before the start of rolling, otherwise roll will get crack. Overheating of roll can be avoided by establishing adequate cooling measures.

(*b*) Sudden change of rolling speed during rolling is not at all desirable, as it causes temperature fluctuations on the surface of rolls.

(*c*) When rolling is to be stopped, cooling should not be stopped suddenly. Rolls should revolve slowly, so that it can cool evenly on the surface of rolls.

(*d*) Rolling of excessive tonnage from the specified tonnage should be avoided.

(*e*) Improper handling of rolls may cause damage of the roll.

(*f*) Water cooling nozzles cleaning on daily basis.

(*g*) Water cooling headers cleaning once in a week.

(*h*) Power tripping/power failure should be nil.

10.0 PROPER COOLING OF ROLLS

10.1 Methodology of Wear-Out

1. During hot rolling, high temperature of the stock and the heat of deformation raises the temperature of the rolls in the contact zone (Fig. 3.6) adversely affects the tribological conditions of the working surface.

10 Proper Cooling of Rolls

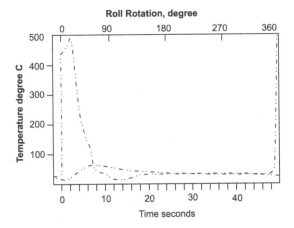

Fig. 3.6 Variation of roll temperature during the rolling of first slab through the stand.

Roll surface layer is subjected to cyclic temperature variation, which leads to development of tensile and compressive stresses on the surface.

2. When the stock leaves the contact zone and air/water cooling of roll surfaces commences, temperature of the surface gets lowered. The cyclic variation of stresses, due to the thermal fatigue, a network (Fig. 3.7) leads to the formation of a network of micro cracks, which is known as 'Fire cracks' and flakes of metal get torn-off under high pressure and temperature variations.

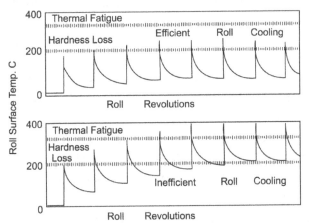

Fig. 3.7 Roll surface temperature cycle and comparison of characteristics- Efficient vis-à-vis Inefficient Roll Cooling.

3. Even if roll temperature is insufficient to cause thermal fatigue, higher roll temperature decreases the hardness of the rolls, resulting in excessive abrasive wear. In view of this it is imperative to retard the flow of heat in the rolls by effective and efficient roll cooling techniques.

4. Oxidation of the surface layer also takes place due to high temperature.

10.2 Factors Affecting Wear-Out of Rolls

(*a*) Surface Condition of Roll.

(*b*) Hardness and Toughness of Roll.

(*c*) Chemical Composition and Structure of Rolls and metal to be rolled.

10.3 What is Proper Roll Cooling

1. The roll cooling should be such that the surface temperature of rolls should be around 80°C. It means that, at any time during rolling one can touch rolls with bare hands.

2. As the bar touches the rolls, at that contact surface of rolls, the temperature get immediately rises to that of bar temperature, but it stands only for the split of second. Therefore, water, which comes in contact with surface of the roll, is not able to do the cooling, as a layer of steam is formed on the roll surface. Problem of water-cooling has become more acute in case of slow rolling, since bar remains in contacts with the rolls for a longer period of time, compared to the high speed rolling.

3. Hence, to start a proper cooling, a high pressure jet of water should be impinged on the rolls immediately at the point where bar leaves the rolls *i.e.*, at the exit side of rolls. This will take away the steam formed on the surface of rolls and by that surface of roll get exposed to the cooling.

4. The cooling of the bottom roll is very important. Water jet has to work against the gravity in case of bottom roll, and because to this, water is not able to reach to the full peripheral surface of the roll, contrary to top roll. Proper care should be taken during the rolling process for checking regularly for water jets installed for bottom rolls, which generally gets blocked by scales generated during rolling process.

5. Cooling water connection to the lower roll should be so set, so that water will not come into the entry side of the pass, as slightly cold front will then lead to the more spread and bar will not able to enter in to the next stand, causing cobbles.

6. Whatever effective is the cooling process, the surface temperature of rolls gets fluctuates from 1000°C to 80°C during the every rotation of roll and by this phenomenon, fire cracks get develops on the roll surface. If these fire cracks will be allowed to go deep inside the rolls, then it will become the cause of roll breakage. The Fig. 3.8, shows that how an effective rolls cooling arrangement is required for a hot rolling mill.

10 Proper Cooling of Rolls

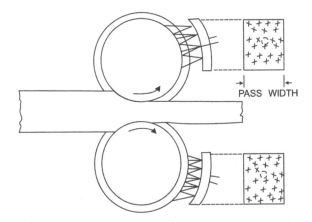

Fig. 3.8 Effective Roll Cooling System for Hot Rolling Mill.

7. Roll should not be suddenly get heated or cooled. It is a general practice to run the roll idle for sometimes before the start of rolling to avoid the sudden heating or cooling of the rolls and *vice versa*, not to stop just after the rolling.

10.4 Conventional Roll Cooling

The conventional cooling of work rolls reduces its life and causes production delays due to the frequent roll changes. It also results in production of the poor quality finished products, due to uneven wear out of rolls due to poor roll cooling. It is seen that generally in long products rolling, roll passes are invariably cooled with the help of water hoses, may be placed at some convenient place towards the exit side of the pass, whereas in the flat product mills, roll cooling is carried out by perforated pipes/nozzles on the entry as well as exit side without much of consideration for location and water distribution.

10.5 Effective Roll Cooling System

To have the effective use of water and to avoid frequent roll changes and breakages, the roll cooling system should be designed upon the principle of rapid heat extraction from the roll surface. Cooling should start immediately after the hot rolled metal loses its contact with the roll. Cooling system should be covered to the all segments for the shaped profile. Spray density and water pressure should be selected such to suite the pass configurations, rolling rate, roll material and stock temperature. Significant improvement in roll life has seen in the wide hot strip mill by only optimizing the distribution and proper orientation of nozzles and water pressure.

In structural mills, wear of the rolls on side walls of the passes plays an important role in achieving dimensional tolerances as well as to regain the pass profile after roll turning. Therefore, cooling headers should take care of effective cooling of full pass width as shown in Fig. 3.8. Cooling of rolls from entry side is not generally recommended for mills producing long products. Roll cooling headers for high speed mills like Bar and Rod, Wire Rod and Merchant Mill, should not only cover the full width of the pass, but also cover the maximum possible arc of the rolls on exit side. However, in case of hot strip mills, the cooling should be carried out from the exit as well as from the entry side also, due to more heat input to the rolls given by the rolled stock as shown in Fig. 3.9. But it should be noted that water on exit side should be substantially more than the entry side that for the effective roll cooling.

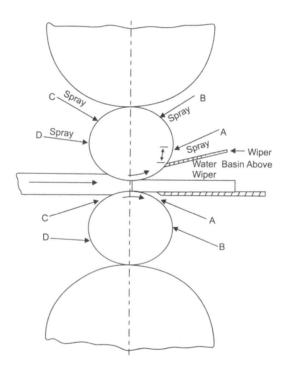

Fig. 3.9 Effective roll cooling for Hot Strip Mill.

10.6 Roll Cooling Parameter

Details of roll cooling parameters for each stand of different mills are given in the Table 3.24 below.

10 Proper Cooling of Rolls   115

Table 3.24 Distribution of Water for Effective Cooling in Different Mills

Type of mill	Type of Cooling	Water flow rate (m³/hr)	Water pressure (kg/cm²)
Heavy Structural and rail mill	Segmental 70° on exit side	6.0–10.0	3.5–5.0
Wire rod and Bar mill	Segmental 90°–100° on exit side	1.0–2.0	2.5–4.0
Medium section mill	Segmental 70°–80° on exit side	5.0–7.0	2.5–4.0
Narrow hot strip mill	Concentrated water jets	15.0–20.0	5.0–7.0
Wide hot strip mill	Concentrated water jets	400–750	5.0–7.0

10.7 Basic Requirements of Effective Roll Cooling System

Following are the basic requirement for effective roll cooling.

- Flow rate,
- Pressure,
- Temperature,
- Design of roll cooling pipe,
- Quality of water,
- Mill speed,
- Control valves,
- Mill scale removal,

10.7.1 Flow Rate

Recommended flow rate for roll cooling is shown in above Table 4.24.

Water consumption in a rolling mill depends upon water circulation rate, atmospheric temperature and relative humidity, depending upon these factors. It varies between 1 to 3% of the circulation rate.

Example: For a bar mill of say 400,000 tons per annum. Water circulation rate is expected to be 650 m³/hr and the evaporation loss shall be about 13 m³/hour. That means mill needs about 260 m³ of makeup water per day.

10.7.2 Water Pressure

Recommended water pressure for roll cooling is shown in Table 4.24. Water pressure depends upon quality of roll. It is desired to have double the pressure in case of tungsten carbide rolls.

10.7.3 Water Temperature

Recommended water temperature for roll cooling at the point of application shall not exceed 80° C.

Fig. 3.10 Steam Formation during Rolling.

When hot metal (at a rolling temperature exceeding 1000° C) comes in contact with roll surface then there is a direct transfer of heat by conduction.

But when cooling water temperature is higher, its adverse effect will be seen. It will have a greater tendency of reaching the boiling point and of formation of steam layer. This steam acts as an insulating layer between roll surface and the water stream. The dispitation of heat from the roll surface is thus badly affected.

10.7.4 Design of cooling pipe

Followings points are to be noted, while designing the cooling pipe:
- Pressure Gauge on supply header.
- Maximum flow at delivery point.
- Fixing the pipe on delivery guide box will automatically sets the alignment.
- Spray nozzle to deliver 30% of the water on the groove.

10.7.5 Quality of Water for Roll Cooling

Following quality norms to be adhered, while maintaining the quality of water:
- Total suspended solids (TSS), should not exceeds 200 mg/l for normal rolls and 80 mg/l for tungsten carbide rolls.
- Maximum particle size at application should not exceed 200 micron.
- Total dissolved solids (TDS) should not exceeds 1500 mg/l.
- Oil and grease content should not exceeds 40 mg/l.
- pH of water should be between 8 to 8.5 in case of Tungsten carbide rolls.
- Install oil skimmer in the sedimentation tank so that oil/grease do not clog the fills and spray nozzles.

10 Proper Cooling of Rolls

10.7.6 Peripheral Speed of Roll

In no case, pheripheral speed should go below 0.2 m/sec, as due to longer period of contact of metal with rolls, fire crack get developed as shown in Fig. 3.11.

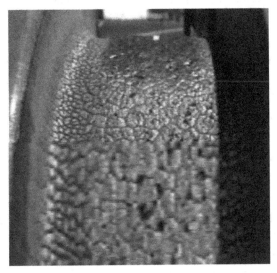

Fig. 3.11 Fire crack Formation.

10.8 Use of Additives and Effective Roll Cooling System for Improvement of Roll Life

1. Advantages derived with the introduction of roll lubrication technology, alongwith effective Roll cooling system are:
   - Improvement in the surface quality of product.
   - Improvement in the dimensional tolerance of finished product.
   - Improvement in the operational environment in the mill due to less pass/roll changes.
   - Mill loading also get increased.
2. The roll pass lubrication system is unlikely to give positive impact, in the absence of efficient roll pass cooling; rather it will have the detrimental effect. The efficient roll cooling system will cause the following:
   - The roll temperature should not exceed beyond 80°C.
   - Roll cooling headers designed to cool the pass groove, to be set as close at the exit of the pass and also to be kept with proper orientation and distance from the roll.

- Cooling water pressure should be between 2.5-4.0 kg/cm^2 with water flow rate of 5-7 M^3/hr in section mill, having rolling speed <10 m/sec. It keeps the temperature of the roll within the desired limit by controlling the burning across the roll groove, generation of fire cracks; thermal shocks. It will also break the oxide film which formed on the surface of roll.

3. Another aspect which needs attention while designing of a roll lubrication system is the logic to control duration of lubricant application. The continuous application of roll lubrication, during when there is no bar in the pass, may lead to biting problem due to loss of friction co-efficient across of roll groove. It will also lead to wastage of lubricant oil. Automation of system through the monitoring of current loading of drive motor is not feasible because of the generation of fault signal generation will be very high. Therefore, photo sensors should be installed to sense the presence of hot bar.

4. It also needs utmost care in deciding the concentration of lubricant in the emulsion. Lubricant is chosen on the basis of their ability to burn off during the pause. But in case of high speed of rolling mill, the time gap between the stoppage of emulsion with exit of bar and energisation of lubricant application with biting of next bar is very small. The lubricant gets very little time to burn off or to get washed away. It leads to burning of lubricants, when it comes in contact with the hot bar. Roll surface then becomes black and it gets roughen in very short time. Thus, it adversely affects the roll life. Therefore, use of lubricants should begin with oil concentration should be as low as possible.

11 ROLL MANAGEMENT

As roll is the most costly material in rolling, the efficient management system helps in reduction in roll inventory and efficient campaign planning of section to be rolled. In addition it also helps in regular monitoring and to take better purchase decision. It includes:
- Estimate the Requirement of rolls,
- Procurement of rolls,
- Receipt of Rolls,
- Stacking of Rolls–Mill wise,
- Turning of Rolls,
- Supply Rolls to Mills,

11 Roll Management

- Redressing of Rolls and its resupply,
- Claim Settlement,
- Reclamation/Salvaging of Rolls,
- Scrapping/ Recycling the Rolls for sale.

11.1 Objective of Roll Management

- To supply suitable roll to rolling mill to produce good quality rolled product,
- To guarantee supply of safe roll to the mill avoiding cracks, spalling,
- To minimize consumption of roll through minimization of dressings,
- To choose most suitable quality of roll for different application,
- To minimize storing of rolls in roll shop,
- To make required modification to the roll drawing in order to minimize production cost,
- To achieve reduction of roll cost target.

11.2 Roll Inventory

The minimum number of three sets is required for normal rolling. One in rolling, one hot spare and one in redressing. Since they are critical items, judicious planning and procurement method is to be employed for continuous rolling. For this future-rolling plan is a must so that rolls are available on time.

11.3 Proper Layout of the Shop

Objective of Roll Shop is as follows:

- To supply suitable roll to rolling mill to produce good quality rolled product as concerning profile and surface aspect.
- to guarantee supply of safe roll to the mill avoiding cracks, spalling and so on.
- to minimize consumption of roll through minimization of dressing.
- to choose most suitable quality of roll for different application.
- to minimize storing of rolls in roll shop.
- to make required modification to the roll drawing in order to minimize production cost.
- to achieve reduction of roll cost target.

To achieve above mentioned objectives, the ideal layout should be such that rolls can be supplied and retrieved from mill at a short notice. A typical layout is shown below:

120 Chapter 3 Rolls, Roll Cooling and Roll Management

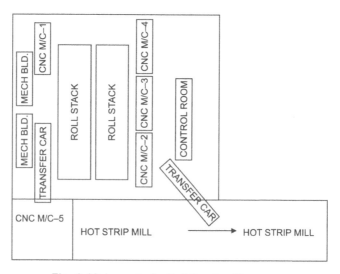

Fig. 3.12 Layout of a Roll Turning Shop.

11.4 Investigation of Roll Failures

Fig. 3.13 Pictorial view of Roll Failure.

Following steps are to be followed up during the investigation

1. Examination of fracture surface

Fig. 3.14 Roll Failure–Brittle Fracture.

11 Roll Management

Fig. 3.15 Roll Failure–Fatigue Fracture.

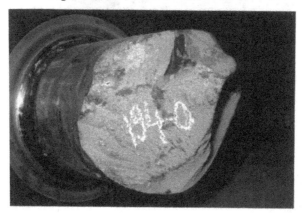

Fig. 3.16. Roll Failure– Torsional Stress.

Fig. 3.17 Roll Failure–Crystalline features.

122   Chapter 3   Rolls, Roll Cooling and Roll Management

Fig. 3.18 Roll Failure-fatigue failure having ratchet marks and brittle fracture.

Investigation of roll failure should be conducted at the site only. It should be done just after the roll breakage. Delay in the examination of fracture surface may lead to the loss of useful information. Aim of investigation is–

(*a*) Whether the roll failed prematurely due to fatigue cracks with ductile fracture or failed in a brittle manner.

(*b*) Whether fracture is coarse or fine.

(*c*) Whether fire cracks are present over the surface and if present, then what is the depth of it.

(*d*) The point at which fatigue cracks get originated.

2. Metallurgical Investigations

Once the fracture surface has been examined, a metallurgical investigation of a roll failure is to be conducted to collect data of fractures such as:

(*a*) History of rolls.

(*b*) What types of heat treatment given to roll before use.

(*c*) Nos. of redressing of roll and its diameter.

(*d*) Amount of draught given, rolling temperature at that time of failure, composition of roll and its hardness.

It is also to know chill depth and uniformity of chill, if it is made of chilled cast iron.

3. Roll investigator has to compare this collected information with the available past data.

4. Photograph of fracture surface should also be taken for investigation report.

5. Sample of fracture portion of rolls has to be collected to assess the microstructure investigation and the hardness of rolls.
6. The establishment of the cause of failure is utmost important, after going through all above mentioned procedure for the investigation of roll failure.
7. Finally to take corrective actions for future rolling and from manufacturer.

12 RECLAMATION OF ROLLS

12.1 Automatic Submerged Arc Welding

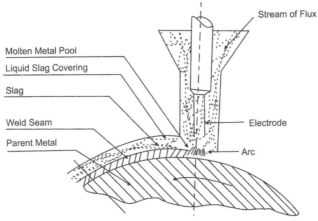

Fig. 3.19 Submerged Arc Welding.

Automatic sub-merged arc welding is used for following:

(*a*) to build up of worn out portion of part

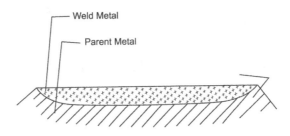

Fig. 3.20 Welding of Simple Shape.

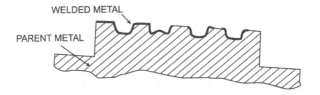

Fig. 3.21 Welding of Complex Shapes.

(b) Reclamation of old and manufacture of new part with improved surface properties such as corrosion resistance, acid resistance, thermal resistance etc. by deposition of metal

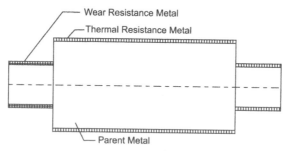

Fig. 3.22 Welding to improve surface properties at different parts.

(c) Rational selection of the blank required to make a part with a view to economize the material and machine hours.

Submerged arc welding is essentially an arc welding process. The working area is shielded by a steam of granular flux. The electrode may be bare or Cu coated and in the form of coils. It is fed into the blanket of flux; the rate of feed is controlled automatically to give correct arc length. DC current supplied by a welding generator to produce the hidden or submerged arc between the electrode and article. Heat generates melts the electrode and base metal. The flux adjacent to the arc gets melted and will float on the surface of the molten metal. This thin film of liquid flux protects the molten metal from the contact of air and also prevents the splashing. Steel rolls of less than 0.8% C content are usually considered for the building up. It is not advisable to build up the full roll to the original size, since thick layer of weld will induce severe internal stresses leading to the formation of cracks. The passes which are built up for the first time, should be first turned in a roll turning lathe and in such a way that all wear out marks and fire cracks get removed from the roll. The thickness of deposited metal must be of a min. thickness of 5 mm at any portion of the working surface of roll after the finish of the turning. For reconditioning, deposit should be made sufficient to restore the dimension of pass. Generally, working passes are built up repeatedly upto 10 times. Further building up of metal is restricted due to the fatigue of the parent metal of the roll. If there are also a deep circular cracks on the surface of the roll or on the roll neck, then welding is not advisable.

The weld building up technology for every work roll should be selected in such a way to ensure the perfect formation of every deposited weld.

12.2 Factors Affecting the Building up of Roll

1. The size of electrode.
2. The RPM of the work roll.
3. The current and the electrode feed speed.
4. The arc voltage.
5. The off set of electrode from the centre and pitch of the built up welds.

12.3 Pre-heating of Rolls

Pre-heating of rolls is to be done at 250°C–400°C depending upon carbon content and size of the roll. Preheating of roll, before welding is required for the advice due to following reasons:

1. To remove internal stresses generated in it as a result of its service in the mill.
2. To avoid martensitic structure which in prone to cracking due to rapid quenching.
3. To reduce the magnitude of shrinkage cracks.

12.4 Post Heating or Stress Relieving of Rolls

Stress relieving is done by heating below critical temperature and then allows it to slow cooling. It is desired for following reasons:

1. Thermal stress–due to temperature change during welding.
2. High internal stress–due to contraction of the weld metal.
3. Hydrogen embrittlement.

12.5 Process of Arcing

An arc is struck between the roll and electrode (normally the wire) when they are connected in an electrical circuit. The arc melts both the wire and to that parts of rolls. The stream of flux, forming a thick layer on the welding surface, protects the metal from the harmful effects of air, maintains the heat of the arc and also prevents splashing of the molten metal. The arc melts and the flux and creates around itself a void encircled by elastic film of liquid flux and thus protects the molten metal. The film of the molten flux does not prevent escaping of the entrapped gases.

12.6 Welding Machine

For mechanized welding, any welding apparatus having mechanical feed of the electrode is used. The most commonly used welding apparatus is which has a constant speed of electrode feed, irrespective of the voltage. This apparatus can be installed on any discarded turning lathe or on any other machine, capable of rotating the roll to be welded at the given speed.

Fig. 3.23 Welding of Rolls.

12.7 Welding Head

They are categorized into three main groups as follows :

1. Carbon steel wires.
2. Alloy steel wires.
3. Highly alloyed steel wires.

12.7.1 Types of Wires

1. CCMS (Copper coated mild steel)
2. Alloy steel wire.

12.7.2 Chemical Composition

Chemical composition of alloy steel wires is as follows:

C-0.25 to 0.35%, Mn-0.8 to 1.1%, Si-0.9 to 1.2%, Cr-0.8 to 1.1%, Ni-0.3%, S-0.25%, P-0.03%.

12.8 Fluxes

Flux is one of the important elements of the welding process; which determine the quality of the weld metal and of the welded seam to a large extent.

12.8.1 Functions of Flux

- Protect the molten metal from the undesirable effects of the atmospheric gases like oxygen, nitrogen etc.
- Ensure good formation of welded head and the stability of the arc.
- Prevent splashing of the molten metal.
- Ensure easy separation of the slag incrustation from the weld.
- Slow down the crystallization of the molten metal and to provide the better condition for the escape of gases from the molten metal.

12 Reclamation of Rolls

12.8.2 Composition of the Flux

Composition of the flux should be such that it should ensure easy removability of slag and to form the uniform shape of the deposited layer. Generally neutral flux is used, alongwith alloyed electrode.

12.8.3 Type of Flux

The fluxes are categorized into two main types, depending upon the method of manufacture:

(a) Fused fluxes–are prepared by melting the charge in the furnace, followed by granulation by pouring in water. Mostly flux used are fused flux. The chemical composition of fused flux is given below:

Si- 41 to 43%, Ca 0-5.5%, Mn 0-34.5 to 37.5%, Mg 0-5.5 to 7.5%, Clay- 3%, Ca 2-35-5.5%, S<0.15%, P<0.12%.

(b) Agglomerated fluxes–are mechanical mixture of ferro alloys with natural minerals. These fluxes are the mechanical mixture of the powders with liquid glass.

12.9 Preparation of the Job for Weld Deposition

The roll received for welding after its service in the mill and is consequent a worn out roll. Selection for a weld deposit depends on the characteristics of the wear out; as explained below:

(a) Rolls having a heavy network of fire cracks and other cracks should be turned down till the fire cracks and other cracks are completely eliminated. Cracks extending to a considerable depth must be removed by turning annular grooves. The sides of the annular grooves will be having a taper of 30-45°.

(b) Roll which has been previously welded and has brought again for rewelding after providing service in the mills. These rolls should be prepared for weld deposition with given consideration to the characteristic of the wear out, as explained above. If blow holes or slag incrustations are noticed on the surface, there these should be removed completely.

(c) The rolls must be thoroughly cleaned and free from grease, graphite, dirt and rust before they are taken for welding.

(d) After cleaning etc., the roll has to be heated before and after welding to relieve the stresses and strains developed during rolling.

12.10 Welding of Rolls

The process of welding starts with the proper setting up of the roll on the machine and proper clamping of it on the face plate and in centers. After the roll has been properly set on machine and checked for proper clamping on the face plate, the welding should be set in accordance with technological chart. The welding head should set for the proper welding regime.

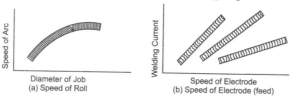

Fig. 3.24 Welding of Rolls.

12.11 Defects of Welding

Causes, prevention and rectification of defects of the welded surface are given below :

12.11.1 Cold Cracks

They are located at right angles to the seams. Sometimes even extending upto the entire depth and even get extended to the parent metal.

Causes: In-sufficient heating prior to welding and rapid cooling after welding.

Preventions: Ensuring proper heating of the entire roll, slow cooling in the controlled cooling pit.

Rectification: Removal of the crack portion by machining and then goes for re-welding.

12.11.2. Hot Cracks

They are generally located circumferentially along the seam with propagation at random angles confined only to the welded seam.

Causes: Insufficient heating of the roll, high welding current for the first layer. Higher welding pitch. Welding wire not corresponding to the parent metal.

Prevention: Ensure the proper heating of the roll. It is not desirable to have high current and low voltage during deposition of the first layer.

Rectification: Removal of the crack portion by machining and re-welding.

12.11.3 Porosity of the Deposited Metal

Along the seam circumferentially as well as across it.

Causes: Higher arc voltage, moist flux contaminating with rust or oil. Insufficient flux coverage over the arc.

Prevention: Measures to ensure that flux not become dirty or moist, wire getting rusted. Avoiding high voltage and open arc.

Rectification: Removal of the crack by machining and re-welding.

12.11.4 Slag Entrapments

At any place, has irregular shapes and size over 1 mm.

Causes: Incomplete removal of the slag incrustation. Bad quality of flux, flux not corresponding to the wire.

Prevention: Preventing over-heating of the roll. Watching for proper welding regime and position of the electrode. Ensuring proper composition and quality of flux and wire.

Rectification: Chipping/drilling and re-welding.

13 ROLL BEARINGS

Roll bearings are required to withstand extreme severe operating conditions such heavy shock loads, varying and reversing speed of rolls and extreme temperature variations. They are also often vulnerable to ingress of scale water and dirt.

The main feature of roll bearing is to withstand very high rolling pressure. Specify pressure in bearing may vary from 210-470 kg/cm^2.

In most of the cases, it has to withstand both the radial roll separating force and the roll end thrust as illustrated in Fig. 3.25 The proposition of end thrust varies considerably being quiet low in flat rolling but high in section rolling.

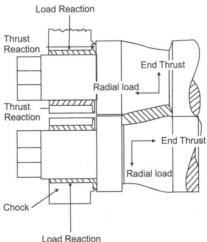

Fig. 3.25 Radial and thrust loading.

There are mainly three main types of bearings:
- Open general bearing with metallic /nonmetallic inserts.
- Roller bearing.
- Enclosed film bearing.

13.1 Open Journal Bearing

13.1.1 Metallic Bearing

At present, open journal bearing with metal inserts like bronze or graphite steel are used in only old sheet mills, where other bearings are not found suitable due to high neck temperature *i.e.*, about 300°C. Draw backs of this bearing is the high co-efficients of friction and of shorter life.

13.1.2 Non-metallic Bearing

Open bearing with non-metallic inserts were introduced in rolls first time in 1930, since then it is being used, with water lubricants, these bearings can withstand the action of abrasive particles in better way and also it will have less co-efficient of friction. Open bearing with non-metallic lining is used in majority of mills. It is generally used in the primary or intermediate stands of the section and plate mill, however the use of these bearings could not be extended to the strip mill, merchant mill, wire rod mill and cold rolling mill, owing to the finer degree of tolerances required for rolling of finished products

Most common material used for non-metallic lining for roll is the fabric impregnated in the resin at a high specific pressure (about 250–300 kg/cm^2) is "Textolite "and its substitute is "Lignofol".

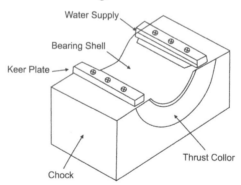

Fig. 3.26 Fabric Bearing.

Since, thermal conductivity of fabric bearing is less than that metal, that's why, it requires increased intensive cooling, as water act as a lubricant and coolant to keep these bearing cool. Sufficient quantities of water are required to be sprayed on roll necks, in order to have very quick heat dissipation and to

13.0 Roll Bearings

avoid any rise of bearing temperature. Critical temperature for the working of textolite bearing is 80°C. If temperature exceeds the critical temperature *i.e.*, 80°C, then water film between surrounding the roll neck surface and bearing gets break. At high temperature, water does not retain its lubricant properties. Water should be free from any acidity and from abrasives solids.

Empirical formula for quantity of water required can be derived from the following formula:

Quantity of water required = D.L.U.N/1600 gallons /hr

$$D = \text{Roll neck diameter}$$

$$L = \text{Load on bearing}$$

$$U = \text{co-efficients of friction}$$

$$N = \text{rpm of rolls}$$

Small amount of grease is sometime added in water to improve the efficiency of coolant. For all reversing mills, it is necessary to supplement the water lubrication with intermittent grease lubrication. It is an acceptable fact that there will be neck to bearing contact, when the roll is stationery. This condition also prevails when roll get reversal. The roll is bound to be stationery at one stage of reversal.

Fabric bearings are 10 to 40 times more resistant to wear than the metallic bearings. This can be explained by ability of bearing to absorb abrasive particles.

Other advantage of fabric bearings are:

1. Economical.
2. Longer life of bearing.
3. Good running in properties *i.e.*, easy confirm to roll neck.
4. Coefficient of friction is 10 to 20 times less than bronze or babbit bearing.

13.2 Roller Bearing

Roller bearings are used for rolls in cold rolling mills, strip, billet and other type of mills. Roller bearing is used when frequent starts and reversal under load is very high, the high-speed roller bearing will be used, whereas for oil film bearing it is not desirable owing to considerable force required to overcome friction in starting under load.

Why is roller bearing are used and not ball bearing? The main reason is that among antifriction bearing, roller bearing with linear contacts between rolling elements and rolling track has a greater load carrying capacity than the ball bearings of same size, whereas in ball bearing, contact between rolling elements and rolling track is only at one point.

Roller bearing with tapered, cylindrical or sometime self-longing spherical rollers are used. Fitting is very much important, whatever type of roller bearing being used. The fitting of bearing to roll neck depends upon the type of duty *i.e.,* loose fit bearing are used for slow speed application and shrink fit on tapered neck are used for high speed application.

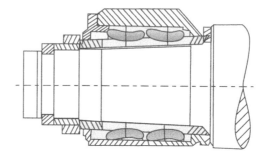

(*a*) Taper Roller Bearing for Radial Load and Thrust.

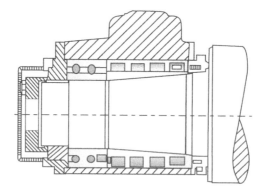

(*b*) Parallel Roller for Radial Load and Angular Contact Ball Bearing for Thrust.

Fig. 3.27 Roller Bearing.

13.0 Roll Bearings

Other advantage of roller bearing is as follow :

1. Very long life.
2. No adjustment is required to take care of bearing wear.
3. Large tonnage can be rolled without any variation in size of section being rolled.
4. Low cost per ton of product rolled.
5. Minimum maintenance.
6. Greater cleanness possible with automatic grease lubricants.

13.3 Enclosed Oil Film Bearing

The permissible speed of rolling mills, using roller bearings is strictly limited and their efficiency falls sharply with increase in speed and load .The permissible speed of mills, using oil film bearings is limited in practice only by its efficiency of cooling. Their load capacity does not decrease, rather it improves with increase in speed, subject only to adequate cooling. The fabric bearings cannot be used in Bar and rod mill, strip and cold rolling mill due to requirement of finer tolerance of finished product. For this reason only, oil film bearings are used in higher speed mill and finds its use in bar and rod and high-speed strip mills. The use of oil film bearing becomes wide spread due to its greater life and to provide rolling accuracy.

Oil film bearing has low co-efficient of friction, high rigidity and long life. The use of such bearing with low friction helps to reduce the power consumption. These bearings are termed as Morgan oil bearing as having been patented by Morgan Construction Co. USA.

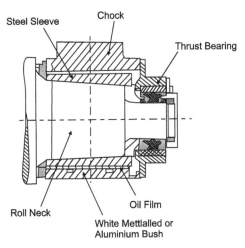

Fig. 3.28 Oil Film Bearing.

The main part of bearing is sleeve and bush. The sleeve is an alloy steel forging, heat treated and ground to mirror finish. The sleeve has a top red bore which fits on the tapered roll neck thus become the journal bush surround sleeve, which is made of aluminium or steel shell lined with about 5 mm thick centrifugally cast babbit material.

The bush and sleeves combination works on the principle of hydrodynamic lubrication. The rotating journal work as a pump and force lubricant in the gap between bush and sleeves. At high speed the oil film thus formed and prevents any form of metal to metal contact.

Freedom from wear and fatigue of working ensures a long life. The oil film is maintained by re-circulating oil under controlled temperature and pressure in the bush.

These bearing are used in high speed mills and have a high load carrying capacity. Due to high load carrying capacity, these bearing are also used for backup rolls of wide strip mills and heavy plate mills.

4
Rolling of Blooms and Slabs

1 INTRODUCTION

Blooms and slabs are products of primary rolling process. It is rolled from an ingot in a Blooming or Slabbing mill. These blooms and slabs are further rolled into various finished products and in various types of finishing rolling mills.

Blooming Mill

Slabbing Mill

Fig. 4.1 Type of Primary Rolling Mill.

Bloom is generally a square section, produced in a Blooming mill. The minimum and maximum cross section of a bloom is usually varied from 150 mm × 150 mm to 400 × 400 mm. Bloom is a primary material for rolling of long products.

Blooming mill also produces shaped beam blanks requires for rolling heavy beams and channels sections.

Slab is a rectangular product with thickness varies from 50 to 300 mm and width from 500 to 1800 mm or even more. Slabs are rolled from ingots either in a Blooming or Slabbing mill. Slab is an initial material for rolling of flat products.

2 INGOT

A steel ingot is heterogeneous in its density, chemical composition and in distribution of non-metallic inclusions. The physical non-homogeneity includes defects such as piping, blowholes and slag inclusions.

With the development of steel industry, there is always the aim of roll pass designer to increase the ingot weight and to achieve the highest productivity. Ten or more tones ingots are used for rolling into blooms and forty or even more tones are the weight of ingots, which can be used for rolling slabs.

2.1 Classification of Ingot

Ingots may be of any shape *viz.* square, rectangle, round or polygon. The parameter of ingot (*i.e.*, the ratio of cross section dimension, height, taper, head size, shape of bottom and configuration of side face) is choosen on the basis of the casting, heating and rolling and it is also based on the requirement of finished product.

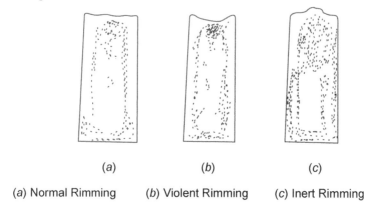

(a) Normal Rimming (b) Violent Rimming (c) Inert Rimming

Fig. 4.2 Rimming Steel Ingot.

According to the character of deoxidation, steel is classified into rimming, killed or semi killed steel. Killed steel is completely deoxidized steel, while gas evolution will be at maximum in the rimming steel. Semi killed steel holds an intermediate position between rimming and killed steel. Rimming steel is used as electrode quality steel, due to pure iron rim. Killing is required for all special quality steel; while semi killed steel is widely used for the production of rail, structural and wire rods.

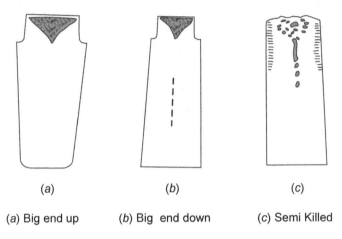

(a) Big end up   (b) Big end down   (c) Semi Killed

Fig. 4.3 Killed (*a* & *b*) and Semi-killed (*c*) Steel Ingot.

2.2 Selection of Dimensions and Design of Ingot

The selection of weight and form of an ingot depends upon the grade of steel, the characteristics of rolling mills where it is to be rolled and type of product rolled. Heavier ingots are used to produce slabs. Selection of a particular ingot weight is based upon the roll diameter of the mill and available power of drive motor. The ingot weight should be selected in such a way to ensure the maximum output of rolling mill and to produce the proper quality of product with minimum rejections. Presently, the weight of ingot for bloom varies from 5 to 10 tones and for slab it may weigh upto 40 tones.

Ingot is generally square or rectangle in cross section. Firstly, rectangular shaped ingots are preferred for rolling in blooming mill, because if ingot will be rolled in uneven number of passings in blooming mills, then rectangular shape of ingot will facilitate to have one more pass of draught on one side than on other side of ingot to form square bloom, as it has to proceed towards to billet mill which is in line with the blooming mill. Secondly, a rectangular ingot will have a better heating and crystallization conditions, while heating in a soaking pit. Rectangular shaped ingot facilitates the better holding of ingot between tongs of the soaking pit crane and on the roller table before the blooming mill.

3 MILL PROPER

3.1 Layout

Single stand two high reversing mill with roll diameter varies from 1000 to 1350 mm. They are widely used blooming mills, all over the world. Equipments of mills are arranged in three parallel bays. The first bay comprises walled premises of motor room. The second bay comprises of all the mechanism of mills and third bay is a scrap bay.

A layout of a modern 1150 mm, single stand two high reversing blooming mills with longitudinal soaking pit arrangement is shown in Fig. 4.4 below. Such a layout ensures convenient transportation of the ingot from steel making shop and it also enables rail road tracks to be laid in all bays of the mill building.

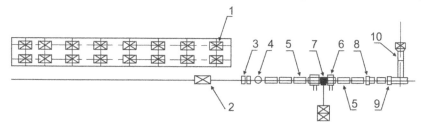

1. Soaking pit, 2. Ingot car, 3. Receiving Roll table with a stationary tilter, 4. Turntable, 5. Front and back mill roll table, 6. Manipulator, 7. Roll Stand, 8. Flame Scarfing Machine, 9. Shear, 10. Scrap conveyer.

Fig. 4.4 Layout of a 1150 mm Single –stand Two –high Reversing Blomming Mill.

Fig. 4.5 Illustrates the main line of the 1,150 mm blooming mill comprising the roll stand, universal spindles and the two motors for the individual drives of the rolls.

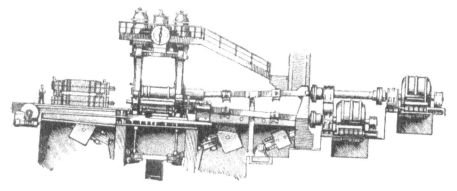

Fig. 4.5 Main Line of 1150 mm Blooming Mill.

3.2 Roll Stand

The roll stand is the main part of the blooming mill, because actual process of rolling is performed in it. It comprises of steel close–topped housings, roll with chocks and roll setting or adjusting device, breast rollers and roll changing gear.

The stand of a blooming mill comprises several elements as shown in Fig 4.6. Rolls are mounted on the bearings and roll necks, having a fork shape and serving to link the roll to the universal spindle. The most rational length to diameter ratio of roll barrel varies from 2.2 to 2.7. The roll neck diameter is generally taken as 0.55 to 0.63 of roll barrel diameter. The length of neck is usually taken equal to neck diameter.

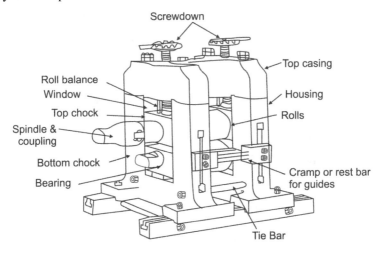

Fig. 4.6 A conventional Blooming Mill.

On contrary to blooming mill, slabbing mills are of universal two high reversing type of mill. The horizontal rolls have plain barrels (without grooves), 1100 to 1350 mm in diameter and 2150 to 2250 mm long. The stand with vertical rolls (diameter around 900-1000 mm), consists of two housings secured above and below through pads to horizontal roll housing.

The roll stand is the main part of a blooming or slabbing mill because in it the actual process of rolling is performed. It comprises two steel closed-topped housings, rolls with chocks and roll setting or adjusting devices, first and breast rolls and roll changing gear.

3.3 Stand Housing

The housings of blooming mill stand are heavy steel castings. They are mounted on bed plates and are tied together by heavy cast separators (cross members) of tubular cross-section through which the tie rods pass.

Due to their complex form, housings of rolling mills cannot be exactly calculated. Such calculations, however, even though they are only approximate, must always be made to check their strength and rigidity.

Housings are calculated to withstand the maximum vertical forces acting during rolling on the roll necks. Horizontal forces to which the rolls and housings are subjected at the moment of bar entry are usually ignored as they are very small in comparison with the vertical forces.

Since, the housings are the most import part of the rolling mill, they should be designed with a large factor of safety. There must be no permanent set in the housings when the rolls are broken.

While a safety factor of about 5 is usually accepted for roll calculations, a factor of approximately 10 is usual practice for housings. Thus, the permissible stress will equal :

$$R_b = \frac{eb}{10} \times 100 \text{ kg per sq. cm}$$

Where, eb is the tensile strength of the housing material in kg per sq. mm. If the tensile strength of the cast steel housings is

eb = 50 to 60 kg per sq mm, then the permissible stress will be

R_b = 500 to 600 kg/cm²

3.4 Rolls for Primary Mill

The roll of a blooming or slabbing mill comprises several elements (Fig. 4.7 below): 1-barrel diameter D and length L; 2–necks, diameter d and length l, on both sides of the barrel and mounted in the bearings and 3 – roll neck, having a fork shape and serving to link the roll to the universal spindle.

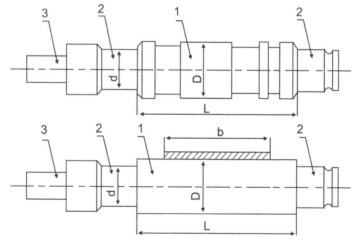

Fig. 4.7 Roll of a Blooming (*a*) and of a Slabbing Mill (*b*).

3 Mill Proper

The other end of the roll is either cylindrical or notched as a wobbler for rigging the hoisting rope when handling with a crane or for driving the roll during redressing in a roll lathe.

The chief dimensions of the roll (diameter and barrel length) are calculated on the basis of the resistance of the roll to bending stresses.

The most rational length to diameter ratio of the roll barrel, established by practice, varies in the range: $L/D = 2.2$ to 2.7.

The roll neck diameters for open type sleeve bearings are usually taken in accordance with the roll barrel diameter. For blooming and slabbing mills, $d = (0.55$ to $0.63) D$.

The length of the neck is usually taken equal to its diameter. The neck of roll is mounted on antifriction or sleeve bearings operating with fluid friction bearing.

As mentioned above, in designing and calculating rolls, a safety factor of 5 atleast is to be considered. Thus, for :

(1) Forged rolls of carbon steel

$$eb = 60 \text{ to } 65 \text{ kg/mm}^2$$
$$R_b = 1200 \text{ to } 1300 \text{ kg/cm}^2$$

(2) Cast rolls of carbon steel

$$eb = 50 \text{ to } 60 \text{ kg/mm}^2$$
$$R_b = 1000 \text{ to } 1200 \text{ kg/cm}^2$$

The permissible amount of redressing is limited only by the strength of the rolls. As a rule, blooming and slabbing rolls may be turned down from 10 to 12 percent of their nominal diameter.

Hot rolling on a blooming mill entails high pressures and temperatures and the rolls must possess high strength (their hardness is of no particular importance); therefore they are made of forged or cast steel with a high ductility. Such rolls have a high bending strength.

3.5 Selection of Rolls for Primary Mill

Blooming and slabbing rolls are made of forged alloy carbon steel with a carbon content of 0.6 to 0.8 percent, with the addition of alloying elements lying chromium nickel etc.

Surface hardening is frequently applied to increase the wear resistance of roll necks which running in textolite bearings. Recently, the service life of roll get prolonged by electrical hard facing.

These measures enable the number of redressing of blooming mill can be increased to 14 from the normal no of redressing i.e., 8-10. And 1.2 to 1.5 million tons of ingots can be rolled from one pair of rolls. In rolling low carbon steels the barrels of rolls should be amply cooled by streams of water.

3.6 Bearing for Primary Mill

Textolite bearings are mainly used in the rolls of blooming and slabbing mills. In some countries, antifriction and fluid friction bearings have been applied.

Textolite is not only a substitute for bronze but possess a number of advantages when used as a bearing material.

Advantages are:

- It is lubricated by water.
- Has good running in properties (conforms to the roll neck).
- Has a very low co-efficient of friction.
- Substantially reducing the energy consumption etc.

The low heat conductivity of textolite bearings necessitates the quick disposal of heat generated by friction. The maximum permissible load on the bearing depends on the heat disposal facilities and on the quality of the lubricant supplied to the bearing.

In accordance with their operating conditions, textolite bearings are lubricated either with water or with emulsion (soluble oil). Bearings must be cooled; their temperature should never exceed 60°-80°C since charring occurs at higher temperature.

Soluble oil is often used when water (for example, sea water) cannot be used due to its corrosive properties (acid or salt content).

The co-efficient of friction of textolite bearings is about 10 to 20 times less than that of bronze bearings and it is almost as low as that of roller bearings at high speeds.

> The co-efficient of friction
> for bronze = 0.06 to 0.11; roller bearings = 0.002 to 0.005
> and for textolite bearings = 0.004 to 0.006.

3.7 Top Roll Adjustment System

In blooming and slabbing mills, during the interval between passes, the top roll is lowered by an amount equal to the draught of the next pass. Before starting to roll an ingot, the top roll should be in its extreme upper position so that there is maximum clearance between the rolls.

The top roll is adjusted in a vertical plane by the roll adjusting mechanism or screw down. In such cases, the chocks of the bottom roll are stationary in the housings.

3 Mill Proper

The rate of travel should be high in top roll adjustment to ensure a high mill output. As a rule, the rate of adjustment (which equals the rate of housing screw travel) is taken from the following ranges:

Table 4.1

Sl. No.	Type of Mill	Lift of top roll, mm	Rate of adjustment mm sec
1.	1150-1200 mm Blooming Mill	1500 to 1800	125 to 250
2.	1150 to 1200 mm Blooming Mill	1000 to 1300	100 to 200
3.	1000 mm Blooming Mill	900 to 1200	90 to 180
4.	1150 mm Slabbing Mill	Upto 2000	125 to 250

Two types of motor driven roll adjusting mechanisms are used on modern blooming and slabbing mills:

(1) Horizontally arranged with the drive through worm reducing gears, and

(2) Vertically arranged with the drive through cylindrical gear reducing gears.

Horizontally arranged with the drive through worm reducing gears first type is illustrated in Fig. 4.8.

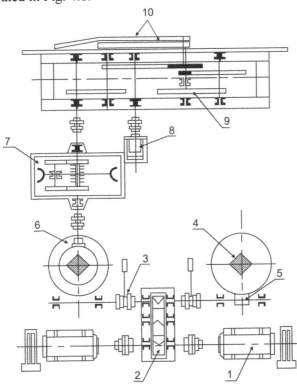

1-Motor, 2-Reducing Gear, 3- Clutch, 4-Housing Screw, 5-Worm Gear, 6-Bevel Gear, 7-Differential Reducing Gear, 8-Selsyn, 9-Kinematic reducing gear and 10-Hand of Draught Control

Fig. 4.8 Top Roll Adjusting Mechanism of Blooming Mill.

In the first system, the housing screws are powered from 2, 150 hp motors arranged for a 1:2 up-speed adjustment by means of a Ward-Leonard system incorporating rotary amplifier controls. This arrangement provides for the required acceleration of the whole mechanism during frequent starts and consequently reduces the period between passes. Power is transmitted from motors to the screws through an intermediate two stage herringbone reducing gear and worm reducing gears. The maximum rate of top roll adjustment is 184 mm per sec.

3.8 Spindles for Primary Mill

Universal joint spindles are used to transmit power to rolls, where centre to centre distance may vary in a wide range. The design of these spindles is based on the principle of universal (Hooke's) joint, so that they can transmit rotation when they are at inclination upto 10° of alignment with the rolls and motor shaft.

Universal-joint spindles can effectively transmit heavy torques at inclination upto 10° and therefore, the spindle length must be selected with this maximum angle.

The tang on the roll end usually has the form of a fork with an open slot for the sliding block. Due to this feature, the roll can be readily uncoupled during roll changing. The joint at the other end of spindle has a round hole in the tang for a pin. Since, motor shaft or pinions linked to the spindles are often not required to be changed and are not uncoupled in an axial direction as are the rolls.

3.9 Roll Speed

The speed of the rolls varies during the pass when heavy ingots are rolled in reversing blooming or slabbing mills. A typical diagram showing the variation

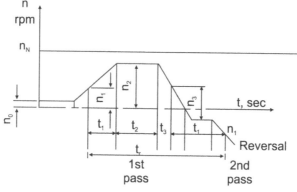

Fig. 4.9 Variation in Roll Speed and Torque in a Blooming Mill during Rolling.

3 Mill Proper

in roll speed for a two-high reversing mill is shown in Fig. 4.9. The rolls bite the bar at the speed n_1 which is then increased to the maximum value n_2 and, at the end of the pass, is reduced to n_3. After the bar leaves the rolls, the motor is stopped and reversed for a pass in the opposite direction.

The following period are distinguished in the diagram :

(a) Acceleration of the mill without the ingot, in which the roll speed increases from 0 to n_1;

(b) Acceleration of the mill with the ingot, in which the roll speed increases from n_1 to n_2;

(c) constant speed, in which the rolls maintain a constant speed of n_2;

(d) deceleration of the mill with the ingot in which the roll speed drops from n_2 to n_3; and

(e) deceleration of the mill without the ingot, in which the roll speed drops from n_3 to 0. If t_1 denotes the time for mill acceleration with the ingot, t_2 – the time for the constant speed period and t_3- the time for mill deceleration with the ingot, then the time for the pass with equal :

$$t_T = t_1 + t_2 + t_3.$$

The time for the period of mill acceleration and deceleration without the ingot is part of the interval between passes.

3.10 Main Drives of Motor

The main drive motor of a Blooming or Slabbing should have the following:

(1) Reversible duty with frequent starts and stops (upto 600-1000 reversals per hour).

(2) Minimum reversal time (from rated speed in one direction to rated speed in the other direction) which will reduce the operative time per bar rolled and increase the mill output.

(3) Large overloads when metal is entered and is rolled with high accelerations. It leads to considerable dynamic current loads in the armature and high torques.

(4) Speed variation in a wide range.

These requirements are fully satisfied with a DC drive motor.

Modern Blooming mill with individually driven rolls 1100 to 1200 mm in diameter are powered with motors of 4000-8000 kW. Blooming mill with rolls of 1100 mm in diameter or less are also equipped with more powerful motors. At present, drive through pinion stand are also used in many cases.

Recent practice is to increase the acceleration and deceleration of blooming and slabbing mill motors, in addition to increasing their capacity.

The acceleration of modern blooming and slabbing mill designing at the present time ranges from 80 to 120 rpm per second and deceleration –from 80 to 130 rpm per sec.

3.11 Roll Tables, Manupalaters of Mill

Blooming and Slabbing mills are equipped at front and back with roll tables, whose purpose is to advance ingots to the rolls. It is in the process of rolling, where convey the bar to the shear and from there to the yard. In accordance with operations performed by them, roll tables are classified as receiving, mill and conveying roll tables.

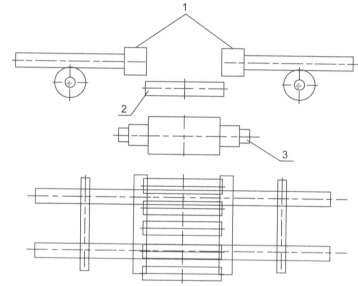

With Two Side Drive 1-Pusher Strip of the Manipulator, 2- Table Roll,3- Mill Roll

Fig. 4.10 Principle of Blooming Mill Manipulator.

The bar moves on the mill roll tables from pass to pass along roll barrel on either side of the stand by manipulators. The manipulator can also do the straighten bars, if they become bent in rolling.

3.12 Shears for Primary Mill

Shears (with parallel blades) are used to cut hot bar rolled in blooming or slabbing mills. Cutting force decides size of cross-section to be cut. The blade length is taken equal to 2 to 2.5 times of the maximum width of bloom for blooming mills or 100 to 200 mm more than maximum slab width for slabbing mills.

3.0 Mill Proper

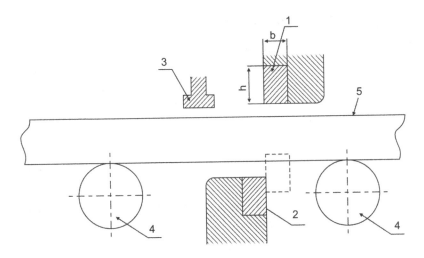

1-Upper Blade, 2-Lower Blade, 3-Hold-Down, 4-Table Roll, 5-Bar.
Fig. 4.11 Cutting Principle of a Parallel–Blade Shear.

The size of cross-section cut by such shears depends upon their designation and the maximum cutting force. The blade stroke is selected so that a bar of maximum cross-section passes freely under the power hold-down. The blade length is taken equal to:

(a) 2 to 2.5 times the maximum bloom width for Blooming mill shear.

(b) 100 to 200 mm more than the maximum slab width for Slabbing mill.

The cross-section of blade is usually based on the ratio = h/b = 2.5 to 3.

In selecting the size of the shear, the maximum cutting force can be determined from the formula;

$$P = K_1 K_2\, eb\, \frac{F}{1000} \simeq 0.8\, eb\, \frac{F}{1000}$$

Where K_1 – a coefficient accounting for the increase in the cutting force upon dulling of the blade and an increase in the clearance between them.

K_2 – ratio of the shear strength to the tensile strength

F – cross–sectional area of the bar being cut, sq. mm

eb – Tensile strength of the metal at the temperature at which, it is cut, kg/mm^2.

Blooms and slabs are delivered to the storage yard through cooling bed. It has the provision of stamping, marking and weighing facilities.

148 Chapter 4 Rolling of Blooms and Slabs

3.13 Scale Disposal for Mill

Much attention is to be paid to the scale disposal in Blooming and Slabbing mill. Out of the various methods of scale disposal, the most popular is hydraulic method, in which the scale is washed away with the flowing water. Generally, flight conveyors are deployed below the roll table, towards the direction of scrap way for the disposal of both fine as well as large pieces of scrap.

Large pieces of scraps slide down to inclined grates onto the flight conveyors, which deliver them into the boxes placed in pits of scrap bay and dumped on rail road cars.

Fine scale drops through the slots in the inclined grates and falls into inclined channels through the continuously flows of water. From channels, the scale is washed into large settling pit in the scrap way. From here it is removed by a clamshell crane.

4 COMPARATIVE STUDY OF BLOOMING MILLS

A comparative study of six different types of Blooming Mill, operated at different parts of world shown in Table 4.2, below:

Table 4.2

Equipment & Facilities	1	2	3	4	5	6
Soaking pits arrangement	At right angle to mill bay	Parallel to mill bay	At right angle to mill bay	At right angle to mill bay	Longitudinal	Longitudinal
Stand	Single stand (blooming & slabbing)	Single stand (Blooming)	Single stand (Blooming)	Twin stand (blooming)	Single stand (Blooming)	Single stand (Blooming)
Roll diameter in mm	1350	1350	1300	1400/1060	1150	1300
Shear in tonne	2300	1100	1250	1200	1000	1700
PERFORMANCE DATA						
Ingot weight	11 T	8/10 Tones	13 T	7/12/14 T	8.5T	7/10/14 T
Products Bloom Size in mm	200×200–350×350	200×200 –400×400	300×300 370×370	230×230–280×280	280×280–320×320	280×280–300×300
Slab Size in mm	200 mm thick 600 mm width		100-200 thick & 700–1000 mm width	100–50 mm thick & Upto 450 mm width		250-50 mm thick & 400.00 mm width

Contd...

5 Roll Pass Design for Blooming Mill

Equipment & Facilities	1	2	3	4	5	6
Capacity/ year	3 millions ton	3 millions ton	6 millions ton	3 millions ton	2.5 millions ton	3 millions ton
Yield (approx)	90%	N A	N A	91 %		89 %

5 ROLL PASS DESIGN FOR BLOOMING MILL

5.1 Pattern of Reduction

The ingot as cast is tapered from one end to other for ease of stripping from ingot mould. The cast structure may be very coarse with marked trans crystalline zone and it may be porous due to blow holes.

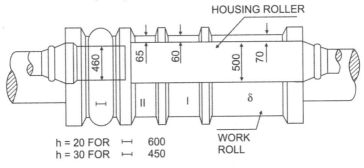

Fig. 4.12 Shows the placement of roller table vis- a-vis working roll of Blooming Mill.

The function of early passes is to eliminate the taper of ingot and help to break-down the cast structure into more workable one and densities the material by closing up of internal blow holes. In majority of cases, during early passings, when cross-section of work piece is more and temperature of metal is also high, a considerable high degree of deformation is recommended, which is only be limited by angle of bite and strength of rolls. But in case of high carbon or special steel, high deformation at early stage may lead to surface defects due to the presence of sub-cutaneous blow holes, near to the surface and will get exposed due to high reduction. Once it appears, these defects can only be removed by dressing the bloom by scarfing or chipping. Dressing of bloom will reduce the yield and increase the cost of production.

With initial draughting, cast structure become stronger, as it gets refined by working, though the resistance to deformation which may not necessarily increases at this stage. Higher draught is always advisable at intermediate passings; there also, it is limited by strength characteristics of roll and capacity of main drive motor.

In final passings, the draught has to be reduced extensively, since resistance to deformation increase with the drops of metal temperature.

5.2 Arrangement of Grooves in Blooming Mills Rolls

Usually three to five grooves are made in blooming mills rolls depending upon the character of manufacture of finished products and local conditions of prevailing. Generally, following two types of arrangements are used.

5.2.1 Series or Russian System

The bullhead pass in the form of shallow box pass is located not at the center of rolls, but at the end of barrel as shown in Fig. 4.13 (a). Passes are arranged in the order of rolling. This design leads to reduction of rolling time and corresponding increase of the output of Blooming mill, as stock move along the barrel of rolls.

Other advantages are:

- In shallow bullhead pass, groove is cut in roll upto certain depth. Such depth of pass even in insignificant amount, makes the easier rolling for operator. It reduces the height as the top roll is required to be raised or lowered. This may help in increasing the output of blooming mill by at least by 5-10%.
- The shallow pass in place of barrel pass prevents the requirement of stepped entry roller before Blooming mill.

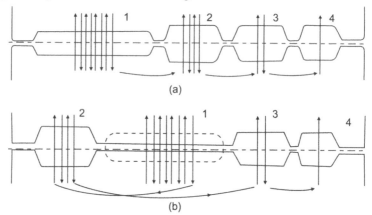

(a) Series or Russian System (b) Symmetrical or American System

Fig. 4.13 Pass Arrangements on Rolls of Blooming Mill.

5.2.2 Symmetrical or American System

The first bullhead pass, having maximum effective diameter is at the center of roll body and other passes are arranged on the both side of bullhead pass. This system provides symmetry in distribution of passes and rolling load around center of roll. Hence, this system is called as symmetrical or American system [Fig. 4.13(*b*)] above. The main characteristics of this system are as follows:

- Bullhead pass is used for breaking down pass for the ingot or slabs in initial stage.
- The roll load during rolling in bullhead pass has equally distributed on roll necks; this reduces the non-uniform wear of roll bearings.
- Rolling in bullhead pass can be performed at higher draught.
- Since, bullhead pass does not have any groove cut, it helps in avoiding weakening of roll, as stress concentration at center of roll is highest but any cut at center will make it further weak.
- Scales, which are embedded in ingot, got shedded in initial passings being bullhead pass is at center. This design ensures less possibility of scales getting into the roll neck.
- Absence of collars in bullhead pass, allows length of roll barrel to be used more effectively.

Major disadvantage of this system is that after earlier passing in bullhead pass, the stock has to be shifted from left side of bullhead pass to right of it. This increases idle time of rolling, in turn total rolling time gets expanded. In addition, absence of collars in bullhead pass affects on the stability of stock, especially after tilting.

6 TYPES OF PASSES USED IN BLOOMING MILLS

Box passes are used in Blooming mills for following reasons.

Box pass can be used for several passings of stock by screwing down the top roll after each passing. This reduces number of grooves required on rolls and enables a large number of passings can be accomplished in one blooming mill stand. Besides, mill can also accommodate various initial and final cross sections of ingots and blooms.

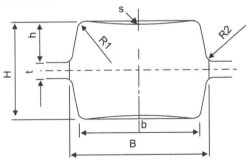

Fig. 4.14 Construction of Box pass for Blooming Mill.

152 Chapter 4 Rolling of Blooms and Slabs

Other advantages to using Box poss are:

- Roll with box passes is stronger than those with diagonal or square passes, as later sunk diagonally into roll and weaken the roll to a greater extend.
- Rolling in box pass ensures uniform draft along width of stock and deformation may proceed with various degree of spread restriction.
- Scale is easily removed from side of stock during rolling inbox passes.

7 SIZE AND SHAPE OF PASS

7.1 Height of Pass (H)

Height of pass (H) is determined by number of passing required and total amount of reduction takes place in that pass. Depth of cut should be half of the smallest size of stock to be rolled through that pass. Deeper the cut of pass will have the more rolling stability of work piece and better will be the quality and dimensional accuracy of finished product.

The depth of groove for blooming mill fluctuates between 0 to 150 mm.

7.2 Width of Pass

The width of the Bullhead pass (B) is taken approximately 200 mm more than the maximum width of slab to be rolled, subject in case both bloom and slab are to be rolled from the same mill. The width of Bullhead pass is kept slightly more than the maximum width of ingot, if only bloom is to be rolled in Blooming mill.

Width of pass at highest depth (b) should be so chosen that it should not be less than minimum width of stock by 0-10 mm. It helps in protecting stock from twisting, especially during course of rolling.

Width of pass at separation line of rolling should be more than maximum width of stock to avoid fin formation due to spread.

7.3 Taper

The taper of sides of blooming mill passes amounts to 5-30 %, generally taper of around 20 % is provided to avoid collaring of stock and formation of fine seams lap on side surface.

7.4 Corner Radii

$$R_1 = \{0.12\text{-}0.20\}\ B$$
$$\text{And, } R_2 = \{0.08\text{-}0.12\}\ B$$

7.5 Pass Convexity

The aim of pass convexity (s) is to provide room for spread in succeeding pass. The pass convexity is generally provided to approximately 5% of pass height. Pass convexity is given in intermediate passes, it is not recommended for final pass. Pass convexity also helps in stability in movement of work piece on roll table.

7.6 Roll Collars

Generally, alloy forged steel rolls are used in Blooming mills. The widths of collars are taken atleast 50% of the depth of groove in a roll.

8 BOTTOM PRESSURE

To prevent hitting of work piece to the roll table, which is about 7 to 10 times heavier than the roller of roll table. Bottom pressure is given in Blooming mills, by making difference in size of roll diameter to the extend of 5-10 mm. In case of Blooming mill, wherein each roll is driven separately by individual motor, the rpm of bottom motor is generally kept more than the top roll to get the desired amount of difference in roll peripheral speed of top and bottom roll.

9 TURNING OF THE WORKPIECE

From the point of view of increasing output, work piece should be turned to least possible number of times during rolling. But from the point of view of better quality and high surface finish of the finished product, the stock should be turned to maximum number of times.

Repetitive reduction in one direction causes tension to setup in side surface due to spreading and due to the differential elongation between surface and center. It may ultimately lead to cracking on side surface

It is advisable to turn work piece after every two passing, particularly at the beginning and later turning will depend upon size of groove available and tilting facilities of mill. It is to note that width to thickness ratio of work piece should not exceed 1:3 when rolling in bullhead pass and 1:1.7, when rolling in other passes in case of rolling of blooms.

The bar should be turned after every two passings, in case of rolling of alloy or high carbon steel.

10 ROLL PASS DESIGN DATA SHEET OF BLOOMING MILLS

Based on the principle discussed above, "Roll Pass Design" of blooming mills can be evolved, from where, there certain basic information are required as given in Table 6.3 below:

Table 4.3

	Necessary Information	Example
A	The cross sectional dimension of the ingot which are to be rolled together with weights. This is required to fix the size of the bullhead or first pass width.	$\dfrac{810\times770}{790\times750} \times 2.5$ mt
B	The smallest bloom, which is required to be rolled and hence the stock length from ingot weight	285×285
C	Any intermediate bloom sizes which may be required	325×325

NOTE: Point B and C control the widths of the edging grooves, It is also desirable to know what are sizes of blooms are required to roll and with what accuracy they must be produced with regard to flatness of sides and corner radii

D	The maximum and minimum width of any slabs required to be rolled. This is required for designing of the width of first pass or of the bullhead pass. The length of roll barrel available	Max. width – 700 mm Min. Width– 400 mm 2800 mm
E	(a) The diameter of rolls *i.e.* maximum and minimum diameter (b) Roll Neck Diameter	1180 mm 1080 mm 690 mm
F	The manipulation facilities available at each side of the mill The rolling sequence must make the best use of tilting to have both quality and quantity in production. Availability of tilter one side or both the side Nos of tilters	Tilter is only available at entry side 1
G	Distance from roll axis to nearest finger of tilter Manipulator Length of guard Max working opening of guards	2550 mm 8750 mm 2750 mm

Steel quality to be rolled which together with mill power determine the max draft	Mild steel Rail steel
Roll material	Forged steel
Top roll lift	1120 mm
Motor Motor capacity in kW Motor rpm	2×8000
Initial temperature of rolling	1200°C
Shear Max. blade pressure Blade length Max. size of bloom cut Max. size of slab cut	1000 T 1200 mm 400×400 mm 200×1000 mm

10.1 Rolling Schemes for Rolling Different Size of Blooms/ Beam Blanks

Table 4.4

MS Blooms-280×285, 260×300, 265×340

Pass	No. of Passing	Section 810×770 / 790×750	Reduction
δ	1	715	95/75
	2	620×800	95
	TILT 90°		
	3	730	70
	4	660	70
	5	590	70
	6	520	70
	7	450	70
	8	380×660	70
	TILT 90°		
I	9	575	85
	10	480	95
	11	385	95
	12	295×420	90
	TILT 90°		
II	13	350	70
	14	295	55
	15	265×305	30
	TILT 90°		
II	13	380	40
	14	335×315	45
	TILT 90°		
δ	15	270×345	45
	TILT 90°		
II	13	345	75
	14	275×325	70
	TILT 90°		
II	15	285×290	40

Beam Blank-450

Pass	No. of Passing	Section 810×770 / 790×750	Reduction
δ	1	715	95/75
	2	620×800	95
	TILT 90°		
	3	710	90
	4	620	90
	5	550	70
	6	490	60
	7	430	60
	8	370×660	60
	TILT 90°		
B 450	9	580	80
	10	510	70
	11	440	70
	12	370	70
	13	310	60
	14	255×415	55
	TILT 90°		
δ	15	375	40
	16	335×270	40
	TILT 90°		
B 450	17	195	75
	18	145	50
	19	95×415	50

Note: Sizes shown in the schemes are hot section, *i.e.*, 285×290 mm hot section will produce a bloom size of 280×285 mm cold section

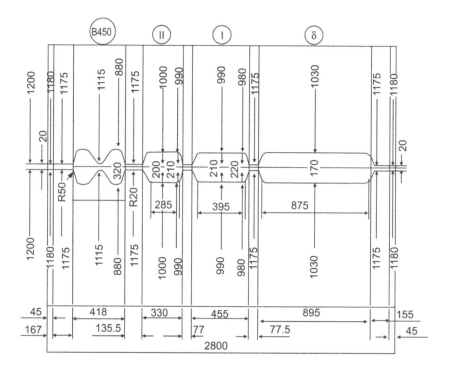

Fig. 4.15 Design of Pass arrangement in 1150 mm Blooming Mill.

11 PASS DESIGN OF BLOOMING MILL

11.1 Barrel or Bullhead Pass

The minimum width of barrel pass is determined by maximum width of ingot and slab. Suppose, ingot width is 785 mm, then the minimum width of barrel pass should be 100 mm more than the maximum width of ingot. The width of barrel pass also depends upon the maximum size of slabs to be rolled. Keeping these into account, the minimum and maximum width of barrel pass is kept as 875 mm and 895 mm respectively as shown in Fig. 4.15, above. The maximum size of slab which can be rolled from this pass will be around 700 mm.

The depth of pass is determined by max lift and strength of the roll. The depth of pass should be made as low as possible, otherwise increases the rolling time, as roll has to travel more. The higher depth of barrel pass will affect the strength of roll, which is the most important criteria for deciding the draft. Angle of bite will also get increases with the depth of the pass. Generally depth of barrel pass varies from 0-200 mm. It is taken as 170 mm as shown in Fig. 6.12 above. The max lift of the top roll is 1120 mm. The radius of 30 mm is given at corners in the barrel pass.

11 Pass Design of Blooming Mill

Roughening of blooming mill rolls by ragging or knurling are made for increasing friction and may raise angle of bite upto 34°.

11.2 First Pass

The width of first pass (Fig. 4.16) at the highest depth should be almost equal entry width of work piece to avoid twisting and wear out. Max width is determined by considering total spread of work piece in that pass after reversal passings that is why, if the width of work piece entering in pass is around 380 mm, then the minimum and maximum width of pass will be taken as 395 and 455 mm respectively, after considering total spread of work piece in that pass.

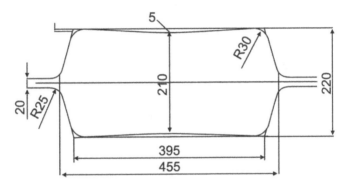

Fig. 4.16 First Pass of Blooming Mill.

The height of first pass is taken as 220 mm, which is higher than the barrel pass by 50 mm. It will not only provide better stability of work piece, but also with this depth, roll will have sufficient strength and also in turn higher angle of bite that will ensure higher draught in this pass.

The convexity of first pass is kept to 5 mm, to give better stability of work piece on roll table and to reduce the chances of fin.

The radius of 25 mm is given at corners of finishing pass, taken into consideration of all product sizes to be rolled from this pass.

11.3 Second or Finishing Pass

The second or finishing pass (Fig. 4.17) width of 285 mm is chosen on the basis of the width of incoming work piece. The max width of 330 mm is selected on the basis of final cross section of bloom. The height of pass is almost kept as same of pass 1 *i.e.,* 210 mm and convexity of pass is also taken as 5 mm.

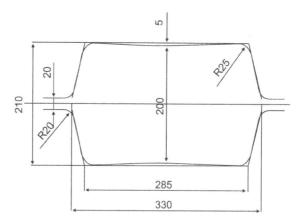

Fig. 4.17 Second of Finishing Pass.

11.4 Beam Blank

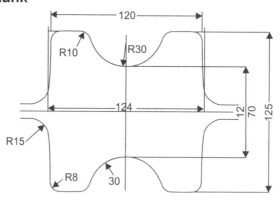

Fig. 4.18 Beam blank pass for Beam 450 mm.

The size of beam blank is determined by the input requirement of the sizes of beams to be rolled in subsequent structural rolling mills. The depth and flanges of beam blank are accordingly designed. In the Fig. 4.18 above, Beam blank pass for Beam 450 is shown.

12 DEFECTS IN BLOOM ROLLING

12.1 Defects due to Rolling

Generally following defects during rolling of blooms may be observed.

1. Skewed or diamonded blooms [Fig. 4.19(*a*)]

It is due to misalignment of the roll grooves and non-uniform heating of the ingot, the rolling in a pass wider than the entering bloom or when edging of a bar with a ratio of the sides in excess of 1.5 (especially for small cross sections).

12 Defects In Bloom Rolling

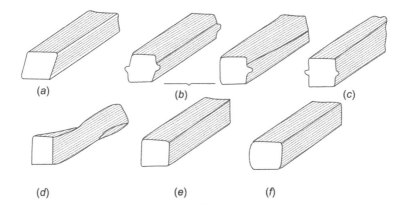

Fig. 4.19 Defects due to Rolling.

2. Collaring [Fig. 4.19(*b*)]

It is mainly due to groove misalignment or bar rolled on collars.

3. Fins [Fig. 4.19(*c*)] on opposite sides as a result of deviations from the draughting schedule. It may also be due to insufficient inclination of the sides of the grooves of a box pass when several passes are made in the same hole only.

4. Twisted blooms [Fig. 4.19(*d*)] are with the result of pass misalignment. non-parallel rolls, entering the bar on the collars and also, due to, non-uniform heating. It also depends upon the amount of pass wall inclination and amount of draught. It may be also caused by excessive draught, faulty setting of entry and delivery guides or lack of uniform temperature across billet cross section.

The twisting of blooms [Fig. 4.19(*d*)] is prevented by properly performed roll turning and in particular, precise alignment of the grooves in the top and bottom rolls, by properly entry of the bar into the pass; by proper setting of the rolls in reference to each other and by uniform heating of the ingot. Besides this, the bar should never be rolled in excessively worn passes, and the rolling speed should be reduced, especially when the front end of the bar enters the rolls.

5. The reason for lopsidedness of bloom [Fig. 4.19(*e*)] is because of different clearances between the end collars of both sides of the rolls.

6. Bellying of bloom [Fig. 4.19(*f*)] is if the walls of the pass are excessively worn out, either due to higher reduction or metal is not hot enough for that pass.

7. Under filling: This condition arises, when pass doesn't get filled up completely. The metal from previous stand is to be enhanced.

12.2 Defects due to Steel

1. Crack

These may be classified into three groups.

(*a*) Cracks on the sides.

(*b*) Cracks at the corners.

(*c*) Hair cracks on the surface of Bar.

2. Scab

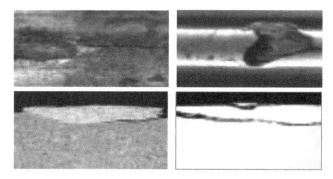

Fig. 4.20 Different forms of Scab on surface.

Relatively thin flakes or tongues of metal imperfectly attached to the surface of the billet, resulting from ingot defects *i.e.*, 'double skin', 'splash', flash, fin, scab or from the exposure and oxidation of large sub-cutaneous blow holes near to the skin in the soaking pits. The defects are sometimes termed as shells, Silver or Spills.

3. Seams

It is a shallow groove formation on the surface of bloom or billet. They are generally present in the ingot and also formed by the elongation of oxidized surface or subsurface blow holes. They may arise from a badly rippled surface or from bad teeming practices.

4. Spongy

A porous and cracked surface caused by oxidized sub-cutaneous blow holes.

Surface defects of blooms, slabs include :

1. Lap, which has the appearance of fine longitudinal cracks, is formed by rolling down a fin; if projections of random form are rolled down, the lap usually resembles a scab.
2. Scabs are thin flat seams due to splashes of metal in top teeming of the ingots, sub-cutaneous blowholes and poor deoxidizing of the metal. Small scabs are usually burned out while large scabs have to be removed.

12 Defects In Bloom Rolling

3. Slag, sand and other nonmetallic inclusions are the basis for establishing that the steel was unsatisfactorily made or cast. However, the cause of these nonmetallic inclusions may be due to contamination of the metal not only at the steelmaking stage, but also during the rolling, especially during the heating of ingot. During heating, fluid or semi fluid slag from the bottom of the heating furnace, or pit, get stick to the surface of the ingot.
4. Scratches and scoring of the surface of the billet are due to guides which must be immediately replaced, sometimes also may be due to various surface irregularities on the rolls.
5. Certain defects of blooms, slabs are arises due to bad shearing. These include a lack of squareness of the cut when the bar is incorrectly positioned in reference to the blades; bending of the bar end may be due to an improperly set shear (large clearance between the blades, dull blades etc.) and the formation of a lip on the end of the bar due to dull blades or an incorrectly set shear. If the ends of blooms are insufficiently cropped, then traces of the pipe and shrinkage of porosity may remain in the bloom.

12.3 Defects Due to Bad Heating Practice

1. Overheating /Burnt Blooms

Fig. 4.21 Burnt Blooms.

Overheating/Burnt are either due to heating of steel either at a very high temperature or due to local impingement of flame on the surface of ingot. Primarily, it is due to negligent heating practice by the furnace saff.

2. **Clinks:** Rupture in ingots, when the ingots are stripped early and then exposed to a cold atmosphere. This defect generally occurs at relatively low temperature and is caused by the formation of high thermal stresses (tensile) in the outer skin, producing external rupture.

162 Chapter 4 Rolling of Blooms and Slabs

Clinks also occur during subsequent processing of the ingot, either by too rapid heating of cold material again in the soaking pit, resulting in internal cracks.

3. **Molten Core:** Molten core is the Liquid formed in the ingot core, after soaking in soaking pits. This liquid may come out while rolling in blooming mill or during shearing of the bloom ends.

Molten core may be due to some segregation of nonmetallic inclusions, which remain at liquid state as they have low melting point, too high soaking temperature and also holding of ingots for longer duration.

4. **Cold or Quenching Crack:** External cracks are formed on blooms in the course of cooling but observed, after the metal has been cooled (after several hours or even days). For this reason they are called "cold" or quenching cracks since metal with these cracks usually has an increased hardness, *i.e.,* hardened by rapid cooling.

Sometimes these cracks are shallow and resemble to hair cracks. As a rule, they stretch in intermittent lines in the direction of rolling. Cold cracks may, however, penetrate to the centre of the billet, while in steels of low plasticity large through cracks may be seen cutting across the whole billet at its ends.

The reasons for the formation of cold cracks are internal residual stresses generated during rolling process, as well as during cooling of the metal after rolling (thermal and structural stresses). The higher the rate of cooling after rolling, the more stresses will appear and there is more probability of the formation of cold cracks.

However, all steels are not susceptible to cold cracking upon rapid cooling after rolling. For example, blooms, slabs and billets of low carbon steel may be cooled at any rate whatsoever without the danger of external cracks being formed of the any deterioration of the quality of the steel.

As a rule, the rate of cooling is controlled after rolling blooms, slabs and billets of alloy and high alloy steels. Some of these steels are more susceptible to cold cracks. Various methods are used to prevent cold cracks depending on the type of steel being rolled.

The highest tendency to form cold cracks is found in high-speed steels, high–alloy tool steels, alloy tool steels, high chromium, chromium-nickel, chromium-nickel-tungsten and other steels.

Flakes are internal cracks observed in macro and micro sections. Chromium-nickel, chromium-nickel-molybdenum, chromium, manganese and other steels (steels of the martensite and pearlitic classes) are most susceptible to the formation of flakes. Flakes are also found in carbon steels with manganese content from 0.7 to 1.0 percent.

12 Defects In Bloom Rolling

No flakes are observed in steels of the ledeburitic (eutectic) and austenitic classes (high speed, stainless, heat-resistant, high-manganese and other steels).

Flakes are most frequently found in blooms, slabs and billets as well as in finished products of the larger sizes (shapes rolled in heavy and medium section mills).

Retarded cooling and heat treatment of the semi finished product are used to prevent the formation of cold cracks and flakes. Retarded cooling is carried out in stacks, in heat-insulating materials and in heated or unheated pits. Heat-insulating materials include blast furnace slag, slag wool, sand and others.

More recently, cooling in reheating furnaces is widely applied (especially for flake-sensitive steels). This method enables the metal to be held isothermally in the range of most favourable temperatures, thus reducing the time required for heat treatment.

One measure for preventing flakes in the rolling is the proper heating of the ingots. This entails charging ingots to a definite temperature, slow heating and holding at 1100°-1150°C for a longer period to allow the escape of hydrogen from the metal by diffusion.

5

Rolling of Billets

1 INTRODUCTION

Semi-finished product of a square or almost square cross-section, of size less than 150×150 mm in size is termed as "BILLET".

Fig. 5.1 Rolled Billet.

It is used as input material for further rolling into Bars, Rods etc. It may also be directly used for forging or machining.

2 TYPES OF MILL AND GENERAL LAYOUT

To avoid the installation of a separate Billet Mill, it is a seldom practice for rolling ingot directly to sizes within billet range in a Blooming Mill. It is established as a non-economic propositions, as with large diameter rolls and small size of finished stock, the rate of reduction get reduced with increase

166 Chapter 5 Rolling of Billets

in spread, in turn; output got seriously affected due to slow speed of rolling in the Blooming Mill. It also requires a large number of passings. Therefore, separate Billet Mill is to be installed to roll billet with smaller rolls and with higher output speed.

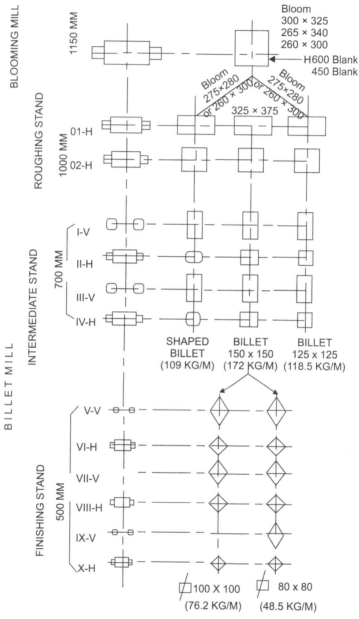

Fig. 5.2. Flow chart of Rolling from Ingot to Billet.

2 Types of Mill and General Layout

The selection of the proper size of input bloom to be rolled in Billet Mill. It depends upon the size of output required, number of stands in the mill, speed and power of motors and capacity of shear to cut the minimum and maximum cross section. Generally, input bloom size for present day billet mill is taken around 300 mm × 300 mm.

Generally; billet mills can be categorized into three main types.

2.1 Open Train Billet Mill

Open train billet mill may consist of 2 high or 3 high stands. The 3 high reversing stands are not generally preferred for billet mill, as there will be difficulties in manipulation of stock in a 3 high stand, which will compel to use a relatively shorter length input bloom. Blooms are to be cut in several small pieces in bloom yard, before rolling in it in 3 high Billet Mill. In addition, with the rolling of these small length blooms at different temperatures, there will be difficulties to control the dimensions of finished billet section, as with fluctuating temperature conditions, there will be change in spread characteristics.

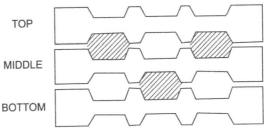

Fig. 5.3 A 3-High Billet Mill

Advantage of using 3 high stands in Billet mill is that the initial cost of installation of mill will be lower and such design is only favoured, when limited tonnage is required to be rolled.

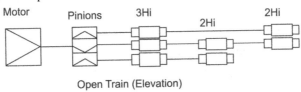

Open Train (Elevation)

Fig. 5.4 Open Train Billet Mill.

Production of 3–high mill can be enhanced by installing two additional 2–high stands in this open train mill (Fig. 5.4), using the same drive and with dividing work between two stands. A single 2 high reversing stand is capable of rolling the bloom in one length only, first stand is used as roughing stand to reduce the Bloom drastically, while the second stand serve as finishing stand for bigger sizes of billets and intermediate stand for smaller sizes. Sometimes, the third additional stand is also installed, which will serve as a finishing stand for rolling of smaller size Billets.

2.2 Cross Country Billet Mill

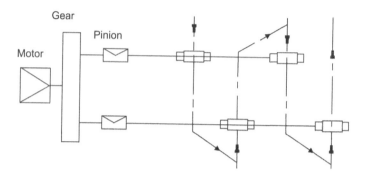

Fig. 5.5. Cross Country Billet Mill.

As shown in Fig. 5.5 above, the cross country mill consists of several 2 high and 3 high stands, following the blooming mill. The speed of last stand governs the speed of flow required and all other stands should be designed in such a way to provide a continuous supply to the last stand. As explained above, due to 3 high mills, the bloom is to be cut in several lengths to facilitate manipulation and to satisfy the limitation of the maximum finished length. Although operation of the cross country mill is faster than the open train, but there will be variation in rolling temperature and a low yield due to extra end cropping requirements.

2.3 Continuous Billet Mill

In a 2 high continuous train, bloom delivers from blooming mill, comes straight to billet mill without reheating and shearing into smaller length. It gives a high yield output and high uniform finishing temperature. The control of finishing product is much better in continuous mill than in other type of billet mill. Here, billet is cut to length by a flying shear, as it leaves the last stand.

The mill train is subdivided into roughing and finishing train. The low speed of roughing train lessens difficulties in biting. The stands of each train must be balanced for the speed and reduction, in such a manner, there will be minimum of pulling and pushing between stands. Stands are either individually or group driven.

Continuous mill often incorporates vertical stands. Vertical stands are used as edging stands and with the use of these stands, twist guide and tilters

2 Types of Mill and General Layout

between horizontal stands get eliminated. It also improves the quality of finished billet as the chances of guide mark and twisting of stock get reduced.

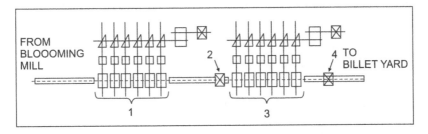

1.-First Continuous Group, 2- Pendulum Shear,
3-Second Continuous Group, 4- Flying Shear

Fig. 5.6 Layout of Continuous Billet Mill (With Group Drive).

The most widely used layout of a continuous billet mill with group drive, which was used until recently is shown in Fig. 7.4. The mill comprises of two groups of six stands. Each stand arranged consecutively in line with the delivery roll table of the bloom shear.

Ingots rolled in the blooming mill, to a cross-section of 200×200 to 250×250 mm and they were further reduced in a continuous billet mill.

The first group of stands produced billets 100×100 mm in cross-section and while the second group rolled the billets to a size of upto 50×50 mm. It is a well-known fact that increasing the final cross-section of the blooms considerably raises the production capacity of a blooming mill.

In this arrangement, the continuous billet mill also consists of two groups of six stands each with rolls from 720 to 800 mm in diameter (first group) and from 500 to 550 mm in diameter (second group). All the stands of each group are driven from a single 5,000 hp motor through a reducing gear and pinion stands.

Automatic turning of the bar is of great importance in these mills which do not have vertical roll stands.

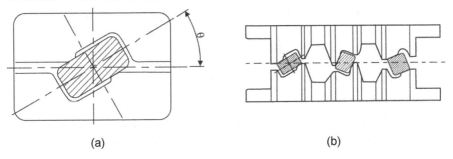

(a) (b)

Fig. 5.7 Bar Turning Devices.

Bars are turned in these mills either by helical twist guides [Fig. 5.7(a)] or by tilting rolls [Fig. 5.7(b)]. It is not advisable to use helical guide as most of surface defects of the metal (scratches, scoring etc.) comes due to the use of helical guide.

Roller twist guides, wherein bar slides along rotating rollers, can be used to reduce friction and guide wear.

The most appropriate solution of the problem of turning in these continuous billet mills was found in the use of tilting rolls.

These rolls have inclined passes. Upon entering these rolls, the end of the bar is twisted through a certain angle. In its further movement, the front end of the bar continues to twist about its axis and when it approaches the pass in the next stand the bar has been turned to the required angle. Tilting rolls are entirely free of the disadvantages of helical twist guides. They have approximately of the same diameter and barrel length that the working rolls have and are mounted in a special frame secured to the roll stand housings.

Characteristic features of modern continuous billet mills (Fig. 5.8) are:

Alternating stands with horizontal and vertical (edging) rolls and individual drive for each stand with speed variation facilities. The total available power of the motors driving each group will be higher than for the older mills with group drives.

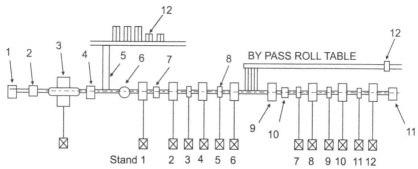

1. Receiving roll table, 2. Manipulator Tilter, 3. 1150 mm Blooming Mill, 4. 1000T Shear 5. Chain Transfer, 6. Turn Table, 7. 90° Tilter, 8. Rope transfer Table, 9. 250 T Pendulum Shear 10.45° Tilter 11. Flying Shear 12. Bloom Yard

Fig. 5.8 Layout of combined Blooming and Continuous Billet Mill.

The bar is reduced in these mills alternatively in the horizontal and vertical directions without turning. Therefore, the formation of cracks and scratches from the action of twist guides is excluded. Individual drive provides for proper speed adjustment; to simplifies mill settings and roll pass design of a continuous mill.

2 Types of Mill and General Layout

The first group comprises of six stands of which the 1st, 3rd and 5th are vertical rolls while the rest i.e., 2^{nd,} 4th and 6th are horizontal rolls. Each stand is driven by a separate 1,800 hp motor running at a speed in the range from 250 to 500 rpm.

The barrel length of the vertical rolls is of 1,000 mm; that of the horizontal rolls is 1,200 mm.

The second group also consists of six stands of which the 7th, 9th and 11th have vertical rolls while the rest i.e., 8^{th,}, 10th, 12th are of the horizontal roll type. The rolls of all the stands are of the same diameter i.e., 530 mm. The vertical rolls have a barrel length of 600 mm; and horizontal rolls are of 800 mm. Each stand of this group is also driven from a separate 1,800 hp motor with a speed range from 250 to 500 rpm.

The rolling speed in each successive stand of a continuous mill increases with the elongation of the bar.

The distance from the axis of the last stand in the first group and to the axis of the first stand in the second group is to be designed in such a manner, so that the longest bar can be accommodated on the roll table between these two groups. This will also facilitates, pushing out of the bar sideways after passing through the first group.

Transfers are provided to deliver bars from the first group sideways to a bypass roll table which, in turn, delivers the bar to the shear for cutting it into measured lengths.

The rolls of all the stands, are manufactured of cast steel or SGCI and run in four row roller bearings. The chocks of the rolls are mounted in two internal horizontal frames of rectangular form. The upper roller bearing is mounted in the upper chock, which has guide tongues at the sides. Through these tongues, the weight of the roll and chocks and any axial load in rolling, is transmitted to the upper internal frame. Through vertical side members, the upper and lower internal horizontal frames are linked together. The lower frame carries the chocks of the lower bearings. The lower frame rests on two housing screws while devices for balancing the vertical rolls are mounted in the side members. Thus, the vertical rolls are mounted in an ordinary stand of two housings (frames) having two lateral screw down devices. This stand, in turn, is mounted in two vertical housings on top of which the combination reducing gear is arranged for roll drive.

The outer vertical housings are erected with their feet on the bed plate and are linked together by separators of welded design (one above and two below). Housing screws, mounted in the lower cross-piece of these housings, adjust the height of the internal roll stand.

In the horizontal direction, the vertical rolls are adjusted by lateral screw downs mounted in the internal housings.

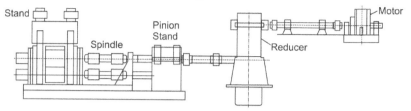

Fig. 5.9 Main line of a Horizontal stand in the first group of 730/530 mm Continuous Billet Mill.

The horizontal roll stand with its drive is illustrated in Fig. 5.9. The rolls are mounted in four row roller bearings; the housings are of the closed top type. Two roll adjusting devices–upper and lower–are incorporated to precisely adjust the rolls in height. The upper roll is balanced by springs. Since the line of rolling is constant for all stands of the mill. It is necessary to shift the rolls axially when a pass is worn-out excessively. This roll adjustment is accomplished by shifting the whole roll stand to the right or left. For this purpose, fastening of the housing to the bed plates is released, and the stand is to be moved by means of a lateral screw mechanism.

The universal joint spindles are designed so that they are extended from the pinion stand and follow the roll stand together with their spring over balancing units.

DC motors are used to drive the individual stands of a continuous billet mill. Speed variation simplifies roll pass design and ensures precise setting up of the mill. These motors are supplied by converters (motor generator sets) or mercury arc rectifiers.

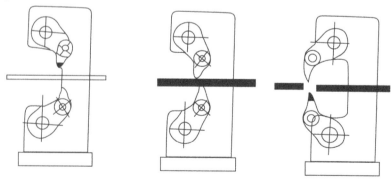

Fig. 5.10 Working of Flying Shear.

A tilter and pendulum shear are arranged before the second group of stands. These shear cuts off the front end of the bar when it has defects (Split end etc.)

3 Specific Features of Continuous Mill

which prevent it from being entered into the rolls of the first stand. The tilter turns the bar when rolling is performed in a diamond-square sequence.

A flying shear is (Fig. 5.10), designed to cut billets while they are in motion, to provided to cut long bars delivered from the second group into measured lengths, usually upto 12 m. The installation of this shear eliminated restrictions on ingot weight, facilitated an increase in mill output and enabled a blooming mill to be combined with a continuous billet mill.

In present day continuous mill, the delivery speed of bar after the finishing stand may reach upto 6 m/sec, flying shear works on eccentric crank principle.

This shear cuts off the front end of the bar of a length from 150 to 200 mm. The rear ends are of random length. These cut-off ends are removed as follows: the short front end drops beyond the blades on to an inclined chute from where it slides down to a side conveyor which takes it to a scrap pit. Here the scrap is removed by a magnet crane to railroad cars. All cut-off rear ends also drop on the chute and are removed by the conveyor.

A machine for stamping the side surface of the billet is installed after the flying shear. The marking is stamped on the side of each billet from 50 to 100 mm behind the front end.

Cooling beds are provided to allow the billets to cool before they are delivered to the billet yard.

All billets produced in billet mill are delivered to general store house, which is also called as billet yard. Billets are delivered to the hot racks of cooling beds. The billet drops in to loading cradle are removed by the crane and stacked in to piles. The billet is to be inspected and are placed by crane on the rack, where they are inspected. The billets are turned over by means of tilting conveyer.

If billets are rolled with alloy and highly alloy steels, the billet storage are equipped with pits for controlled cooling, as well as furnace for heat treatment of billets.

3 SPECIFIC FEATURES OF CONTINUOUS MILL

In case of continuous mill, the work piece is rolled simultaneously in two or more stands at a time *i.e.*, volume/second passing through each stand of a continuous mill is a constant.

As shown in Fig. 5.11

$$V_1 \times F_1 = V_2 \times F_2 = \text{........} = V_{n-1} \times F_{n-2} = V_n \times F_n$$

Where V and F are peripheral speed of roll and area of metal of various stands of a continuous group.

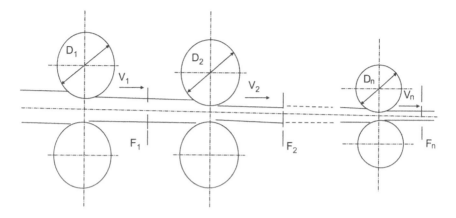

Fig. 5.11 Principle of a Continuous Mill

$$\frac{\pi \times D \times N(1+S) \times F}{60} = \text{Constant}$$

or, $F \times D \times N = C$, if forward slip is negligible.

Here, C = Rolling constant D = Rolling diameter of roll and N = Roll rpm = n/i, where n is motor rpm and i is a reducer ratio of pinion stand, F = Area of metal and S is forward slip.

3.1 -Advantages of Individually Driven Stand Over Group Combined Driven Stand

In a combined drive rolling mill, the selection of the metal section and rolling diameter of rolls at different stands of a continuous group is a very difficult task. The constant i.e., volume per second passing through each stand can only be varied either with roll diameter or with roll speed. In a combined drive of a group driven mill, the flexibility of doing speed maneuvering to change the rolling constant will not be possible. For, only remedy available, if roll diameters can suitably altered to meet the requirement. Roll diameter, in fact can only be varied within a limit of close tolerance.

Thus, group drives imposes limitation on product mix of the mill, while individual driven stand of a group can offer a wide range of speed regulation. Group driven continuous mills are still in use, where mill is required to roll only one specific section, thus considerably reduces the mill manufacturing cost.

In an individual driven continuous mill, loop growth and tension are completely eliminated by the adjustment of the speed of individual stand of the continuous group. This makes the simplification of technology and production of better quality products within close tolerance range.

4 PASS DESIGN DETAILS OF BILLET MILL

4.1 Open Box Pass

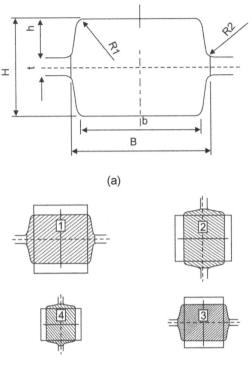

(a)

(b) Box-Box System

Fig. 5.12 Open Box System.

Box passes are used in the roughing group of billet mill, but it is seldom used in finishing train of billet mill because of its limitation of rolling finished billets within the close tolerance.

4.2 Diamond-Diamond System

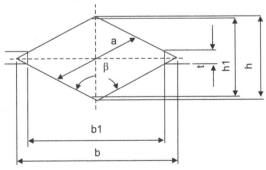

Fig. 5.13 Diamond Pass With Rounded Corners.

The diamond-diamond system sequence is applied directly after the box passes in open train billet mills, continuous billet mill and roughing stands of medium section mills. This sequence is generally used for rolling high quality steels.

The distinctive advantages of this system are:

- The possibility of obtaining a square cross-section from each diamond pass, by turning 90 degree between two passes of one pair of rolling stands.
- The possibility of obtaining square billets of several sizes from the same pass by varying the clearances between rolls.
- Simple and easy roll adjustments, when changing over of rolling to a different section or steel of different grades.
- Stability of stock in diamond pass is higher than other passes. Stock is seldom twisted during entry and rolling in diamond pass.
- No complex guides are required in rolling in diamond pass. This simplifies the working of rolling crew, as only side guides are used.

The diamond-diamond sequence has the following disadvantages.

- In comparison with other breakdown passes, diamond passes are sunk deeper into the rolls, therefore, weakens it to a greater extend.
- A square billets rolled in diagonal pass is of octagonal in shape. It has a detrimental effect, when such billets are loaded in and moved along in heating furnace.

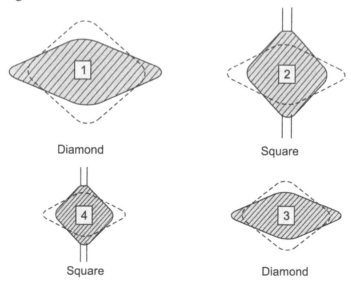

Fig. 5.14 Diamond – Square Sequence.

- Furnace scales are not properly removed from the bar during rolling in diamond pass, leading to formation of various defects. Diamond passes are never employed for initial breakdown passes for this reason. A combination of box and diamond pass is applied and is proven more effective.

4.3 Diamond-Square Sequence

In this sequence, diamond passes are placed alternate with square passes. After being delivered from a large square pass, the bar is turned to 90° and entered into a diamond pass. After delivery from the diamond pass, the bar is turned to 90° again and so forth. The diamond–square system usually applied for rolling square of comparatively angle varies from 100° to 125° and square angle varies from 90° to 96°, depending upon the placement of stand.

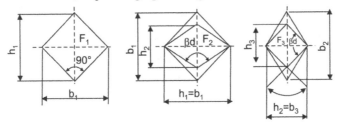

(a) Pass Designed on Principle- "Height of One Pass equals the width of next pass"

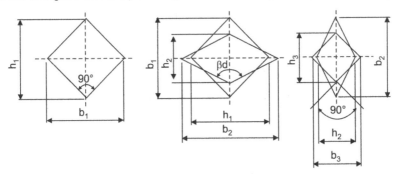

(b) Pass Designed on Principle- "Height of One Pass is less than the width of next pass"

Fig. 5.15 Different Pass design in Diamond-Square System.

The sequence is designed with following principles:

(a) The sequence is designed so that width (horizontal diagonals) of any given pass is equal to the height (vertical diagonals) of the preceding pass [Fig. 5.15 (a)]. In this case diamond pass is intermediate between two square passes.

178   Chapter 5   Rolling of Billets

With angle at the apex of diamond $\beta d = 120°$, the maximum co-efficient of elongation of diamond-square sequence will be

$$\mu_d = \mu_{sq} = \tan \beta d /2 = \tan 120°/2 = 1.732$$

(b) The sequence is designed so that the width (horizontal diagonals) of any given pass is larger than the height of preceding pass. (Fig. 5.15(b)).

In this sequence the coefficient of reduction will be a function of apex of diamond βd and co-efficient k, where k will be

$$K = (b_2 - h_1)/(b_1 - h_2)$$

4.3.1 Advantages of Diamond-square System

- Possibility of rolling squares of several sizes due to presence of different square in rolling sequence.
- Possibility of getting geometrically exact square with correct angle.
- Approximately even deformation over the whole width of cross-section.
- Shape of pass protects against formation of cracks, especially at corners, as corners of bar, the whole cross-section cools less extensively than the diamond-diamond system.

4.3.2 Disadvantages of Diamond-Square Sequence

- The use of diamond-square system makes the roll weaker, as this sequence has more depth of cut, in comparison with box passes; This sequence is always advised to be used in finishing train, with the use of the box sequence in roughing train.
- Quicker wear out of pass due to wide difference in the peripheral speed along the width of the pass.

4.4 Oval-Square System

In this system, the stock after its delivery from preceding square pass enters flat down into the following oval pass. After leaving the oval pass, the stock is turned to 90° and then enters to the next succeeding smaller square pass. The oval-square pass sequence is the most effective and heavy reduction sequence of the break down sequences. It is preferred mainly for rolling smaller square sections and wire rod mill. This is rarely used for billet production. However, this system can be recommended only for billet rolling, when there is a shortage of stands in Billet mill and there is no limitation from motor power side for giving higher reduction.

5. Selection of a Pass Sequence

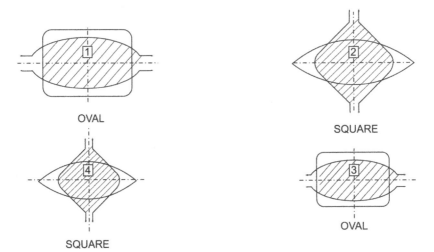

Fig. 5.16 Oval –Square System

4.4.1 Advantages of Oval-square System

As shown in Fig. 5.16, due to the continuous renewal of the corners and turning of bar in each pass of oval-square system, it makes to ensure the uniform distribution of temperature across the cross-section of the stock and it enables to achieve uniform properties of finished billets. Besides this, corner renewal avoids tensile stress concentration at the corner of stock.

4.4.2 Disadvantages of Oval-square System

- The shape of pass combined with the chances for scale to enter at the roll opening. It gives rise to conditions, where scale breaks out from oval and got pinched in the bottom fillet of square pass and causing impression along the length of finished billet.
- Rapid and uneven wear out of passes, especially in oval pass in this system.

5 SELECTION OF A PASS SEQUENCE

The selection of a pass sequence for billet rolling depends upon –

(a) Overall reduction required from bloom A to finished billet B.

(b) Number of passes available i.e., N, then overall reduction will be $\dfrac{A}{B} \times 100\%$ and, overall elongation will be $E = \dfrac{A}{B}$

- If, alternate passes are square in this system, then square to square reduction will be $\dfrac{N/2}{A/B}$.
- If, it is less than 1.5, the open square system may be considered.

- If, it is not greater than 1.86, a diamond-diamond sequence may be considered.
- If, it is more than 1.86, the oval-square system may be considered.
- The reduction between two passes of square to square system may be proportioned as mentioned below:

Square-oval : Oval-square = 1.0 : 0.70

Square-diamond : Diamond-square = 1.0 : 0.83

6 COMPARATIVE STUDY OF BILLET MILLS

A comparative study of six different types of billet mill is placed at Table 5.1

Table 5.1

Equipment & Facilities	1	2	3	4	5	6
Stand	5 Stand (semi-continuous)	12 Stands (continuous)	14 Stands (continuous)	10 Stands (continuous)	12 Stands (continuous)	10 Stands (continuous)
Roll Dia. in mm	960 / 700	950/750/650	900/700/500	1000/850/650	1000/700/500	1000/700/500
Input size in mm	240 × 350	300×300 - 360×360	< 370×370	280 × 280	320 × 320	280 × 280 - 320 × 320
Output size	150 x 150	70 x 70 - 240-240 sq.	80 x 80 - 200-200 sq	75 x 75 - 125-125 sq.	85 x 85 - 150-150 sq.	75 x 75 - 130-130 sq.
Capacity per year	2.0 million tons	1.7 million tons	5.5 million tons	2.7 million tons	1.26 million tons	2.0 million tons

7 LAYOUT OF CONTINUOUS BILLET MILL

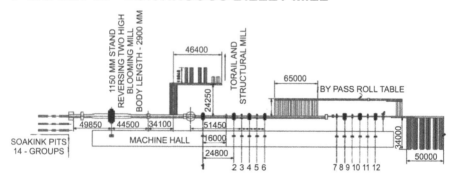

Fig. 5.17 Layout of Continuous Billet Mill (With Individual drive for each stand).

As shown in Fig. 5.17 above, there are 12 stands in continuous Billet Mill. Stand 01 and 02 are 1000 mm horizontal Stands. There is a 90° tilter between these two stands, to tilt the work piece to 90°, Bloom 325×325 mm is directly fed from 1150 mm blooming mill to stand 01 of the Billet mill.

All these stands are having box passes only. Maximum reduction in these two stands *i.e.*, 01 and 02 are limited only by angle of bite and motor capacity.

Intermediate group consists of four stands having equal numbers of horizontal and vertical stands (two each). This group is called as 700 mm group (stand 3 to 6), whereas 700 mm is the distance between the pinion centers. This group is having box passes and the use of horizontal and vertical stands are in sequence to avoid the use of tilter between stands.

The 500 mm finishing group (stand 6 to 10) consists of six stands (equal numbers of horizontal and vertical stands). There is a transfer bed, which separates intermediate group from the finishing group. There is a tilter before the first stand of finishing group, to turn work piece by 45°, to facilitate the entry of work piece in diamond-square sequence of finishing group. The finishing billet moves to the cooling bed of billet mill, before being cut at the required length by flying shear placed after the finishing stand.

In case of breakdown/shutdown in finishing group, flexibility in the layout has been provided by keeping provision for the diversion of square coming out of intermediate group to roll-out sideways into a bypass roll table for time being, until the finishing group get ready. By this rolling in first group will continue without the interruption of production. A separate line with roll tables is installed in parallel to finishing group. A down cut shear of 800 T capacity is placed to cut this square to any length.

8 ROLL PASS DESIGN OF BILLET 90 mm × 90 mm

Once, data related to the size of input bloom and finished billet section are firmed up, (decided on the basis of input billet size requirement of the down line finishing mills *viz*, Bar and rod mill ,light structural and Wire rod mill), Roll pass designer has to then firm up the reduction schedule of the billet mill.

Reduction in each stand of roughing, when intermediate and finishing group of mill is based upon the capabilities of mill (Fig. 5.18) like:

- What are the strength and weakness of the mill, in which this product is planned to roll.
- The type and layout of the mill, production capacity of furnace, mill proper and finishing section of the mill.
- Size of billet available to roll this section.
- Temperature of input billet and finished product temperature.
- Type of layout of mill *i.e.*, whether it is an open, semi continuous or continuous mill.

The details of mill facilities *viz.*, mill configuration *i.e.*, numbers of rolling stands, numbers of vertical stands available in each group. The distance between stands and in between each group, stand size, type of stand to be used for rolling *i.e.*, whether it is open or close type of stands, which is to determine the amount of mill spring and ease of roll changing. The type of roll used to specified rolling load and stand motor capacity and numbers of tillers in the mill and its spacing from preceding and succeeding stand. Shifting facilities to transfer the bar from main line bypass line and its length decides the maximum length of bloom, which can be taken up for rolling.

All above mentioned parameters decide the type of roll pass design schedule, optimum numbers of passes to be used in the sequence and also to determine the type of reduction pattern to be adopted.

Following are details of information, which should be available with roll pass designer to design a billet section:

8.1 Details Regarding Input Bloom & Finished Product

(*a*) The cross sectional of input material, *i.e.*, Bloom size, length and billet weight with positive tolerance.

(*b*) The smallest billet size to be rolled in the billet mill *i.e.*, cross-section size, length and billet weight with negative tolerance.

(*c*) Any intermediate sizes, which may be required as input material for rolling in finishing mill.

8.2 Details Regarding Mill

Following details are required to be collected for the billet mill (Fig. 5.18)

• Maximum diameter of rolls.

• Minimum diameter of rolls.

• Length of roll body.

• Maximum opening of rolls.

• Minimum opening of rolls.

• Roll neck diameter.

• Maximum rolling load in tons.

• Maximum rolling torque in Tons-Meters.

• Motor capacity in kW.

• Other information like.

(*a*) Length of roll table between stand 1 and 2, to have free tilting of work piece by 90°, maximum length of stock which can be tilted, between stand 1 and 2, in above cited layout, is 14 meter.

(*b*) Length of transfer table between intermediate and finishing group, to accommodate the stock freely before entering into finishing group of billet mill the length is 65 m.

8 Roll Pass Design of Billet 90 mm × 90 mm 183

(*c*) Details of Tilter before 500 mm group
- Max. section of turned billet: 150 mm×150 mm
- Max. length of turned billet: 58 m
- Max. weight of turned billet: 6.3 Tons.
- Angle of turning: 45°- 53.5°

(*d*) Details of Flying shear
- Max. section to cut: 106 mm×106 mm
- Min. section to cut: 50 mm×50 mm
- Accuracy of cut: ± 0.5%
- Rated speed of cut with section
  100 mm×100 mm: 1.8 m/sec.
  75 mm×75 mm: 3.13 m/sec.
  60 mm×60 mm: 4.36 m/sec.
- Length of cut section: 5 m / 6 m / 7.5 m / 9 m / 10 m / 12 m

(*e*) 800 T Shear
  Max. Blade pressure: 800 T
  Max. section to be cut: 150 mm×150 mm.

8.3 Tolerance of Billet

Table 5.2

Size of Square mm	Tolerance on sides mm	Tolerance on diagonal ± mm	Sectional weight kg/m
80 × 80	± 2	5	48.5
90 × 90	± 2	5	61.5
100 × 100	± 2	5	76.2
125 × 125	± 3	6	118.5
150 × 150	± 4	8	172.0

Lastly, the size of intermediate section *i.e.*, the final section coming out from the intermediate group is to be firmed up.

With the consideration of above stated factors, the size 150 mm×150 mm is selected as the best intermediate section for this mill.

8.4 Computation of Reduction

A. Maximum reduction:

It is computed as follows:

1. Total reduction for intermediate group

H–325 mm B–325 mm

Where H and B are height and width of input bloom

$$h - 150 \text{ mm } b - 150 \text{ mm}$$

Where h and b are height and width of output billet

$$\sum \Delta h = 1.15 \, [(H - h) + (B - b)]$$
$$= 1.15 \, [(325 - 150) + (325 - 150)]$$
$$= 1.15 \times 350$$
$$\sim 400 \text{ mm}$$

2. Total reduction for finishing group

$$H - 150 \text{ mm } B - 150 \text{ mm}$$

Where and B are height and width of input square, $h - 90$ mm $b - 90$ mm, Where and b are the height and width of finished billet

$$\sum \Delta h = 1.15 \, [(150 - 90) + (150 - 90)]$$
$$= 1.15 \times 120$$
$$\sim 138 \text{ mm}$$

B. Maximum Reduction

$$\Delta h \text{ max.} = D \, (1 - 1/ \sqrt{1+f^2})$$

Where, $f = K_1 \times K_2 \times K_3 \, (1.05 - 0.0005 \, t°C)$

Co-efficient $K_1 = 1$, for steel roll $= 1$,

$K_2 = 1$, for rolling speed < 1 m/sec.

$K_3 = 1$, for rolling material of plain steel composition

(*i*) For stand 1 & 2

$$= 1060 \left(1 - \frac{1}{\sqrt{1 + (1.05 - 0.0085 \times 1000)^2}} \right)$$

Where, 1060 mm is maximum diameter and 1000°C is the rolling temperature.

~ 132 mm

(*ii*) For 700 mm Group (Roughing)

~ 90 mm

(*iii*) For 500 mm Group (Finishing)

~ 65 mm

C. Average Reduction

$$\Delta_{av} = (0.8 \sim 0.9) \, \Delta h \text{ max.}$$

(*i*) For 1000 mm stand

$$= 0.8 \times 132 = 105 \text{ mm}$$

(*ii*) For 700 mm Group

$$= 0.8 \times 90 = 72 \text{ mm}$$

(*iii*) For 500 Group
$$= 0.8 \times 65 = 52 \text{ mm}$$

Keeping in view of the above parameters, average reduction of each individual stand can be decided, which may be limited again by the capacity of motor of each stand.

8.5 Roll Pass Design of Finishing Group

8.5.1 Stand 12 (Finishing)

As per the basic principle, roll pass design starts with finishing pass only.

Here, size of the finished billet is 90 mm × 90 mm

Hot size of billet will be, size of billet × (1 + 0.000012 × *t*°C), *i.e.* the finishing temperature of billet, coming out from finishing stand, the desirable finishing temperature is around 850°C.

Then hot section size = cold section × factor for expansion at 850°C

$$= 90 \times 1.011$$

$$\sim 91.0 \text{ mm}$$

The final dimension of finishing pass is to be determined on the basis of tolerances specified on the finished product. Here, in this case, the size tolerance on 90 × 90 mm size billet is given as ± 2 mm. Designer should select the size in such a way, so that even after the cooling of the billet, it should not become less than 88 mm or should not go beyond 92 mm.

Generally, designer recommend for hot size is to be taken same as equivalent to cold size of billet, that's why 90 mm is selected as hot size of billet, which will give finished product within tolerance *i.e.*, in between 88–92 mm.

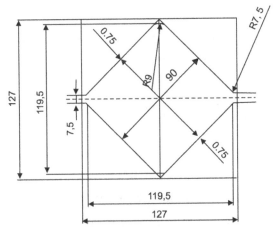

Fig. 5.18 Pass Design of Stand 12.

$$F_{12} = 90 \times 90 = 8100 \text{ mm}^2$$

$$B = H = \sqrt{2} \times 90 \text{ mm}$$

where B & H are width and height of the finishing pass.

$$= 1.414 \times 90 \sim 127 \text{ mm}$$

Radius $\qquad r = (0.1 - 0.15) \times A_{12}$

$$= 0.01 \times 90 = 9 \text{ mm}$$

$$H_K = H - 0.83 \times r$$

$$= 127 - 0.83 \times 9 \sim 119.5 \text{ mm}$$

$$t = (0.1 - 0.2) \times D$$

where t is the roll gap and D is the diameter of finishing stand.

$$= 0.15 \times 500 = 7.5 \text{ mm}$$

$$B_K = B - t,$$

$$= 127 - 7.5 = 119.5 \text{mm}$$

where B_K is effective width of the pass

$$F_{K12} = (F - \text{Area of four corners})$$

$$= 8100 - 70 = 8030 \text{ mm}^2$$

Rolling Dia, $\quad D_K = D_c - \dfrac{F_{k12}}{B_k}$

$$= 540 - \dfrac{8030}{119.5}$$

$$\sim 472.5 \text{ mm}$$

Rolling constant $= \dfrac{(F_{12} \times D_K \times N)}{i}$

Where N and i are the motor rpm and reducer ratio respectively.

$$= \dfrac{8030 \times 472.5 \times 200}{1.605} = 4.733 \times 10^8$$

Velocity $= \dfrac{(\pi \times DK \times N)}{60 \times i}$

$$= \dfrac{3.14 \times 472.5 \times 200}{60 \times 1.605 \times 1000}$$

~3.1 M/ sec, where speed is selected as 200 rpm,

$S = 0.75$ mm, Concavity of 0.75 mm is provided on the all sides of the square to take care of

1.5 Sulphur Dioxide

(*a*) Wear out in subsequent rolling in same pass.

(*b*) It also helps in movement of concave billets on the roll tables of cooling bed, while a convex billets will have no stability on roll table.

(*c*) To avoid fins in subsequence passes.

(*d*) Concave billets will avoids the jumping of billets in reheating furnace of the finishing mill, in place of use of convex/bulge billets.

Angle of bite is also required to be computed to know whether reduction assigned is within the range of angle of bite.

Angle of bite, $\cos \alpha = 1 - \Delta h_{12} / D_k$
$$= 1 - 16.5 / 472.5 = 0.962$$

Where, $\Delta h = 16.5$ mm, $D_K = 472.5$ mm

Angle of bite, $\alpha = 16°$

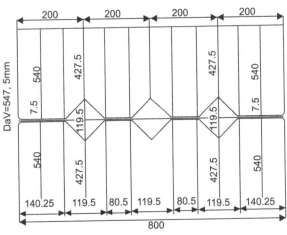

Fig. 5.19 Roll Diagram of Finishing Stand-12 of Billet 90 mm.

Co-efficient of reduction $\mu_{12} = F_{K11}/F_{K12}$
$$= 8763/8030 = 1.09 \text{ or } 9\%$$

which will facilitates the more life of roll due to less wear out of finishing stand, in turn improve the production of mill by fewer pass/roll changing.

8.5.2 Stand 10

$$\mu_{10} = F\,Sq._{12} \times \mu_{12-10}$$

Assumption, $\mu_{12-10} = 1.25$ (it should be taken in the range of 1.20 –1.30)

then, $\quad \mu_{10} = 90^2 \times 1.25 = 10125$ mm^2

Side of square $= \sqrt{10125}$

= approx. 100.6 mm, side of Square for pass-10 is selected as 100 mm.

$$B_{10} = H_{10} = 1.414 \times 100$$
$$= 141.41 \text{ mm or say } 141 \text{ mm}$$

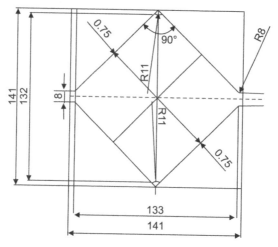

Fig. 5.20 Pass Design of Stand 10

Assumption Roll gap, $t = 8$ mm (Keeping in view of mill spring, tolerance etc.)

$$B_K = B - t = 141 - 8 = 133 \text{ mm}$$

Radius r is taken as 10-12 % of side of pass,

r is assumed as 11 mm, i.e. 11% of the 100 mm side

$$\tan \beta/2 = B/A$$
$$= 141/141 = 1, \text{ i.e. equal to } 45°$$
$$\beta = 90°$$
$$\alpha = 180 - \beta = 90°$$

Actual area

$$F_{K10} = F_{10} - \text{area of four corners}$$
$$= 9992 \text{ mm}^2$$

Rolling Diameter,

$$D_K = D_C - F_{K10}/B_K$$
$$= 540 - 9992/133 = \text{say, } 465 \text{ mm}$$

8 Roll Pass Design of Billet 90 mm × 90 mm

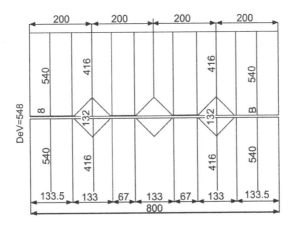

Fig. 5.21 Roll Diagram of Stand-10 (Horizontal).

Angle of bite,
$$\cos \alpha = 1 - \frac{\Delta h_{10}}{D_K}$$

Where, $\Delta h = 37.5$ mm and $D_K = 467$ mm

$$\cos \alpha = 1 - 37.5/465$$
$$= 0.92 = 22.8°$$

Rolling constant $= 4.733 \times 10^8$

Motor speed, $N = \dfrac{C \times i}{F_{10} \times D_K}$

$$= \frac{4.733 \times 10^8 \times 2.808}{9992 \times 465 \times 1000} = 285 \text{ rpm}$$

Rolling speed, $V = \dfrac{(\pi \times D_K \times N)}{60 \times i}$

$$= \frac{3.14 \times 465 \times 285}{60 \times 2.808 \times 1000}$$

~ 2.45 m / sec.

Coefficient of reduction $\mu_{10} = F_{K09}/F_{K10}$
$$= 11984/9992 = 1.20.$$

8.5.3 Stand–11 (Vertical)

After formulating the design parameter of square–square pass, *i.e.*, for stand 10 and 12. It is now require to design the intermediate diamond pass *i.e.*, for stand 11 of the sequence. First of all, appropriate low co-efficient of reduction (μ_{12}) is selected for the finishing pass to derive the area of the diamond pass.

It should be kept as minimum as possible, generally, in the range of 1.10-1.15, as temperature is on the lowest point at this time. With the selection of lower value of reduction in the finishing pass, the wear out will be less. It will also avoid the frequent pass and roll changing and, in turn mill availability will get increased.

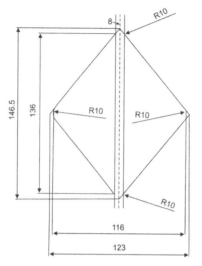

Fig. 5.22 Pass Design of Stand 11 (Vertical)

μ_{12} is assumed as 1.10

$$F_{11} = F_{12} \times \mu_{12}$$
$$F_{11} = 8100 \times 1.10 = 8910 \text{ mm}^2$$
$$A_{10} = 100 \text{ mm}, a_{12} = 90 \text{ mm}$$

Where A_{10} and a_{12} are the sides of pass 8 and 10 respectively

$$B_{11} = 1.95 \times A_{10} - 0.54 \times a_{12}$$
$$= 1.95 \times 100 - 0.54 \times 90$$
$$\sim 146.5 \text{ mm}$$

$$H_{11} = 1.8\, a_{12} - 0.39\, A_{10}$$
$$= 1.8 \times 90 - 0.39 \times 100$$
$$= 123 \text{ mm}$$

b/h ratio of diamond = B_{11} / H_{11}

$$= 146.5 / 123$$
$$= 1.19, \text{ it is o.k. As it is} \leq 1.5$$

8 Roll Pass Design of Billet 90 mm × 90 mm

$$r = r_{12} \times \mu_{12}$$
$$= 9 \times 1.09$$
$$= \text{say, } 10 \text{ mm}$$

Roll gap t is assumed as 8 mm, keeping in view of mill spring and wear out of pass.

$$Bк = (H_{11} - t) \times B_{11}/H_{11}$$
$$= (123 - 8) \times 146.5 / 123$$
$$\sim 136 \text{ mm}$$
$$Hк = H_{11} - 2 \times r \left[\sqrt{[1 + (H_{11}/B_{11})^2]} - 1 \right]$$
$$= 123 - 2 \times 10 \left[\sqrt{[1 + (123/146.5)^2]} - 1 \right]$$
$$\sim \text{say } 116 \text{ mm}$$
$$F_{к11} = 8763 \text{ mm}^2$$

Rolling Dia. $D_к = D_c - F_{к11}/B_к$
$$= 540 - 8763/136$$
$$\sim 475.5 \text{ mm}$$
$$\tan \beta/2 = B/A$$
$$= 123/146.5$$
$$\beta = 79.9°$$
$$\alpha = 180 - \beta$$
$$= 180 - 79.9° = 100.1°$$

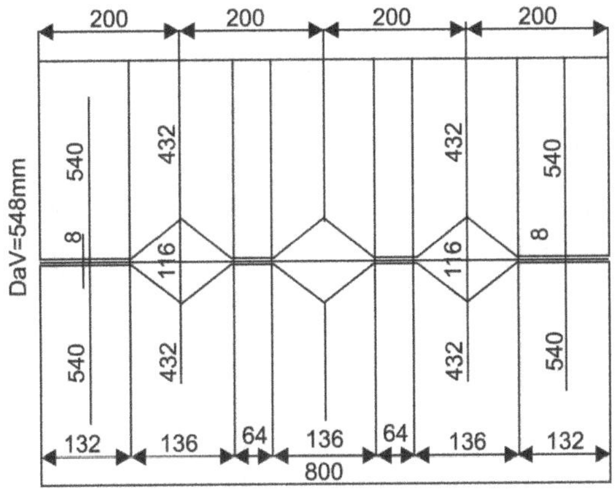

Fig. 5.23 Roll Diagram of Pre-finishing Stand-11 (Vertical)

Rolling constant,

$$C = 4.733 \times 10^8$$

$$N_{11} = \frac{C \times i}{F_9 \times D_k}$$

$$= \frac{4.733 \times 10^8 \times 2.200}{8763 \times 475.5}$$

~ say 250 rpm

$$V_{11} = \frac{\pi \times D_\kappa \times N}{60 \times i}$$

$$= \frac{3.14 \times 475.5 \times 250}{60 \times 2.20 \times 1000}$$

$$= 2.82 \text{ m /sec.}$$

$$\cos \alpha = 1 - \Delta h / D_k$$

Where, $\Delta h = 17$ mm

$$= 1 - 17/475.5 = 15.4°$$

Co-efficient of reduction $\mu_{11} = F_{K10}/F_{K11}$

$$= 9992/8763 = 1.14$$

8.5.4 Stand-8

Stand 8 will be a square pass and parameters of this pass shall be fixed first, before designing the intermediate diamond pass 9. First of all, the reduction from pass 10 to 8 (μ_{10-8}) is to be assumed, based on the same consideration explained above for designing pass 10.

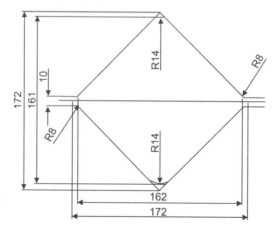

Fig. 5.24 Pass Design of Stand 08 (Horizontal).

In stand 8, depending upon the requirement of the down placed finishing mill, we have to select a square size, in between the size of input *i.e.*, 150 sq.

8 Roll Pass Design of Billet 90 mm × 90 mm

and pass-10 sq *i.e.*, 100 mm². At this stage, higher reduction is suggested, as stock is hotter in pass- 8 than at pass-10. With these considerations, a size 122 × 122 mm is selected for the square pass of stand 8.

$$F_8 = 122 \times 122 = 14884 \text{ mm}^2$$
$$\mu_{10-8} = 14715 / 9992 = 1.47$$
$$B_8 = H_8 = 122 \times \sqrt{2} = 172 \text{ mm}$$

Roll gap, t_8 is assumed as 10 mm

$$B_K = (H_8 - t_8)$$
$$= 172 - 10 = 162 \text{ mm}$$

r_8 is taken as 11% of side *i.e.*, 122 mm

$$= 122 \times 0.11 = \text{approx. } 14 \text{ mm}$$
$$H_K = H_8 - 0.83 \times r_8$$
$$= 172 - (0.83 \times 14)$$
$$\sim 161 \text{ mm}$$

$F_{K8} = (F_8 -$ Area of four corners), where F_{K8} is actual area.
$$= 14715 \text{ mm}^2$$

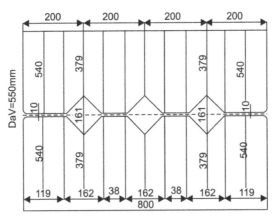

Fig. 5.25 Roll Diagram of Stand-8 (Horizontal).

$D_K = D_c - F_K / B_K$, where D_K is rolling diameter
$$D_K = 540 - 14715 / 162 = 449 \text{ mm}$$
$$\tan \beta/2 = H_K / B_K = 161/162$$
$$= 0.993 \text{ or } 44.8°$$
$$\beta = 44.8 \times 2 = 89.6°$$
$$\alpha = 180 - \beta = 180 - 89.6 = 90.4°$$

Rolling constant, $C = 4.733 \times 10^8$

$$N_8 = \frac{C \times i}{F_K \times D_K}$$

$$= \frac{4.733 \times 10^8 \times 4.00}{14715 \times 449} = 285 \text{ rpm}$$

$$V_8 = \frac{(\pi \times D_K \times N)}{(60 \times i)}$$

$$= \frac{(3.14 \times 449 \times 285)}{60 \times 4.00 \times 1000}$$

~ 1.68 m/sec.

$\cos \alpha = 1 - \Delta h / D_K$

Where, $\Delta h_8 = 50$ mm

$= 1 - 50/449 = 0.889$ or $27.2°$

Co-efficient of reduction $\mu_8 = F_{K7} / F_{K8}$

$= 17742/14715 = 1.20$

8.5.5 Stand-9

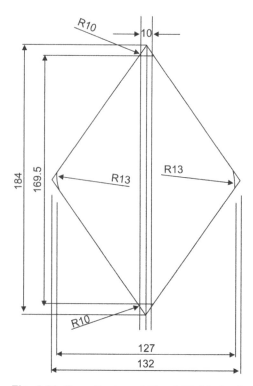

Fig. 5.26 Pass Design of Stand 09 (Vertical)

8 Roll Pass Design of Billet 90 mm × 90 mm

$A_8 = 122$ mm, $a_{10} = 100$ mm

Where A_8 and a_{10} are the sides of pass 8 and 10 respectively

$B_9 = 1.95 \times A_8 - 0.54 \times A_{10}$

$ = 1.95 \times 122 - 0.54 \times 100$

$ \sim 184$ mm

$H_9 = 1.8 \times A_{10} - 0.39 \times A_8$

$ = 1.8 \times 100 - 0.39 \times 122$

$ = 180 - 48 \sim 132$ mm

b/h of diamond pass $= B_9 / H_9$

$\phantom{b/h \text{ of diamond pass }} = 184/ 132 = 1.39$, it is o as ratio is ≤ 1.5

$r_9 = r_{10} \times \mu_{10} = 11 \times 1.2$

$ = $ approx. 13 mm

Roll gap t is assumed as 10 mm

$B_\kappa = (H_9 - t) \times B_9 / H_9$

$ = (132 - 10) \times 1.39$

$ \sim 169.5$ mm

$H_k = H_9 - 2 \times r_9 (\sqrt{1 + (H_9 / B_9)^2} - 1)$

$ = 132 - 2 \times 13 (\sqrt{1 + (132/184)^2} - 1)$

$H_\kappa = $ approx. 127 mm

$F_{\kappa 9} = 11984$ mm^2

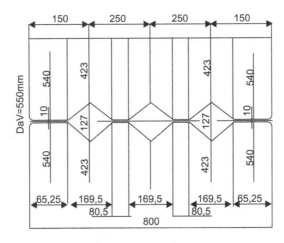

Fig. 5.27 Roll Diagram of Stand-09 (Vertical)

Rolling Dia. $D_\kappa = D_c - F_\kappa / B_\kappa$

$$= 540 - 11984/169.5 = 470 \text{ mm}$$

$$\tan \beta/2 = H_\kappa / B_\kappa$$

$$= 127/169.5$$

$$= 0.749 \text{ or } \beta/2 = 36.8°$$

$$\beta = 73.6°$$

$$\alpha = 180 - \beta = 180 - 73.6°$$

$$= 106.4°$$

Rolling constant, $C = 4.733 \times 10^8$

$$N_9 = \frac{C \times i}{F_9 \times D_\kappa}$$

$$= \frac{4.733 \times 10^8 \times 3.46}{11984 \times 470} = 290 \text{ rpm}$$

$$V_9 = \frac{\pi \times D_\kappa \times N}{60 \times i}$$

$$= \frac{3.14 \times 470 \times 290}{60 \times 3.46 \times 1000}$$

$$\sim 2.07 \text{ m /sec.}$$

$$\cos \alpha = 1 - \Delta h / D_\kappa$$

Where, $\Delta h_8 = 35$ mm

$$= 1 - 35/471 = 0.925 = 22.20°$$

Co-efficient of reduction $\mu_9 = F_{K8}/F_{K9}$

$$= 14715/11984 = 1.22$$

8.5.6 Stand-7

$$A_6 = 150 \text{ mm}, a_8 = 122 \text{ mm}$$

$$B_7 = 1.95 \times A_6 - 0.54 \times a_8$$

$$B_7 = 1.95 \times 150 - 0.54 \times 122$$

$$= 292.5 - 66.0 = 226.5 \text{ mm}$$

$$H_7 = 1.8 \times a_8 - 0.39 \times A_6$$

$$= 1.8 \times 122 - 0.39 \times 150$$

$$= 219.6 - 58.5 \sim \text{say, } 161.0 \text{ mm}$$

8 Roll Pass Design of Billet 90 mm × 90 mm

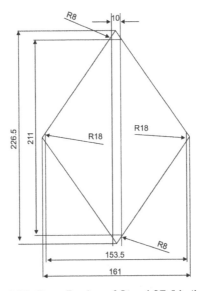

Fig. 5.28 Pass Design of Stand 07 (Vertical).

$B_7/H_7 = 226.5/161.0$

$\qquad = 1.40$, which is ok, as it should be ≤ 1.5,

$r_7 = r_8 \times \mu_8$

$\qquad = 14 \times 1.20$, say 18 mm

Roll gap t is assumed as 10 mm

$B_K = (H_7 - t) \times B_7/H_7$

$\qquad = (161 - 10) \times 1.40$

$\qquad \sim$Say 211 mm

$H_K = H_7 - 2 \times r_7 (\sqrt{1 + (H_7/B_7)^2} - 1)$

$\qquad = 161 - 2 \times 18 (\sqrt{1 + (161/226.5)^2} - 1)$

$\qquad =$ say. 153.5 mm

$F_K = 17742$ mm^2

Rolling dia. $D_K = D_c - F_K/B_K$

$\qquad = 540 - 17934/211 =$ say 455 mm

$\tan \beta/2 = H_K/B_K$

$\qquad = 153.5/211 = 0.726$ or $\beta/2 = 36°$

$\beta = 72°$

$\alpha = 180 - \beta$

$\alpha = 180 - 72° = 108°$

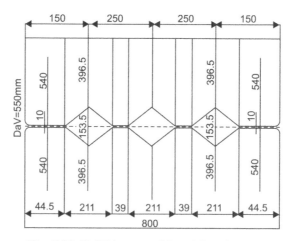

Fig. 5.29 Roll Diagram of Stand-07 (Vertical).

Rolling constant, $C = 4.733 \times 10^8$

$$N_7 = \frac{(C \times i)}{(F_5 \times D_K)}$$

$$= \frac{4.733 \times 10^8 \times 5.43}{17742 \times 455 \times 1000}$$

~ 310 rpm

$$V_7 = \frac{(\pi \times D_K \times N)}{60 \times i}$$

$$= \frac{3.14 \times 455 \times 310}{60 \times 5.43 \times 1000}$$

~ 1.37 m /sec.

The stock from stand 6 enters 45° into the vertical stand-7, with the help of twist guide, the diagonal of metal 150×150 mm is 196 mm as shown in figure 7.30,

$$\Delta h_7 = 196 - 153.5 = 42.5 \text{ mm}$$

$$\cos \alpha = 1 - \Delta h / D_K$$

Where, $\Delta h = 42.5$ mm

$$= 1 - 42.5 / 455 = .907 = 24.9°$$

Co-efficient of reduction $\mu_7 = F_{K6}/F_{K7}$

$$= 21858/17742$$

$$= 1.23$$

8.6 Roll Pass Design of Roughing and Intermediate Group

Roughing group of Billet mill consists of 6 number of stands, out of that 1 and 2 are of 1000 mm horizontal stands and there is a tilter in between 01 and 02 stands to tilt the stock 90°C. Input bloom is of size of 325 × 325 mm (hot dimensions). The intermediate group has 4 number of stands of 700 mm size, having vertical and horizontal stands alternatively. Stand 1 and 3 are vertical and stand 2 and 4 are horizontal stands. The output size of intermediate group is 150 × 150 mm. The maximum reductions for 1000 mm and 700 mm group are arrived by mathematical computation and are 132 mm and 67 mm respectively. For giving of such amount of reduction, the Box–Box sequence is selected. Higher reduction is provided in earlier passes, when metal is very hot, because of that the resistance to deformation will be less. Here, the reduction will be only limited by the angle of bite and the motor capacity. Higher diameter steel rolls are selected for billet mill to improve the angle of bite.

First of all, a rough scheme is to be worked out, taking into consideration of assumption of reduction, based upon the maximum and average reduction work out and spread in individual passes (approximately about 20-30 % of reduction). In this layout, Roll pass design starts with pass-6, which is the last pass of the group.

8.6.1 Stand–6

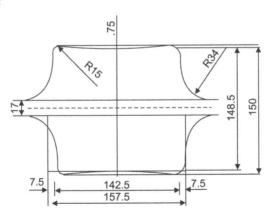

Fig. 5.30 Pass Design of Stand 06 (Horizantal).

H_6 – 150 mm, B_6 – 150 mm

F_6 = 150 × 150

Roll gap, $t = (0.015 - 0.025) \times D_{max}$

Where, D_{max} is max. roll diameter of stand i.e., 730 mm, the roll gap is selected on higher side, depending upon the mill spring, and flexibility of working with higher wear out and to have less pass/roll changing for finishing pass,

$$t = 0.023 \times 730, \text{ or say, } 17 \text{ mm}$$

$$h_6 = \frac{(H_6 - t)}{2} = \frac{(150 - 17)}{2} = 66.5 \text{ mm}$$

$a = h_6 \times \tan \alpha$, taper on the side for this pass, will be in the range of 10-15%

$$= 66.5 \times 11 \% \text{ or say, } 7.5 \text{ mm}$$

$b_6 = B_6 - 2 \times a$, where b_6 is the pass root width.

$$= 150 - 2 \times 7.5 = 142.5 \text{ mm}$$

Concavity at the width of pass ("s") is provided, to facilitate the easy movement of billet on the roll table and it helps to facilitate the easier restoration of pass during redressing of roll, after it use in the mill. "S" is generally given between 0.50 – 1.0 mm (0.75 mm is selected for this design).

Actual area of pass = F_6 – [(area of 4 corners + area of top and bottom concavity)]

$$= 21850 \text{ mm}^2$$

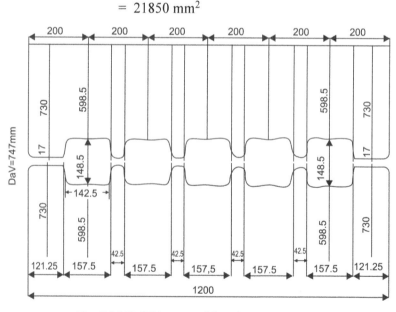

Fig. 5.31 Roll Diagram of Stand-06 (Horizontal).

Rolling diameter, for box pass can be directly computed by the formula given below:

$$D_K = D_c - 2 \times (h_4 - s)$$
$$= 730 - 2 \times (66.5 - 0.75) = 598.5 \text{ mm}$$

8 Roll Pass Design of Billet 90 mm × 90 mm

Radius, r_6 is assumed as 10% of side of pass *i.e.* b_6

$$r_6 = 150 \times 0.10 = 15 \text{ mm}$$

Angle of bite, $\cos \alpha = 1 - \Delta h/D_\kappa$

Δh_6 is assumed as 37 mm, then

Angle of bite $= 1 - 37 / 598.5$
$= 0.938$ or $20.3°$, which is well within the limit

Outer radius R is generally taken to the value of 2-3 times of the roll gap, *i.e.*

$$R = 2 \times 17 = 34 \text{ mm}$$

An angle of bite and rolling load is the main concern during rolling of Billet profile. As the selection of motor speed is of prime importance. Rolling load increases with increase in motor speed at the direct proportional and at the same time, with the increase of motor speed, the angle of bite reduces drastically.

That's why the lowest rpm range of motor is always selected for initial passing. Motor rpm given is $200 - 400$ rpm.

Assumption, $N = 210$ rpm

Rolling constant $= (F_\kappa \times D_\kappa \times N) / i$

Where N and i are the motor rpm and reducer ratio respectively.

Reducer ratio for stand 6 = 5.25

Then, $\qquad C = (21858 \times 598.5 \times 210) / 5.25$
$$= 5.23 \times 10^9$$

Rolling speed, $V = \dfrac{\pi \times D_\kappa \times N}{60 \times i}$

$$= \frac{3.14 \times 598.5 \times 210}{60 \times 5.25 \times 1000} = 1.25 \text{ m/sec.}$$

Co-efficient of reduction, $\mu_6 = F_5/F_6$

$$= 24860/21850 = 1.14$$

8.6.2 Stand-5 (Vertical stand)

Here, the angle of bite and motor capacity are only the criteria for selecting the reduction in Billet mill. The suitable reduction for stand 5 has been already selected while calculating the angle of bite for stand 6.

$$\Delta h_6 = 37 \text{ mm (assumption)}.$$
$$B_{cp5} = H_6 + \Delta h_6, \text{ where } B_{cp5} \text{ is the metal stock width}$$
$$= 150 + 37 = 187 \text{ mm}$$

H_5 which will enter the pass 6 as width of the metal should be made either equal or closer to the root of the pass to avoid the twisting of the incoming stock in the pass. H_5 should be equal to b_6 with a tolerance limit of ± 2.0 mm

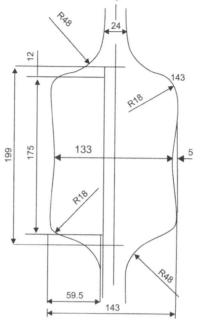

Fig. 5.32 Pass Design of Stand 05

$H_5 = 142.5 + 0.5$, 0.5 mm is selected because it reduces the chances of twisting.

$$= 143.0 \text{ mm}$$

Roll gap, $\quad t = (0.01 - 0.04) \times D_{max}$

Assuming $\quad t = 24$ mm

$$h_5 = \frac{(H_5 - 24)}{2} \quad \text{or} \quad \frac{(143 - 24)}{2}$$

$$= 59.5 \text{ mm}$$

Here, tan α, taper on the side selected should be in the range of 20–25%, as higher % will facilitate the saving of roll, as less turning is required for restoration of pass after the wear out.

$$a = h_5 \times \tan α,$$

$$a = 59.5 \times 0.21, \text{ or say } 12 \text{ mm}$$

b_5 is the pass root width

$$b_5 = B_{cp5} - a$$

$$= 187 - 12 = 175 \text{ mm}$$

$$B_5 = B_{cp5} + a$$
$$= 187 + 12 = 199 \text{ mm}$$

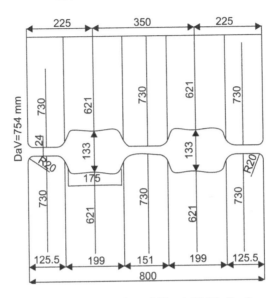

Fig. 5.33 Roll Design of Stand 05 (Vertical).

Rolling dia., for box pass can be directly computed,

$$D_K = D_c - 2 \times (h_4 - s)$$

Concavity, s, can be taken more in this pass to avoid fin formation in pass-6, also because of low temperature during rolling.

Here, S is assumed as 5 mm.

$$= 730 - 2 \times (59.5 - 5)$$
$$= 730 - 109 = 621 \text{ mm}$$

Actual area of stock, $F_{\kappa 5} = F_5 - [(\text{area of 4 corners} + \text{area of top and bottom concavity})]$

$$= 24860 \text{ mm}^2$$

Radius, $\quad r_5 = \mu_6 \times r_6$
$$= 1.14 \times 15$$
$$= 17.2 \text{ mm say } 18 \text{ mm}$$

Δh_5 is assumed as 47 mm, then $\Delta h_5 = 47$ mm

Angle of bite,
$$\cos \alpha = 1 - \Delta h_5 / D_K$$
$$= 1 - 47 / 621 = 0.924 \text{ or } 22.5°$$

Outer radius R, taken as 2-3 times of the roll gap, *i.e.* t
$$R = 2 \times t$$
$$= 2 \times 24 = 48 \text{ mm}$$
Rolling Constant, $C = 5.23 \times 10^9$

$$N_5 = \frac{(C \times i)}{(F_K \times D_K)}$$

$$= \frac{5.23 \times 10^9 \times 7.028}{24860 \times 621}$$

$$= 238, \text{ say } 240 \text{ rpm}$$

$$V_5 = \frac{(\pi \times D_K \times N)}{60 \times i}$$

$$= \frac{3.14 \times 621 \times 240}{60 \times 7.028 \times 1000}$$

$$= 1.11 \text{ m/sec.}$$

Co-efficient of reduction, $\mu_5 = F_{k4}/F_{k5}$
$$= 31344/24860 = 1.26$$

8.6.3 Stand–4 (Horizontal Stand)

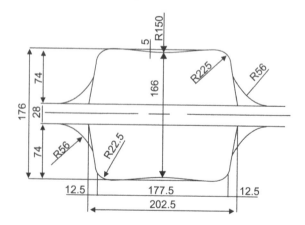

Fig. 5.34 Pass Design of Stand 04.

Δh_5 is assumed as 47 mm.

$B_{cp4} = H_5 + \Delta h_5$, where B_{cp4} is the metal stock width
$$= 143 + 47 = 190 \text{ mm}$$

8 Roll Pass Design of Billet 90 mm × 90 mm

H_4, which will enter the pass 4 as width of the metal, should be closer to the root of the pass to avoid the twisting of the incoming stock. H_2 should be equal to b_4 with a tolerance limit of ± 2.0 mm

$$H_4 = 175 + 1.0,$$
$$= 176.0 \text{ mm}$$

Roll gap, $\quad t = (0.01 - 0.04) \times D_{max}$

Assuming $\quad t = 28$ mm

$$H_4 = (H_2 - t)/2, \text{ or, } (176 - 28)/2$$
$$= 74 \text{ mm}$$

Here, tan α, taper on the side selected should be in the range of 10-25%, as higher % will facilitate the saving of roll, as less turning is required for restoration of pass after wear out.

$$a = h_4 \times \tan \alpha,$$
$$a = 74.0 \times 0.175 \text{ or say } 12.5 \text{ mm}$$
$$b_4 = (B_{cp4} - a), \text{ where, } b_4 \text{ is the pass root width}$$
$$= 190 - 12.5$$
$$= 177.5 \text{ mm}$$
$$B_4 = B_{cp4} + a$$
$$= (190 + 12.5)$$
$$= 202.5 \text{ mm}$$

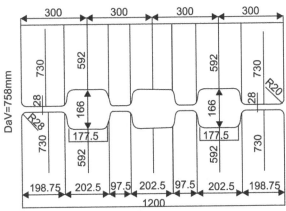

Fig. 5.35 Roll Diagram of Stand 04.

Rolling diameter for box pass can be directly computed,

$$D_K = D_c - 2 \times (h_4 - s)$$

Concavity s, can be taken more in this pass to avoid fin formation in subsequent pass-5,

S is assumed as 5 mm.

$$D_\kappa = 730 - 2 \times (74 - 5) = 592 \text{ mm}$$

Actual area of stock,

$$F_k = F_4 - (\text{area of 4 corners} + \text{area of top and bottom concavity})$$

$$= 31344 \text{ mm}^2$$

Radius, $\quad r_4 = \mu_5 \times r_5,$

$$= 1.26 \times 18$$

$$= \text{say, } 22.5 \text{ mm}$$

Δh_4 is assumed as 69 mm, then $\Delta h_4 = 69$ mm

Angle of bite,

$$\cos \alpha = 1 - \Delta h_4 / D_\kappa$$

$$= 1 - 69 / 592 = 0.883 = 28.0°$$

Outer radius R, taken as 2-3 times of the roll gap, $i.e., t$

$$R = 2 \times t$$

$$= 2 \times 30 = 60 \text{ mm}$$

Rolling constant,

$$C = 5.22 \times 10^9$$

$$N_5 = \frac{(C \times i)}{(F_\kappa \times D_k)}$$

$$= \frac{(5.23 \times 10^9 \times 8.476)}{31344 \times 592}$$

$$= 239 \text{ say, } 240 \text{ rpm}$$

$$V_4 = \frac{\pi \times D_\kappa \times N}{60 \times i}$$

$$= \frac{3.14 \times 592 \times 240}{60 \times 8.476 \times 1000}$$

$$= 0.88 \text{ m/sec.}$$

Co-efficient of reduction, $\mu_4 = F_{K3}/F_{K4}$

$$= 41140/31344 = 1.31$$

8.6.4 Stand-3 (Vertical stand)

The suitable reduction in stand 4(Δh_4) has been already selected while calculating the angle of bite for stand 3. Δh_4 was assumed as 69 mm.

$$B_{cp3} = H_4 + \Delta h_4, \text{ where } B_{cp3} \text{ is the metal stock width}$$
$$= 176 + 69 = 245 \text{ mm}$$

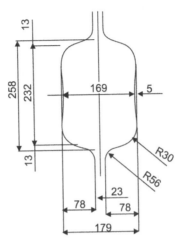

Fig. 5.36 Pass Design of Stand 03

H_3, which will enter the pass 4 as width of the metal, should be closer to the root of the pass to avoid the twisting of the incoming stock.

H_3 should be equal to b_4 with a tolerance limit of ± 2.0 mm

$$H_3 = 177.5 + 1.5, 1.5 \text{ mm is selected because it}$$
$$\text{reduces the chances of twisting.}$$
$$= 179.0 \text{ mm}$$

Roll gap, $\quad t = (0.01 - 0.04) \times D_{max}$

Assuming $\quad t = 23$ mm

$$h_3 = (H_3 - t)/2, \text{ or } (179 - 23)/2$$
$$= 78.0 \text{ mm.}$$

Here, tan α, taper on the side selected should be in the range of 15-25%, as higher % will facilitate the saving of roll, as less turning is required for restoration of pass after wear out.

$$a = h_6 \times \tan α,$$
$$a = 78 \times 0.17, \text{ or say } 13 \text{ mm}$$
$$b_3 = B_{cp3} - a, b_3 \text{ is the pass root width}$$
$$= 245 - 13 = 232 \text{ mm}$$

$$B_3 = B_{cp3} + a$$
$$= 245 + 13$$
$$= 258 \text{ mm}$$

Rolling dia., for box pass can be directly computed,
$$D_K = D_c - 2 \times (h_3 - s)$$
$$= 730 - 2 \times (78 - 5)$$
$$= 584 \text{ mm}$$

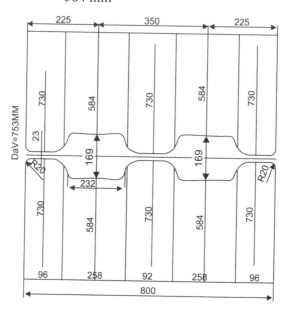

Fig. 5.37 Roll Diagram of Stand 03.

Concavity "s" can be taken more in this pass to avoid fin formation in subsequent pass, because stock is here received at low temperature, Here, S is assumed as 5 mm.

Radius, $r_3 = \mu_4 \times r_4$,
$$= 1.32 \times 22.5$$
$$= \text{say } 30 \text{ mm}$$

Δh_3 for pass-3 is assumed as 70 mm, then $\Delta h_3 = 70$ mm
Angle of bite,
$$\cos \alpha = 1 - \Delta h_3/D_K$$
$$= 1 - 70/584$$
$$= 0.88$$
$$= 28.3°$$

8 Roll Pass Design of Billet 90 mm × 90 mm

Outer radius R, taken as 2-3 times of the roll gap, *i.e.*, t

Assume, $R = 56$ mm

Rolling constant,

$$C = 5.23 \times 10^9$$

$$N_3 = \frac{C \times i}{F_K \times D_K}$$

$$= \frac{5.23 \times 10^9 \times 11.77}{41140 \times 584}$$

$$= \text{say } 260 \text{ rpm}$$

$$V_3 = \frac{\pi \times D_K \times N}{60 \times i}$$

$$= \frac{3.14 \times 584 \times 260}{60 \times 11.77 \times 1000}$$

$$= 0.67 \text{ m/sec.}$$

Co-efficient of reduction, $\mu_3 = F_{K2}/F_{K3}$

$$= 52380/41140 = 1.27$$

8.6.5. Stand–2

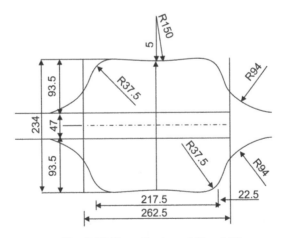

Fig. 5.38 Pass Design of Stand 02.

The suitable reduction in stand 3(Δh_3) has already been computed while calculating the angle of bite for stand-4. Δh_3 is assumed as 70 mm.

$$B_{cp2} = H_3 + \Delta h_3, \text{ where } B_{cp2} \text{ is the metal stock width}$$

$$= 170 + 70 = 240 \text{ mm}$$

H_2, which will enter the pass 3 as width of the metal, its dimension should be taken closer to the root of the pass to avoid the twisting of the incoming stock.

H_2 should be equal to b_1 with a tolerance limit of ± 2.0 mm

$$h_2 = 232 + 2.0,$$
$$= 234 \text{ mm}$$

Roll gap, $\quad t = (0.01 - 0.25) \times D_{max}$

Assuming $\quad t = 47$ mm

$$h_2 = (H_2 - t)/2, (234 - 47)/2$$
$$= 93.5 \text{ mm}.$$

Here, tan α, taper on the side selected should be in the range of 10 – 25%, as higher % will facilitate the saving of roll, as less turning is required for restoration of pass after wear out.

$$a = h_2 \times \tan α,$$
$$a = 93.5 \times 0.24$$
$$= \text{say } 22.5 \text{ mm}$$
$$b_2 = (B_{cp2} - a), \text{ where, } b_2 \text{ is the pass root width}$$
$$= 240 - 22.5 = 217.5 \text{ mm}$$
$$B_2 = B_{cp2} + a$$
$$= (240 + 22.5) = 262.5 \text{ mm}$$

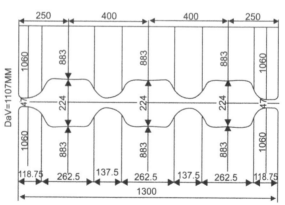

Fig. 5.39 Roll Diagram of Stand 2

Rolling diameter, for box pass can be directly computed,

$$D_K = D_c - 2 \times (h_2 - s)$$

8 Roll Pass Design of Billet 90 mm × 90 mm

Concavity s, can be taken more in this pass to avoid fin formation in subsequent pass, s is assumed as 5 mm.

$$D_\kappa = 1060 - 2 \times (93.5 - 5) = 883 \text{ mm}$$

Actual area of stock, $F_\kappa = F_2 - [(\text{area of 4 corners} + \text{area of top and bottom concavity})]$

$$= 56160 - (2046 + 1740)$$
$$= 52380 \text{ mm}^2$$

Radius, $r_3 = \mu_3 \times r_3,$
$$= 1.27 \times 28$$
$$= 36 \text{ say } 37.5 \text{ mm}$$

Δh_2 is assumed as 111 mm, then $\Delta h = 111$ mm

Angle of bite,

$$\cos \alpha = 1 - \Delta h_2 / D_\kappa$$
$$= 1 - 111 / 883$$
$$= 0.874 \text{ or } 29°$$

Outer radius R, taken as 2-3 times of roll gap, *i.e. t*

Assumption $R = 2 \times t$
$$= 2 \times 47 \text{ mm} = 94 \text{ mm}$$

Rolling constant,

$$C = 5.23 \times 10^9$$

$$N_3 = \frac{(C \times i)}{F_\kappa \times D_\kappa}$$

$$= \frac{5.23 \times 10^9 \times 16.75}{52380 \times 883}$$

$$= \text{say}, 190 \text{ rpm}$$

$$V_1 = \frac{\pi \times D_\kappa \times N}{60 \times i}$$

$$= \frac{3.14 \times 883 \times 190}{60 \times 16.75 \times 1000} = 0.52 \text{ m/sec.}$$

Co-efficient of reduction, $\mu_2 = F_{K1}/F_{K2}$

$$= 72040/52380 = 1.37$$

8.5.6 Stand-1

The suitable reduction in stand 2 (Δh_2) has been already selected while calculating the angle of bite for stand 3. A tilter is provided between stand-1 and 2 to tilt the stock by 90°. Length of bar after stand-1 is required to be computed, before the finalization of the reduction in stand-1, it is required because the bar should leave freely the stand-1, before entering the tilter. That means length of bar should be less than the distance between stand-1 and the entering point of tilter. Δh_2 was assumed as 111 mm.

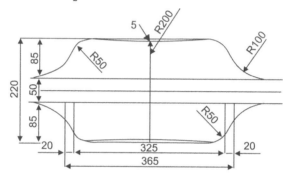

Fig. 6.40 Pass Design of Stand-1.

$B_{cp1} = H_2 + \Delta h_2$, where B_{cp1} is the metal stock width
$= 234 + 111 = 345$ mm

H_1, after tilting, will enter the pass-2 as width of the metal. H_1 should be equal to b_2 with a tolerance limit of ± 2.0 mm

$H_1 = 217.5 + 2.0$,
$= 219.5$ say, 220 mm

Roll gap, $t = (0.01 - 0.05) \times D_{max}$
Assuming $t = 50$ mm
$h_1 = (H_1 - t)/2$, or $(220 - 50)/2 = 85$ mm.

Here, tan α, taper on the side selected should be in the range of 10-25%, as higher % will facilitate the saving of roll, as less turning is required for restoration of pass after wear out.

$a = h_1 \times \tan α$,
$a = 85 \times 0.23 =$ say 20 mm
$b_1 = (B_{cp_1} - a)$, where, b_1 is the pass root width
$= 345 - 20 = 325$ mm
$B_1 = (B_{cp_1} + a)$
$= (345 + 20) = 365$ mm

8 Roll Pass Design of Billet 90 mm × 90 mm

Rolling dia., for box pass can be directly computed,

$$D_K = D_c - 2 \times (h_1 - s)$$

Concavity s, can be taken more in this pass, to avoid fin formation in subsequent pass, S here is assumed as 5 mm.

$$D_K = 1060 - 2 \times (85 - 5) = 900 \text{ mm}$$

Actual area of stock, $F_K = F_1 -$ (area of 4 corners + area of top and bottom concavity)

$$= 75900 - (1700 + 2160) = 72040 \text{ mm}^2$$

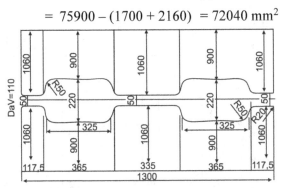

Fig. 5.41 Roll Diagram of Stand 01.

Radius, $\quad r_1 = \mu_2 \times r_2 = 1.37 \times 37.5 = 51.3$ mm, say 50 mm

$\Delta h_1 = H$ of bloom $- H_{01}$

$\quad\quad = 350 - 220 = 130$ mm

Angle of bite,

$$\cos \alpha = 1 - \Delta h_1 / D_K = 1 - 130/900 = 0.855 = 31.3°$$

Outer radius R, generally taken as 2-3 times of roll gap, *i.e.* t

$$R = 2 \times t$$
$$= 2 \times 50 \text{ mm} = 100 \text{ mm}$$

As stand-1 does not part of the continuous group, here rolling constant is not required to be computed. Stand-1 of this group run with AC drive with a constant speed of 100 rpm.

$$V_1 = \frac{(\pi \times D_K \times N)}{60 \times i}$$

$$= \frac{3.14 \times 900 \times 100}{60 \times 13.45 \times 1000} = 0.350 \text{ m/sec.}$$

Co-efficient of reduction, $\mu_1 = F_{Kbloom}/F_{K1}$

$$= 100340/72040 = 1.39$$

8.7 Rolling Scheme of Billet 90 × 90 mm From Bloom 325 × 325 mm With Intermediate Size of 150 × 150 mm

Table 5.3 Roll Pass Design Sheet For Rolling Billet 90×90 Mm Billet Size : 325 X 325 Mm

Stand	Metal Size H × B		Roll Gap t	Draught ΔH	Area	Rolling dia.	Spread ΔB	Angle of bite α	Radius of Pass	Coeff. of Redⁿ μ	Reducer Ratio	Speed of the motor	Rolling Velocity
	mm	mm	mm	mm	mm²	mm	mm	Degree	mm		i	n	m/sec
1H	220 Tilter	345 90°	50	105	72040	900	20.0	31.3°	50	1.39	13.45	100	–
2H	234	240	47	111	52380	883	20.0	29°	37.5	1.37	16.75	190	0.52
3V	179	245	23	61	41140	584	11.0	28.3°	30	1.27	11.77	260	0.67
4H	176	190	28	69	31344	592	11.0	28°	22.5	1.31	8.476	240	0.88
5V	143	187	24	47	24860	621	15.0	22.5°	18.0	1.26	7.028	240	1.11
6H	150	150	17	37	21850	598.5	7.0	20.3°	15.0	1.14	5.25	210	1.25
7V	153.5	211	10	42.5	17742	455	15.0	24.9°	18	1.23	5.43	310	1.37
8H	161	162	10	50.0	14715	449	8.5	27.2°	14	1.20	4.00	285	1.68
9V	127	169.5	10	35.0	11984	470	8.5	22.2°	12	1.22	3.46	290	2.07
10H	132	133	8	37.5	9992	465	6.0	22.8°	11	1.20	2.808	285	2.45
11V	116	136	8	17.0	8763	475.5	4.0	15.4°	10	1.14	2.2	250	2.82
12H	119.5	119.5	7.5	16.5	8030	472.5	3.5	16°	9	1.09	1.605	200	3.10

Note: (1) There is a tilter between Stand 1 & 2.H and V are horizontal and Vertical Stand

(2) Stand 1 is an independent stand having driven by constant speed *i.e.*, 100 rpm.
(3) Stand 2 to 6 form intermediate group of Billet mill, having the intermediate output as 150×150 mm
(4) Stand 7 to 12 form the finishing group, having output billet sizes from 80 × 80 mm to 100 ×100 mm

9 DEFECTS IN BILLET ROLLING
9.1 Defects Due to Bad Rolling

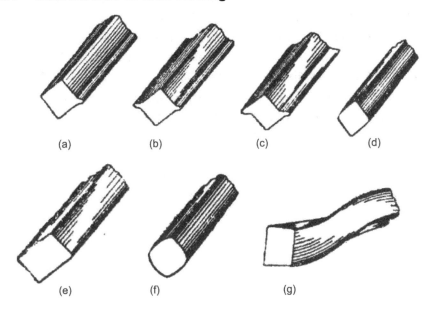

Fig. 5.42 Defects due to Bad Rolling.

1. **A one sided fin** (Fig. 5.42(*a*)): It is the result of, if the entry guides are not set symmetrical to the vertical axis of the pass.
2. **Two sided fins** (Fig. 5.42(*b*)): It arises due to higher reduction given to particular pass than the calibrated reduction as per rolling scheme or metal is not hot enough as it was required for rolling.
3. **Collaring of billets** (Fig. 5.42(*c*)): It is due to misalignment of the grooves in a pass and also due to insufficient turning of the bar before it entered the pass.
4. **Under-filling of the horizontal corners** (Fig. 5.42(*d*)) of the billet is due to insufficient metal entering in the given pass.
5. **A rectangular instead of a square shape** (Fig. 5.42(*e*)): It is obtained when the billet is rolled in a square pass having misaligned grooves.

216   Chapter 5   Rolling of Billets

6. **Bellied sides are obtained on the billet** (Fig. 5.42 (*f*)) : if the walls of the pass are excessively worn, then defect may be corrected by going over to a new square pass.

7. **Twisting of billet** (Fig. 5.42 (*g*)): A condition wherein the ends of a rolled billet have been forced to rotate in relatively opposite directions about its longitudinal axis, may be caused by excessive draft, faulty setting of entry and delivery guides or lack of uniform temperature across billet cross section.

8. **Bent:** The rolled products bend on the cooling bed due to improper pushing or when they are pushed in hot condition.

9. **Fish-tail:** When the surface or skin of the billet is hotter than the core and get elongates more than the core, A V–notch may then be developed at the back end of the length of billet, somewhat resembling to the tail of a fish.

10. **Lap:** A defect appearance on the surface of the billet, caused by a portion of steel being folded over on itself and it did't gets welded on further rolling. Lap may arise due to careless working, fins on the previous metal and defective pass in rolling.

11. **Under filling:** This condition arises, when pass doesn't get filled up completely. The metal from previous stand is to be enhanced.

□□□

6

Rolling of Rounds and TMTS Bars

1 INTRODUCTION

Rounds are the most common shape produced in a rolling mill. These are also the most difficult symmetrical shape to roll and to maintain a good shape and within the accurate tolerance limit while rolling in a conventional rolling Mill.

Generally rounds are classified in two main categories:

(*a*) **Merchant bar quality**: Rounds, that are produced with standard commercial specifications, on tolerances, steel grade and surface requirements.

Fig. 6.1 Merchant Bar.

(*b*) **Special bar quality:** Rounds that are produced only for special requirements with respect to either size, tolerance, steel grade or surface condition. One such type is Rock Bolts, which are embedded into the mine roof and support the wire mesh, to protect the miners from injury, during rock fall.

Fig. 6.2 Rock bolt rounds.

2 SELECTION OF ROLLING SEQUENCE

In selection of rolling sequence for rolling steel rounds, it is necessary to take into the consideration the size and end use of rounds, grade of steel, type of mill and extend to which mill is mechanized.

According, to standards, rounds are rolled in diameter from 5 to 200 mm. The classification on size and type and size of mill requires to roll these sizes are explained below:

Table 6.1

Size mm	Nomenclature	Type of mill	Size of mill mm
> 5 – up to 12	Rods, mostly rolled in coils	Wire rod mill	250
> 12 – up to 25	Bar, mostly rolled in straight length.	Bar & rod mill or Light rod / structural mill	250–350
> 25 – < 75	Medium Rounds rolled in straight length.	Light structural mill/Merchant mill	350–450
> 75 – 200	Heavy Rounds rolled in straight length.	Heavy structural mill	500–800

However, with the advent of Bar and Rod Mill, having a rolling speed of 110 m/sec and with the use of modern technology like housing–less stands, and high hardened rolls; it is possible to roll wire rods upto 22 mm in coils and bar from 8 mm to 40 mm with the annual production upto 1 MT.

There are different types of rolling sequence for rolling rounds. The common sequence into all of these sequences is that the "OVAL" pre finishing pass, which is fed into the round finishing pass. In other words, the shape of pre-finishing pass remains unchanged. This facilitates possibility of working out a common regulation for the behaviour of metal in finishing and pre-finishing pass in all methods of rolling rounds. There are about 10 different types of method of rolling round.

Details of each of these methods are given below:

2 Selection of Rolling Sequence

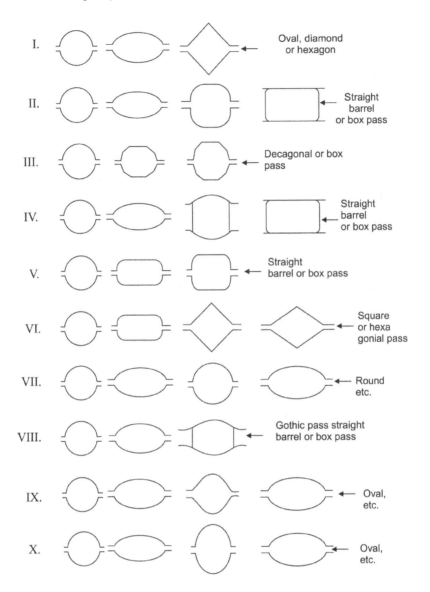

Fig. 6.3 Different types of rolling sequence for rolling Rounds.

2.1 First Method

The peculiar features of this method is the presence of leader square pass, which cut diagonally. This square can be produced out of any system *i.e.*, hexagonal square, rhombus-square or oval-square. Elongation in this system is high and hence number of passes required are less compare to other system. The drawbacks of this system are uneven deformation in oval pass and deep cut of rolls for square pass, especially, when a higher size of rounds

220   Chapter 6   Rolling of Rounds and TMTS Bars

are rolled. This method can be applied effectively for rounds of size between 5 mm–20 mm.

2.2 Second Method

This system is distinguished by producing different size of rounds from the same pre-finishing pass. Second, advantage of this system is the even deformation in pre-finishing oval and leader edged oval pass. It also makes possible for easy and quick removal of scale from the surface of rolled products. This method is generally used for rolling rounds of size 20 mm–100 mm of plain and as well as for alloyed steel. It is highly recommended for rolling mill, which are working with the principle of one pass for one stand, mainly for continuous mill.

2.3 Third Method

In this method, the leader oval is replaced by a decagon pass. This formation decreases the unequal deformation in the pre-finishing passes and help to secure the stability of rolling, without application of twist guides. This method is generally used for rolling rounds above 150 mm and above.

2.4 Fourth Method

The fourth method is similar to the second one, only have difference in the shape of the edging pass, which does not restrict the metal spreading. The absence of side wall in the pass helps for better scale removal. The disadvantage of this sequence is the deteriotating condition of side surface of the strip, which may get crack, when metal is not sufficiently hot. Metal may also get twisted, due to poor holding of incoming square in edging pass. This version is used for rolling rounds of various sizes, but it is not desirable to use, when metal ductility is low.

2.5 Fifth and Sixth Method

These versions envisage the application of flat oval passes, which offer considerable draught and ensure for better vertical stability of rolled section. Besides, the condition of bite is easier and wear out of finishing pass comparatively will be more uniform. But rolling of round in this system requires more accurate mill setting, as with the small excess of metal, overfilling may occur with the formation of fins.

2.6 Seventh, Eighth, Ninth and Tenth System

These versions are based on the use of oval-round series. The disadvantage of this system is the correct holding of oval strip in the guide, while feeding it into

4 Design of Passes

the round passes. It is a very complicated process. Uniformity of deformation in this system permits smoother working of products, thus giving a good surface finish and highly accurate dimension without any surface defects. This system is widely accepted system for rolling high quality steel.

3 ROLLING TOLERANCES OF ROUNDS AS PER BIS SPECIFICATION

3.1 On Size of Round

Table 6.2

Sizes		Tolerance mm
Over mm	Upto including mm	
-	25	± 0.5
5	35	± 0.5
35	50	± 0.8
50	80	± 1.0
80	100	± 1.3
100	-	± 1.6% of diameter

3.2 Ovality or Out of Square

The permissible ovality of rounds is measured as the difference between the maximum and minimum diameters. It should be less than 75% of the total tolerance (plus and minus) specified on the size.

3.3 Weight

The tolerances on weight per meter for round bar will be as follows:

Table 6.3

Sizes		Tolerance percentage
Over mm	Upto and including mm	
–	10	± 7
10	16	± 5
above 16	–	± 3

4 DESIGN OF PASSES

4.1 Design of Finishing Pass

The reduction in finishing pass is recommended from 5 to 15%. The reduction varies according to shape of pre-finishing pass, surface finish desired, size of round, and tolerance of the finished product. Smaller reduction is recommended for bigger sizes rounds. Reduction is generally fixed around 12%.

It is also to be noted that before, designing the finishing pass; tolerances on size specified by the customer or as specified in reputed specification are to be taken into consideration for designing finishing pass.

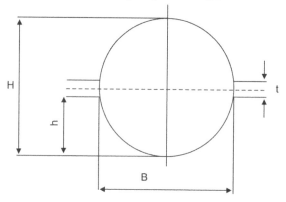

Fig. 6.4 Design of Finishing Passes.

Considering tolerances on size and ovality for finished product, the elements of finished pass is designed as follows:

4.1.1 Height of Finishing Pass

Roll pass design is always made on hot dimensions. As metal will get shrink in cold condition, a correction for shrinkage is to be made in the design, taken into the account of the co-efficient of expansion of rolled product and then depending upon the tolerances assigned, a part of tolerance value is subtracted from the obtained value.

D hot $= (1.010 - 1.015) D$ cold, as finishing temperature varies from 850-950°C. Roll pass designer generally takes height of finishing pass equals to the size of the finished round.

4.1.2 Width of Pass

Horizontal diameter of finishing pass should be taken little more than the vertical one. It necessitate because if pass is made of correct round size, then even the smallest overfilling connected with the continual change of condition of widening (may be changed due change of co-efficient of friction), leads to formation of fins. It may cause rejection of material or production of inferior quality products.

Other reasons are:

- Shrinkage in horizontal direction is always more than the vertical direction.
- After wear out of the pass, it will remains as a good shaped round.
- It will also take care of the widening of the width of metal, due to change in tension in the continuous group.

4 Design of Passes

Following are the methods for computing width of round pass.

(a) $B = 1.015 \times D_c$, for round 6–30 mm

(b) $B = (1.005 - 1.025) \times Dc$

(c) $B = Dc +$ Tolerance on size + co-efficient of contraction

4.1.3 Roll Gap

It is generally taken equivalent to 1% of the mill size. The computation is based upon

Roll Gap or $t =$ Nominal roll gap + spring of mill + Wear of pass allowed + Roll turning defect + Safety allowance

4.1.4 Radius of Finished Pass

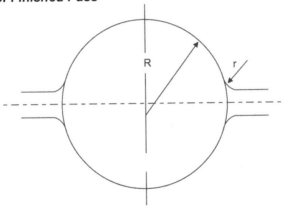

Fig 6.5 Radius of Finished Pass.

There are three methods, called as single plug, two plugs and three plug methods depending upon number of radius used for designing round. Application of one of these methods depends upon how much accuracy desired in finished products and need for longer run of pass without the dressing of the rolls. The three plug method is not generally used, unless high precision rounds are required.

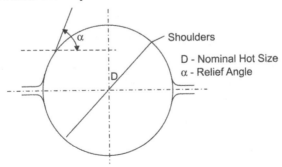

Fig. 6.6 Single Plug Method for Designing Rounds.

For single plug method

$$R = (1.01 - 1.02) \times Dc / 2$$

It is recommended that R should be taken as $1.01\ Dc/2$ for rounds of size of 16 – 36 mm, $(1.012 - 1.015) \times Dc/2$ for size 40 – 45 mm and $(1.015 - 1.017)\ Dc/2$ for rounds 50 – 75 mm.

4.2 Design of Pre-finishing Oval Pass

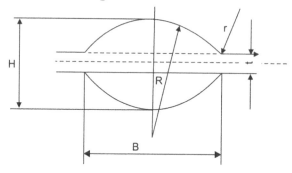

Fig. 6.7 Single Radius Oval.

Selection of the form and dimensions of pre-finishing oval pass is based on the following considerations:

The more nearly the shape of oval pass to the round pass, the better will be the quality and will have close control on the size of finished product, as less draught will give less spread and hence there will be less chance of incorrect filling of round pass.

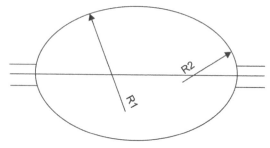

Fig. 6.8 Double Radius Plum Oval.

Plum ovals are used for higher diameter of rounds and slim oval are used for smaller diameter of rounds. This will facilitate to avoid the turning of work piece in pass, lead to twisting of the bar. The co-efficient of elongation in the oval pass is generally recommended to 1.13 -1.50.

4.2.1 Height of Oval

It is determined on the basis of spread in the finishing pass.

4 Design of Passes

H oval $= Dc$ – spread in the finishing pass

There are other formulas to calculate the spread in the finishing pass, the most commonly used formula is

Spread in finishing pass $= (0.03 - 0.04) \times \sqrt{D} \times \sqrt{d}$

Where D and d are diameter of roll and size of round desired. Higher value is used for smaller size rounds and vice versa.

4.2.2 Width/Height Ratio of Oval Pass

B/H, varies with the size of oval, it is recommended to use such derived ratio for oval pass :

Size of Rounds	B/H Ratio
16–28 mm	1.80–2.20
32–40 mm	1.50–1.70
45–75 mm	1.40–1.50

4.2.3 Radius of oval

It is suggested to use single radius i.e., R_1 for the size upto 36 mm. Double radius i.e., R_1 and R_2 are recommended for size above 36 mm.

Computation of R_1 will be as follows:

$$R_1 = \frac{B^2 + (H - t)}{4 \times (H - t)}$$

Calculation for R_2 radius for double radius ovals for round size above 32 mm, is derived from the drawing itself, as it will connect outer point of the pass with R_1 radius.

4.2.4 Outer radius r of Oval Pass

It is generally taken equal to the roll gap of oval pass.

Radius $r = t$, where t is roll gap of oval pass.

Roll gap t: Roll gap of pre finishing pass is generally taken 2 mm, more than the finishing pass.

4.2.5 Area of Oval Pass

Area F oval is determined by the formula

$$F \text{ oval} = \frac{2}{3} B \times H$$

Where B and H are the width and height of pass respectively.

4.3 Design of Strand Pass

Strand pass is a pass, which follows the oval pre-finishing pass. It may be square, round or plug /edge oval.

Generally, plug/edge oval is preferred due to following reasons:

1. Uniform wear out of pass due to big radius.
2. Self centralizing of out coming bar from this pass, due to big radii, and whatever position bar comes out of the plug oval; it falls on its bailey, causing smooth entry to the next oval pass. It means no need of tilting device is required.
3. Quicker setting of pass due to its shape.
4. Because of uniform wear out in the plug /edge oval, roll life is increased. It is observed that the plug oval pass gives 1.5 times more life than the square pass. Moreover, the roll life of oval pass also gets increases due to uniform reduction in oval pass.
5. The product obtained by this system will have fewer defects.

Only disadvantage of plug oval–oval system is that plug/edge oval which have not a high co-efficient of elongation. The maximum co-efficient of elongation will be in the range of 1.15 to 1.35 in plug/edge oval.

4.3.1 Strand Plug oval/edge Oval

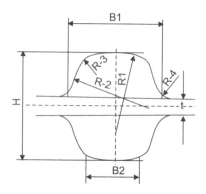

Fig. 6.9 Strand Plug oval/edge oval.

First step is to determine the μ oval, as it is mentioned earlier, the value of μ oval is generally taken between 1.15–1.35. Higher value is selected for lower size and *vice-versa*. Once the value is assigned, then area can be ascertained by the following formula *i.e.*,

F plug oval = μ oval × F oval,

B plug oval = H oval + Δ H oval

4 Design of Passes

ΔH oval can be determined, keeping in view of B/H of the pre-finishing pass. The less reduction is recommended for flatter oval than for plumber oval.

$$H \text{ plug oval} = \frac{F \text{ plug oval}}{B \text{ plug oval}}$$

Generally, the design of plug oval with one radius is used for the round size of $20 - 25$ mm. For bigger rounds *i.e.* from 28 -75 mm, double radius edge oval are used.

4.3.2 Strand Square

Square pass is preferred for strand pass, when a higher reduction is required to be given in strand pass. It is important to note that to avoid the possibility of surface defects, strand square should give a correctly filled oval. The computation of side of square can be related with the size of finishing round.

Generally, side of square is computed by multiplying the size of round by 1.1, thus giving a overall reduction of about 35 %. For size of round above 12 mm, following formula can be used for computation of the side of square

Table 6.4

Round size in mm	Side of square
12	1.125 × Dc
22	1.16 × Dc
25	1.15 × Dc
28	1.13 × Dc

For size above 20 mm, a "Gothic shaped" square is preferred to give a better entry, better riding and less wear out of pass. In the Gothic square pass, the distance across the sides is slightly more than the square pass, but area is slightly less due to its shape.

4.3.3 Strand Round

A round is used as the strand pass, when round–oval–round sequence is applied. This implies a light reduction from strand round to finishing round maybe upto 20–30% only.

4.4 Design of Following Passes in Rolling Sequence

After the design of finishing, pre finishing and strand passes for rolling of particular round, pass design of following passes in rolling sequence are to be selected. Pass sequence can be oval–square, diamond–square, round–oval or box–barrel, depending upon the size of finished round, input billet size, type of mill, finishing temperature of finished bar and other consideration already been discussed.

Generally a square–rectangular sequence is used for round above 20 mm in a structural mill.

5 ROLLING SCHEME OF ROUND–32 MM

Table 6.5 Input Billet Size–100 x 100 x 6 Mtrs

STD	Pass	H	B	t	Dc	Nrpm
1	Barrel	90	103		515	500
Tilt 90						
2	Barrel	76	100		523	320
3	Barrel	66	104		523	310
4	Rd 32 / 40	78	76	20	425	315
5	Barrel	52	92		523	300
Tilt 90						
6	Rd 32- 36	62	62	6	420	340
7	Barrel	37	75		425	335
8	Rd 32-36	46	48	8	420	315
9	Barrel	27	60		400	300
Tilt 90						
10	Rd-32	41	35	6	420	300
11	Rd-32	28	46	6	370	390
12	Rd-32	32	32	3	370	500

Note: Where H and B are the size of metal in the respective passes, t is the roll gap, Dc is the collar diameter of the roll, Nrpm is the motor rpm

6 ROLL PASS DESIGN R-32 MM

6.1 Stand-12 (Finishing Pass)

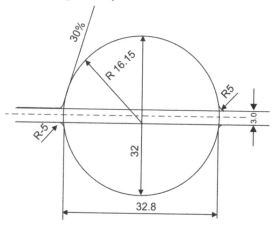

Fig. 6.10 Design of Finishing Pass.

6 Roll Pass Design R-32 mm

6.2 Pass-11 (Pre-finishing Pass)

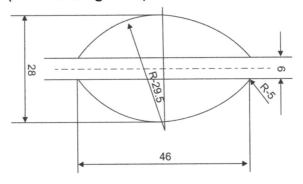

Fig. 6.11 Pre-finishing Pass.

6.3 Pass-10 (Strand Pass)

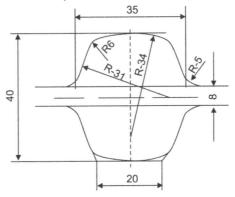

Fig. 6.12 Strand Pass.

6.4 Pass-8 (Rd-32-36 Pass)

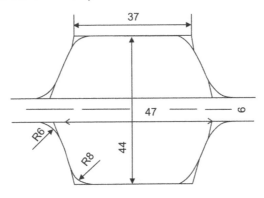

Fig. 6.13 Pass-8

6.5 Pass-6 (Rd-32-36 Pass)

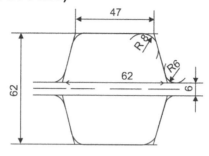

Fig. 6.14 Pass-6.

6.6 Pass-4 (Rd-32-36 Pass)

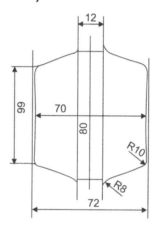

Fig. 6.15 Pass-4.

6.7 Roll Design of Finishing Pass

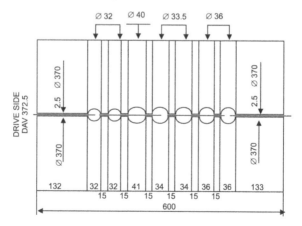

Fig. 6.16 Roll Diagram of Finishing Stand.

6.8 Roll Design of Pre-finishing Pass

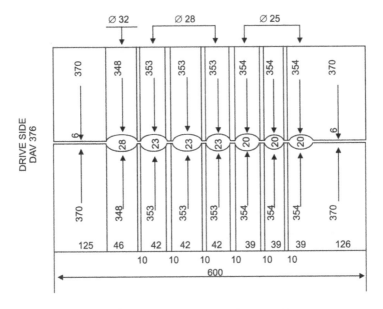

Fig. 6.17 Roll Diagram of Pre-finishing Pass.

6.9 Roll Design of Strand Pass

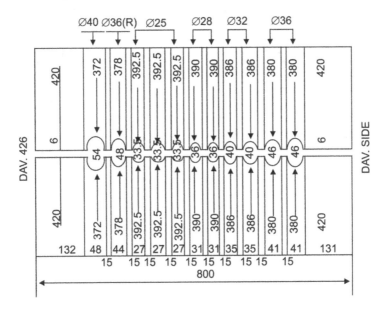

Fig. 6.18 Roll Diagram of Strand Pass.

7 LAYOUT OF MERCHANT MILL

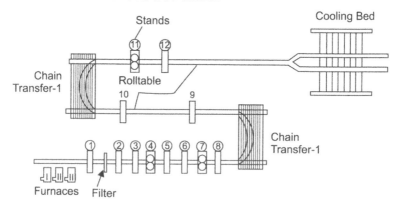

Fig. 6.19 Layout of Merchant Mill.

8 ROLLING OF TMT BARS

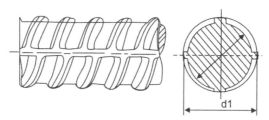

Fig. 6.20 TMT Bar.

TMT stands for "Thermo-Mechanically Treated" steel reinforcement bars.

Type of steel	Yield Strength MPa,min.	Tensile Strength, MPa,min.	Elongation,% min.
IS 2062 Gr A	26	240	23
FE415	410	485	14.5

8.1 Process of TMT

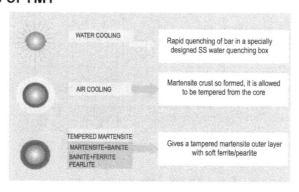

Fig. 6.21 Process of TMT Bar.

8 Rolling of TMT Bars

This process has become very popular, as with the use of process of controlled cooling TMT bars system, yield strength, will have atleast 65% higher than the plain mild steel. At the sametime, bond strength will also be 100% more than the plain round bar. End hooks will no longer required.

The most important point is that though there is a remarkable increase in yield strength, but there will be no major loss of ductility; thereby TMT bars will also have good weld-ability.

The use of TMT bars has not only drastically minimized the quantity of steel requirement. If further reduce the cost of transportation, storage and handling of less quantity of steel, that is with no end hook requirement, will also facilitate easier handling with reduced cost. Laying in Site is simplified and reduces the labour charges.

We can say that TMT bars are having following advantages over the cold twisted ribbed bars:

- High Strength with High Ductility.
- High Corrosion Resistance.
- Very High Weld- ability.
- Very High Bend ability.
- High Strength at Elevated Temperature.
- Bars are Stress Free.
  - Bars have Protective Shield.
  - Pollution Free Process.
  - Lower Manufacturing Cost.
  - International Quality Products.
  - Saving in steel.
  - 50% Less Labour required in Mills.

8.2 Comparison of Different TMT Grades

Table 6.6

Standard	Grade	Yield Stress (min) N/mm^2	UTS (min.) N/mm^2	% Elongation
BIS 1786	Fe415	415	485	14.5
	Fe415 D	415	500	18
	Fe500	500	545	12
	Fe500 D	500	565	16
	Fe550	550	585	10
	Fe550 D	550	600	14.5
	Fe600	600	660	10

Contd...

BS 4449	B500A	500	1.05<UTS/YS	2.5
	B500B	500	1.08<UTS/YS	5.0
	B500C	500	1.15<UTS/YS<1.35	7.5
ASTM 615 M	40	280	420	11-12
	60	420	620	7-8
	75	520	680	6-7

8.3 The Effect of Carbon Equivalent on Weldability

Table 6.7

Carbon equivalent (CE)	Weldability
Upto 0.35	Excellent
0.36 – 0.40	Very Good
0.41-0.45	Good
0.46-0.50	Fair
Over 0.50	Poor

8.4 Rib Design

Purpose of proper rib design is to:

- Increase the bond with concrete.
- Ribs or lugs termed as deformations provide a high degree of interlocking of cements concrete and rebar.
- Minimum requirements for these deformations (spacing, projection, etc.), have been developed by extensive lengthy experimental research and are specified in the standards.

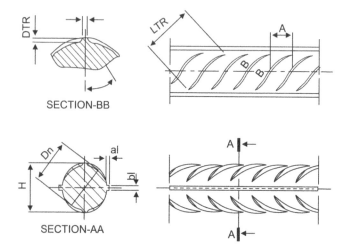

Fig. 6.22 Rib Design.

8 Rolling of TMT Bars

8.5 Control Data for Thermo Mechanically Treated Wire Rods/ Bars

Table 6.8

| Size | Sec-tional Weight kg/M | Core Dia. MM | Max. Dia. H MM | LUGS (Transverse RIB) | | | RIB | | Av. Spacing `A' between lugs Max. MM | Lug Angle "θ" Min. | Proected Area MM^2 |
				DTR Av. Height Min. MM	Max. width `b' MM	Length (Min.) LTR. MM	Thick-ness Max. al MM	Width Max bl MM			
8	.395	7.6	8.72	0.56	0.8	10	0.8	0.8	4.0	55	1.36
10	.617	9.5	10.90	0.70	1.0	12	1.0	1.0	5.3	55	1.70
12	.888	11.4	13.08	0.84	1.2	15	1.2	1.2	6.5	55	2.04
16	1.58	15.2	17.44	1.12	1.6	20	1.6	1.6	7.5	55	2.72
20	2.47	19.0	21.80	1.40	2.0	26	2.0	2.0	9.5	55	3.40
25	3.85	23.75	27.25	1.75	2.5	30	2.5	2.5	11.0	55	4.25
28	4.83	26.6	30.52	1.96	2.8	35	2.8	2.8	11.0	55	4.76
32	6.31	30.4	34.88	2.24	3.2	40	3.2	3.2	11.0	55	5.44
36	7.99	34.2	39.24	2.52	3.6	40	3.6	3.6	13.0	55	6.12
40	9.86	38.0	43.6	2.80	4.0	40	4.0	4.0	13.0	55	6.80

Note: As per BIS, the projected area *i.e.*, 0.17 × D should not be less than the value computed by the following formula based on different elements of TMT product. *i.e.*,

$$= \frac{2/3 \times 2 \text{ Depth of lug (DTR)} \times \text{Length of lug (LTR)} \times \sin \theta}{\text{Spacing between lugs (A)}}$$

Where,

DTR = Depth or Height of Lug (transverse Rib)

LTR = Length Lug (transverse Rib)

A = Average Spacing between two consecutive Lugs

θ = Lug Angle

• Different bar producers use different patterns, all of which should satisfy these requirements.

8.6 Roll Pass Design of TMT-32

Rolling of TMT-32 is same as rolling of plain round 32 mm, except higher metal is required to form lugs and ribs in finishing pass. The area of finishing pass of TMT should be maintained same as that of plain round-32. Metal sizes of subsequent passes are required to be enhanced. To accommodate the metal size, the plug oval pass of Round-36 (one size higher) is taken in stand-10, instead of normal round -32 mm pass of plain round scheme.

Table 6.9 Rolling Scheme for TMT-32 MM
Input Billet Size: 100 × 100 × 6 Mtrs

STD	Pass	H	B	t	Dc	Nr
1	Barrel	90	103		515	500
Tilt 90						
2	Barrel	76	100		523	320
3	Barrel	66	104		523	310
4	Rd32/40	78	76	20	425	315
5	Barrel	52	92		523	300
Tilt 90						
6	Rd 32- 36	62	62	6	405	330
7	Barrel	38	75		436	380
8	Rd 32-36	50	48	12	398	420
9	Barrel	33	59	-	395	450
Tilt 90						
10	Rd-36	43	40	4	385	370
11	Rd-32	31	46	9	360	420
12	TMT-32	32	32	3	375	540

Note: Where H and B are the size of metal in the respective passes t is the roll gap Dc is the collar dia. of the roll, Nr is the motor rpm

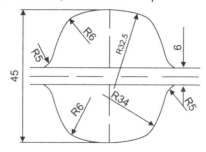

Fig. 6.23 Rd 36 mm Pass for Stand-10 for TMT Rolling.

9 ROLLING DEFECTS AND ITS SOLUTION
9.1 Ovality

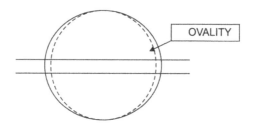

Fig. 6.24 Ovality.

9 Rolling Defects and Its Solution

The sides of pass are not completely filled up. This is due to less metal from previous stand, when height of metal is correct. If the height is more then the top roll of finishing stand should be pressed. Metal from previous stand should be increased if metal height is correct. If, there is one side fin and at other side oval, then entry box of finishing stand should be shifted towards the oval side.

9.2 Diagonal Difference

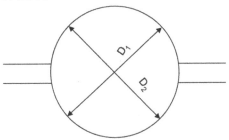

$D_2 - D_1 < 75\%$ of the size tolerance

Fig. 6.25 Diagonal Difference.

Right or left diagonal may be more than the allowable tolerance. Reasons for this cause are the improper setting of the entry guide or rolls. The guide is to be shifted against the opposite direction of the higher shoulder, if it is not adjusted by the shifting of guide, then rolls may be adjusted.

9.3 Fin

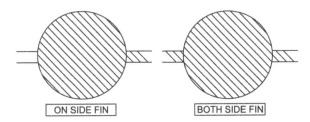

Fig. 6.26 Fin.

- One sided fin indicates the defective fixing of entry box of the finishing stand, which should be shifted opposite to the defect.
- Both side fins may be due to excess metal from the pre-finishing oval pass. The thickness of oval pass is to be reduced and if the oval pass is worn out, then, it should be changed immediately.
- If fins are only at the end of bar, then tension should be removed in continuous group of stands, especially finishing group of the mill.
- If false fins are noticed at the top and bottom of round, then this may be result of oval with fin which wears out the finishing stand pass at top and bottom.

9.4 Under-Filled

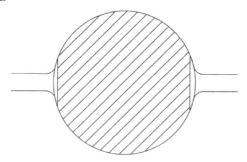

Fig. 6.27 Under-filled.

Under filling is normally caused by

- Insufficient stock in the oval pass,
- Check the tension and if any, then remove it.
- Top/bottom dimension too large, which can be rectified by checking the top and bottom dimensions of finished round and if it is too large then change of pass is recommended.

9.5 Fish-tail

Fish–tail at the end of bar causes jamming inside the delivery rolling tackles or in the cooling pipe of TMT rolling system. Causes are:

- Improper setting of the entry roller box of the finishing stand,
- Jamming of the rollers of the entry box.

9.6 Lap

Avoid overfilling of the pre-finishing oval pass and finishing round pass

9.7 Worn out Mark

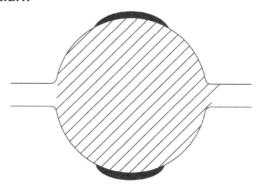

Fig. 6.28 Worn out mark.

9 Rolling Defects and Its Solution

It indicates when diagonals are more than height and width of the pass. Then pass should be immediately changed.

9.8 Pipe Formation in TMT

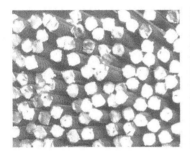
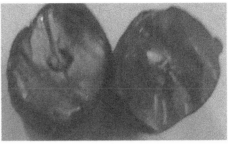

(a) Pin hole Pipe at the End of TMT Bar.

(b) Pipe and Fin   (c) Pipe Exposed

Fig. 6.29 Different Types of Pipe.

Pipe formation is mainly due to faulty steel making practices and can be reduced either by proper cutting of crop ends of bloom/billet at the shear of the blooming or billet mill or by strict inspection of billets at Billet yard, to avoid charging of pipy billets.

9.9 Pass Mark

It is mainly due to the use of worn out pass of pre-finishing or finishing stand or due to wrong setting of pass or guide. If, it is not eliminated by setting, then pass is to be immediately changed.

For good mill setting, the roller and its concerned staff should know about the special features of rolling technologies such as:

(*a*) Twisting.

(*b*) Bending sideways/up or down.

(*c*) Slipping in stands.

(*d*) Rolling schemes for different profiles.

(*e*) Pass positioning and their dimension.

10 MILL SETTING OF ROUNDS/TMTS
10.1 Twisting
1. If rolls are cross.

    Twisting depends upon the type of pass

    (*a*) Box passes with height more than the width-If the top roll is axially displaced to the right, the bar shall twist clockwise and the roll is to be taken to the left *i.e.*, to the opposite to the direction of the twist. It is because lateral deformation is more than the normal deformation.

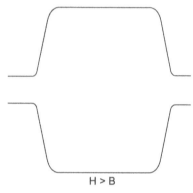

Fig. 6.30.

(*b*) Box passes with height less than the width- If the top roll is axially displaced to the right, the bar will twist anticlockwise and the top roll is to be taken to the left *i.e.*, to the direction of the twist. It is because lateral deformation is less than the normal deformation.

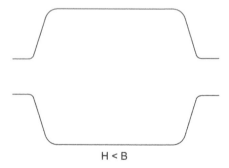

Fig. 6.31.

(*c*) Square box pass
  - Lesser the side draught, lateral deformation will be more than the normal deformation and it will behave as (a) above.
  - More the side draught, lateral deformation will be less than the normal deformation and it will behave as (b) above.

10 Mill Setting of Rounds/TMTS 241

(*d*) Oval and round passes also behaves like square pass as explained in (b) above, (where height of square is less than the width). Roll is to be taken in the direction of the twist.

2. If the feeding bar has unequal diagonal in box, oval and round pass, the out coming bar will twist along the bigger diagonal.

3. If the metal from previous stand is twisted and fed as twisted.

4. If the entry guides held the bar loose. Then the twisting may be in either direction.

5. If the entry guide setting is perfect, but the entry rest bar is inclined, then the metal will twist in the direction of inclination of the rest bar.

6. If the exit rest bar is inclined, then metal will twist in the direction of the inclination of the rest bar.

7. If the pass is not filled properly, the under filling of the pass allows the metal to play in the pass, so twisting may be in any direction.

8. If the entry guides are at one side of the pass, then press the metal of cause the twisting.

9. If the metal with fin enters the pass, the surface of contact will be less. When rolls exert pressure on the metal, then metal losses its balance and get twists in either way.

10. If exit guide of one side is pressing the bar, the bar will twist opposite to the direction of pressure. There will be twisting. Gap between guards should be sufficient and guard should not be too close.

11. If guards are set wrongly, it will act as twist guide and twist the bar. If bad metal sticks on the guard then there will be twisting. Gap between guards should be sufficient and guard should not be too close.

12. If one roll is set inclined to another.

13. If the box roller are inclined or worn out, the bar will be fed inclined, so twisting may take place.

14. If the rolls having axial play.

15. If the entry guide over roller is moving up or down (when roller bearing got damaged), then there will be no proper guiding of the bar and metal get twisted.

16. If temperature of bar is not correct. In that case at higher temperature, the pass may be under filled and piece got twisted. In case of less temperature pass get overfilled and due to fin appearance, metal get twisted.

17. With wider metal, twisting is easier than the narrower metal.

18. With slight cross of rolls, if depth of pass is more, then twisting will be less.

10.2 Bending of Bars (Side Ways)

1. If the left side guide is more open, the piece will bend towards right and if the right guide is too close, then also piece will bend towards right.

2. If the left guide is closed more, piece will bend toward the right and when right guide is closed, the piece will bend towards the left.

3. When rolls are cross, it creates unequal pressure on the sides of metal. If the pressure is from right side, the piece will bend to the left and *vice-versa*.

4. Due to unequal pressure because of roll gap is not equal throughout the length of roll, Bar will bend towards the sides where pressure will be more *i.e.*, where, gap is less in between rolls.

5. If rolls have play, the side pressure will cause bending. If pressure is from right side, the piece will bend to the left and *vice-versa*.

6. If the entry box is loose, the piece may bend either to the left or to the right and finally piece will follow a *zig-zag* path.

7. If during the course of rolling the entry or exit guides get cut deeply by the previous bar, the next bar may bend as per the position of cut.

8. In case of under filling of pass, the piece may bend either way and if the pass is overfilled, the fin will touch exit guide and piece will bend.

9. If the piece is slightly bend and rollers of roll table after stands are not rotating, the bend will become more.

10.3 Bend up and Down

1. Due to top and bottom pressure-If the top roll diameter is more than the bottom roll *i.e.*, as called as "Top Pressure", the piece will go downward and if the bottom roll diameter is more than the top roll *i.e.*, called as "Bottom Pressure" the tendency of piece will go upward.

2. If the top guard is inclined downwards, the piece will also go down. The behavior of the metal is to be carefully judged, if piece is going down after hitting the some stationary elements after coming out of stand, then it can be concluded that piece was actually going up not down.

3. If the top guard is short, the front end of piece will go up. If the bottom guard is short, the front end of piece will down and may hit the roll table and then may go up.

4. If the bad metal sticks on the guard, the piece may go up or down, depending upon where the bad metal is stick.

5. If the top roll jumps, in case of hanging guard, the nose of top guard will also go up, when metal enters and will come down and piece leaves the

10 Mill Setting of Rounds/TMTS 243

roll. Proper care is to be taken for the setting of the guard gap for the tubular guard. Its nose is kept bit higher than the gap.

6. When front end is not having the uniform temperature, the cold surface will not elongate as much as the hot surface and bar will bend towards the cold surface.

7. If one groove in one roll get more worn out than the other, then diameter of that roll will become lesser in worn out roll than the other roll. The metal will have the tendency to go up or down as explained under top or bottom pressure.

8. In case of vertical stand, the metal will go up or down depending on the level of entry or exit traverse. If exit traverse is up, the metal will go up or if the exit traverse is down, the metal will go down.

10.4 Slipping in Stands

1. Slipping of bars in rolls is due to :
   (a) More speed of the stand or group of stands.
   (b) Under filling of the pass.
   (c) More reduction.
   (d) Too much less reduction.
   (e) Twisting of piece from previous stand.
   (f) Cold front end of the bar.
   (g) Too much hot metal.

   The slipping can be avoided by adjusting the above parameters, according to the case.

2. Passes should be ground and burning of individual stand before start of the rolling.

3. Slipping can also be avoided by making the pass rough, either by throwing of sand over the working pass or of ragging or knurling of rolls. Sometimes, welding in steel base rolls are made to reduce the slipping phenomenon.

4. Slipping in stand also occurs if cast iron rolls are used, instead of the steel roll.

10.5 Rolling Scheme

It is very important factor for the roller and all concerned staff to know the all intricacies of the rolling schemes.

(a) These schemes use the following denotation

H – Maximum height of metal at the arrow shown

B – Maximum Width of metal

244 Chapter 6 Rolling of Rounds and TMTS Bars

t – Rolling gap

D_k –Rolling diameter

Dc – Average rolling diameter

N – Speed of motor

i – Reducer ratio of the pinion stand.

(b) Rolling schemes are confirmed by Roll Pass Design Bureau. Any deviation can be allowed with the permission of I/c Opr. Mills in consultation with roll pass design bureau.

(c) Speed of rolls in the rolling scheme is given per the maximum diameter. Actual speed is to be computed as per the actual diameter of rolls.

(d) Roller and his associated staff should know the type of different passes and their behaviour during the rolling.

10.6 Pass Positioning and their Dimensions

Roller should also understand and keep records of the positioning of the working pass in roll and its dimension as per caliber and its condition and dimension after the wear out to make correct plan for rolling.

11 MAJOR POINTS DURING ROLLING

11.1 Setting of Roughing group

Roller should see to the adherence of following points:

- No twisting from any stand
- All passes to be made filled up
- Slipping in stand is to be avoided
- Pass burning of those box passes, which are having maximum reduction in rolling schedule to avoid the slipping.
- Metal dimensions after the roughing group should have the exact dimension as per the rolling scheme.
- Cast iron rolls should not be used in the roughing group.

11.2 Setting of the Intermediate Group

Following points should be kept under consideration, while setting the intermediate group:

- Twisting from the stands,
- The width of the metal of the previous stand, entering in to the box pass, should either be equal or in the range of ±2 mm to the root of the box pass, otherwise twisting may takes place.

11 Major Points During Rolling

- Under filling or overfilling in box passes of the intermediate group are to be adjusted to avoid the twisting.

- The delivery of the metal from stands to be checked and if bending of piece is *zig-zag*, then there may be more gap in the entry guide or entry box may be loose, accordingly adjustment has to be made.

- During initial setting, water should be closed.

- Metal dimensions after the intermediate group should have the exact dimension as per the rolling scheme.

11.3 Setting of the Finishing Group

Hook formation infront of end of piece at strand pass (plug oval), may be due to exit side tubular guard is away from the pass. It should be brought closer to the pass.

- Delivery of the metal fed into the pre-finishing pass should not have twisted front end and the difference of diagonal should not exceed 1 mm. As height of this vertical oval pass is more than the width, the top roll is taken to the opposite direction to the twisting direction.

- Similarly slipping in strand pass, because of under filling can be eliminated either by raising the previous pass or by pressing the strand pass.

- Repeater water should be closed during initial setting.

Twisting of metal in oval pass may be due to :

1. Under filling of the pass. In this case it is necessary to increase the width of the section from previous stand.

2. Difference of diagonals of the section fed into the oval pass. The diagonal of the fed bar should be rectified by adjusting previous stand by shifting rolls, shifting entry guides and with the proper filling of the pass.

3. Due to axial shifting of rolls in oval pass, oval pass behaves as box pass with height less than the width *i.e.*, left roll is to be taken in the direction of twisting of metal until this defect is rectified.

4. After shifting the rolls, it is necessary to check the oval section by putting wooden piece over the bar to find out one sided fins, which indicates unequal diagonals of the section fed to the oval pass.

5. More fins from the oval pass make mark on the bar and gives false mark on the finishing pass also. If fin comes at the bottom side of the oval of the vertical stand and if the roller entry guides of the finishing stand is also set low, then there are chances that metal may go to the collar of the guide and gets cobble.

246   Chapter 6   Rolling of Rounds and TMTS Bars

6. The finishing stand entry box feeds the bar vertically to the round pass. The responsibility of roller is for the correct preparation of the entry box and then the correct setting of the entry box *i.e.,* centre line of roller guide gap and the pass should coincides properly. Gap between the rollers with the metal should not be too tight or loose. The roller should rotate freely.

The jammed rollers causes:

- Mark on the bar
- Cutting in the bar
- Fish-tail formation.

In addition the contours of rolls should be checked with the template of the oval pass and also the smoothness of inserts should be checked.

7. The roller should check the condition of the bar delivered from the finishing stand *i.e.,* the direction of the twist, the fins, ovality on both sides and also the bending of the front end of the finished bar.

8. Roller should check following dimension of rounds in the sample saw and later on the cooling bed, to set the mill further :

(*a*) **Height (top to bottom):** If less, then more metal is taken from the pre finishing and strand pass. If necessary take more metal from the intermediate group.

(*b*) **Width (side to side):** Fin shows that width has come to its limit and roller has to reduce it.

(*c*) **Shoulders (right and left):** If right shoulder is more and twisting is towards right, then it is necessary to move the entry box roller to the left *i.e.,* the direction opposite to the twisting, the top roll should not be shifted.

    • If right shoulder is more and twisting is towards left (anti clock wise). The roll should be shifted in the direction of shift because rounds behaves like box pass with less height than the width.

    • In all cases when the dimensions of shoulders (right and left) exceed the dimension of height and width as per the specification, it becomes necessary to change the pass. This defect may be due to either pass is too much worn out or the turning of pass is not proper.

(*d*) **Fins and ovality:** One sided fin indicates the defective fixing of the entry box of the round pass, which should be shifted opposite to the defect.

    • In case of both sided fin, if observed in the entire length of the bar, it is necessary to reduce the thickness of the oval and if fin is at only one end of the bar, then tension is to be removed in continuous group of stands, especially in the finishing group.

- When false fin is noticed, the pass should be changed. This is the result of oval with fin enters into the finishing pass causes wear out of the finishing pass at top and bottom.

9. Falling of tail ends of the bar and their jamming inside the delivery rolling tackles of finishing stand, indicate that rolling tackles are not set properly (metal loose in entry roller) or the metal fed into finishing stand was of narrow or thin oval section (pass is not getting filled up) or entry box is set far from the rolls. It is true for strand pass also. It causes fish tail formation.

10. Scabs on the rod at regular intervals indicate defect in the pass and pass is to be changed.

11. Marks at regular interval throughout bar may be due to dent mark from the external body at the roll table.

12. Hook formation from finishing stand should be avoided, otherwise It will give trouble on cooling bed, *i.e.*, hitting against the gap between lifting valve. It is more in case of rolling TMT bar.

12 SLIT ROLLING OF BARS

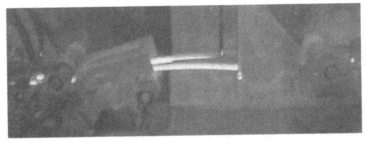

Fig. 6.32 Slit Rolling.

12.1 Why Slitting is Required

Slitting in hot rolling mills started in the beginning of 20th century, when scrapped and rejected rails were slit into web, flange and head by hot rolling process and then it were further rolled to various shapes.

In the late 60's, the slitting concept was adopted by Swedish engineers to enhance the production in bar mills where increasing the speed of mill was either difficult or was not advised due to design limitations of the mill.

Slit rolling is a method of rolling that allows relatively small size bars to be rolled at low finishing speeds, but still match with the higher reheating furnace capacity.

For example, to produce a 10 mm rebar single strand at furnace capacity of 80 T/H,it requires a finishing speed of > 20 m/sec. This speed is not possible on most of Bar Mills (either due to motor capacity or cooling bed deceleration).By slitting the bar into two strands part way through the process, mill can roll at 80 tones per hour and at the sametime, the finishing speed can also be reduced by half to 10 m/sec. Each strand will then produce half of the furnace capacity, *i.e.*,

80 / 2 = 40 t/hr

Advantages of Slit rolling

- Increased production.
- Reduced number of rolling stands. (for a particular product range).
- Lower rolling speeds with increased productivity.
- Increased competitiveness

Disadvantages of slit rolling

- Increased quantity of guide equipment.
- More complex rolling technique.
- Stringent maintenance of rolling mill equipment .
- Increased operator vigilance required

12.2 Sections Produced by Slit Rolling

(*a*) **Re-bars :**   8 mm, 10 mm, 12 mm, 16 mm, 20 mm and 25 mm.
(*b*) **Squares :** 6 mm, 8 mm, 10 mm, 12 mm, 16 mm and 20 mm.
(*c*) **Plain Rounds :** 10 mm, 12 mm, 12,7 mm, 16 mm and 20 mm.
(*d*) **Small Flats :** 12 mm × 3 mm, 16 × 3 mm and 20 × 3 mm.
(*e*) **Small Angles :** 16 × 16 × 2 mm and 20 × 20 × 3 mm

12.3 The Process of Slitting

STAND 1 – OVAL

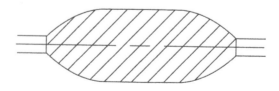

Fig. 6.33 OVAL.

It is to be ensured that the metal should be filled equally on both sides and sharp enough to enter the small radii in the fluted square at stand 2.

12 Slit Rolling of Bars

STAND 2 - FLUTED SQUARE

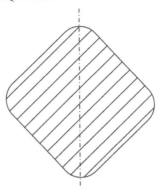

Fig. 6.34 Fluted Square.

It is vital, that this square should be equal, across both the flats end and of diagonals. The entering oval must not be lean.

Stand 3: Dogbone

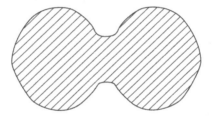

Fig. 6.35 Dog bone.

The dog bone pass must be filled completely, so that 4 roller entry guide should be set centrally on the pass with the front rollers being tight on the bar, and therefore it will facilitates the square to enter absolutely flat in slitting pass. Burn the sides to ensure that the pass should be completely filled, thus ensuring equal stock in each strand.

Stand 4: Slitting Pass

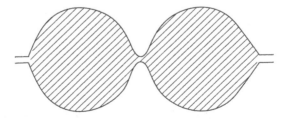

Fig. 6.36 Slitting Pass (Entry).

The pass should have the minimum collar gap. It minimize the amount of steel holding the two bars together, aiding to the slitting. The 4 roller entry guide must be set centrally on the pass with the front rollers holding the dog bone.

To maintain equal stock in the strands, wood burning of the sides should be done.

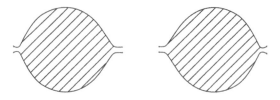

Fig. 6.37 Slitting Pass Guide (Delivery).

The slitting guide must be set centrally on the pass in both the vertical and axial plane, the slitting roller knife points should be touching and be directly above each other. The bearings must be tight, to prevent any axial play in the slitter guide rollers.

Stand 5: Leader Oval

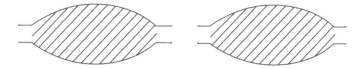

Fig. 6.38 Leader Oval.

Burn the sides of the each oval to ensure that they are filled equally and are not too sharp

Stand 6: Finishing Passes

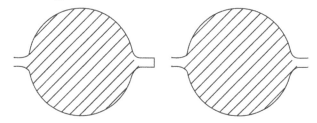

Fig. 6.39 Finishing Pass.

Check both passes for the wear out of pass and steel grade brands.

12.4 Control of Loop Heights between Stands 4 and 5 and 5 and 6

(a) The loop scanner should be scanning down to control the highest loop and to help to prevent speed cobbles.

(b) The loop heights of the two strands should be the same; this will indicate that the two strands will have the same weight/mt.

(c) If there is a small difference between the loop heights of the two strands, then it requires to control the sides and to ensure equal fill with a hint of overfill.

12.5 Method of Slitting

In the conventional slit rolling process, a square is used to form the dog-bone, which is fed into the slitting pass to get two bars of equal cross sections. In the slitting pass the material is torn by the rolling forces into two equal parts. For correct tearing of the material in the slit pass, the settings of pass and guides are very important besides it requires an accurate pass design.

The square used for slitting has the following difficulties:

1. Turning of the square by 45°, to ensure the correct entry of the square into the dog bone pass.
2. Balancing the material equally on both sides of the dog bone.
3. Stability of the guides and rolls.
4. Low pass and guide life.

A typical 12 mm slit rolling conventional slit roll pass design–Using leader square.

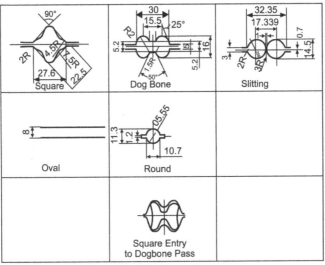

Fig. 6.40 12 mm Slit Rolling Conventional Design Using Leader Square.

12 mm Slit Rolling–Using Leader Round

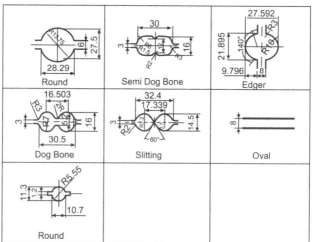

Fig. 6.41 12 mm Slit Rolling Conventional Design Using Leader Round.

The advantage of using round as Leader to form Dog-bone

1. Use of a round to form the Dog-bone would eliminate the twisting required in the case of a square.
2. Better control of equal balancing of the material on both sides of the dog-bone.
3. Maintains the stability of rolls and Guides.
4. Pass and Roll life improves.

3-Strand Slitting

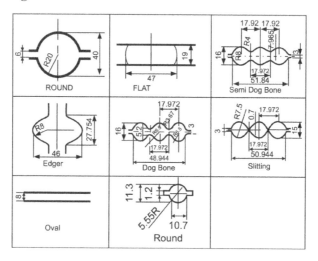

Fig. 6.42 Passes used for 3 Strand Slit Rolling.

12 Slit Rolling of Bars

4-Strand Slitting

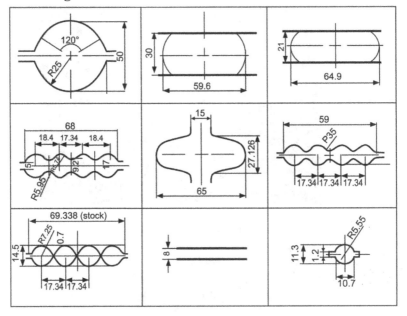

Fig. 6.43 Passes Used For 4 Strand Slit Rolling.

12.6 Slitting and Slit Guide

Slitting of the material is done by the rolls, whereas the guide bars are supposed to act only to guide the individual strands.

However, in actual practice, it has been observed that the slit guide bars help in the tearing process and in addition to their main function to guide the individual strand.

12.6.1 Slitting Force–Slitting Roll Profile

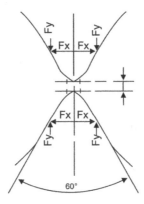

Fig. 6.44 Pass Design for Providing Slitting Roll Force.

12.6.2 Typical Slitting Guide

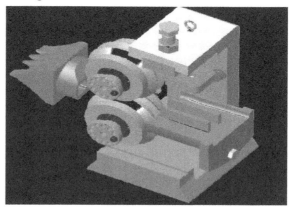

Fig. 6.45 Slitting Guide.

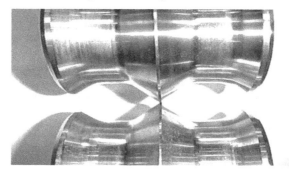

Fig. 6.46 Slitting Guide Point of Contact of Slitter Rollers and Dog bone Points on Slitter Knives, Perfectly Aligned.

The Slitting delivery guide box consists of:

(*a*) A complex nose piece that is accurately cast to fit the dog bone pass.

(*b*) A pair of high grade steel slitting knife rollers.

The nose piece strips the dog bone from the pass and guides it to the center of the slitting rollers. The dog bone is not cutting into two rounds with a slicing or knifing action, but its thrust apart by an axial separating load exerted by the slitting rollers against the inside shoulders of the two rounds.

Therefore, it is important that:

(*a*) The pass should be square, which will be minimizing the material holding the two rounds together.

(*b*) The slitting rollers should be square and their bearings should be tightly fit.

The cassette type guide fits into a guide holder to ensure ease and accuracy of set up during guide changes.

12.7 Problems Faced During Slit Rolling and its Solutions

12.7.1 Fails to Slit and Cobbles

1. Change slitter box, check old one for:

   (*a*) Worn slitter rollers,

   (*b*) Worn or loose slitter box bearings,

   (*c*) Scrap builds up in nose piece or slitter rollers.

2. Worn out dog bone pass or passes.

3. Entry guides on dog bones will ensure that the front rollers are tight on bar.

12.7.2 Excessive Side Rib

To check the excessive side ribs, ensure the following:

1. Size of oval against product guide sheet.

2. Weight, size and collar gap on finished bar.

12.7.3 No Side Rib

To have the no side rib, it is required to check the following:

1. Size of oval against product guide sheet.

2. Weight, size and collar gap on finished bar.

3. Tension between stands.

12.7.4 One Sided Side Rib

To check the one sided rib, ensure the following:

1. Line up of the entry guides (finisher).

2. Fill on the dog bone passes.

12.7.5 Transverse Rib Under Tolerance

To get the transverse rib under tolerance, firstly it is required to determine that whether ribs are on both sides of the bar or only at one side.

(*i*) If It is on one side

   (*a*) Oval is symmetrical in shape.

   (*b*) Oval is entering at an angle due to loop height.

(*ii*) If it is at both sides of bar

   (*a*) Size of oval,

   (*b*) Oval is not too sharp or too blunt,

   (*c*) Oval is leaning in the entry guide.

13 MILL SETTING PROBLEM DURING SLITTING

13.1 Rib Height below Tolerance on Both the Top and Bottom of The Bar

Reasons are:

1. Oval is too sharp.
2. Oval is too blunt.
3. The oval is leaning on entry into the finishing pass.
4. Width of the oval is too narrow causing insufficient draft.
5. Collar gap on the finishing pass too large.

13.2 Rib Height Below Tolerance on Either the Top or Bottom of the Bar

Reasons are:

1. Oval is non symmetrical due to unequal fill.
2. The oval is not entering the finishing pass straight or flat due to the loop height, this will cause the oval to be pushed into the ribs on one roll and held away from the ribs on the other roll. It will also cause a poor delivery.

13.3 Oval With Unequal Fill Due to Misalignment of the Entry Guide

13.4 Small Kinks Every 1 to 1.2 m

These normally occur at the pass brand.

1. Pass brand too deep.
2. Grade brand too deep and situated in the bottom of the pass rather than in the side.
3. The oval not entering the finishing pass straight and flat due to the looper being set too high.
4. Delivery nozzle and tube too large over the finished bar size.

13.5 Lap

1. Worn- out slitter pass.
2. Dogbone slitter pass not square.
3. Worn slitter rollers in guide box.
4. Worn bearings in slitter rollers.
5. Slitter rollers crossed.
6. Overfill on one of the dogbone passes.

13.6 Variations in Weight Between Strands

If the different strands vary significantly in filling of metal or one of the strands is empty or excessively full. The following checks and adjustments should be immediately carried out:

1. Observe, if the highest loop corresponds to the heaviest and/or fuller finished bar.
2. Confirm that the fluted square is equal across both the diagonals and the flats.
3. Wood burn the sides of both dog bones and confirm that the fuller side corresponds to the heaviest finished bar and the strand with the higher loop.
4. Make very small axial adjustments to the entry guide on the first dog bone that exhibits these conditions and observe the change in looper heights and also confirming they are becoming more equal in height.
5. Confirm that the dog bones are filled equally, and the finished rebar bars from all strands are within ± 1/2% on weight.

7
Rolling of Flats and Squares

1 INTRODUCTION

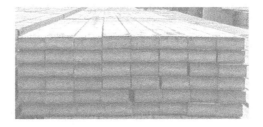

Fig. 7.1. Rolled Flats.

A considerable wide range of rolled products falls into the category of flat steel. This includes:

(*a*) Flats for construction purpose, with widely varied width and thickness.

(*b*) Bus bar section of 40-70 mm width and 5-15 mm in thickness.

(*c*) Specially shaped springs and other type of flat steel, such as brake pedal flat, fire grate bars automotive rim sections and other sizes for automotive and agriculture industries.

2 SELECTION OF ROLLING SEQUENCE FOR ROLLING FLATS

Flats steel are mainly rolled in open and tongue and groove pass design by following two methods in open-train mills:

1. *Open design* (between stepped rolls, using edging passes), intended for working the side edges of flats.

2. *Close design* (tongue and groove) rectangular passes.

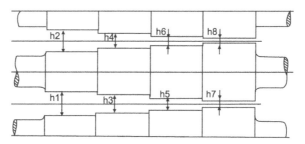

Fig. 7.2 Three-high stepped rolls for the rolling of flats.

Rolling in Open design is widely used design. It is used for producing narrow thick strips or flats of a width upto 100 mm. In this method, a square billet is first rolled between stepped rolls using intermediate edging passes (Fig. 7.2). The number of edging passes may vary. The leader is usually an edging pass (Fig. 7.3), which works the edges and control the width of the flat. The last pass is usually between plain-barrelled (polishing) rolls of a two-high stand. The edging passes are designed in various ways. More often, they are cut to little depth into rolls, this depends upon the design of stand and roll gap adjustment limits. The taper of sides in edging passes amounts to 5-10%.The advantage of this method is the good surface finish of the finished product, as the edging passes of this sequence will remove the scale effectively. Strips of various sizes can be produced in the same stepped rolls by changing the roll adjustment.

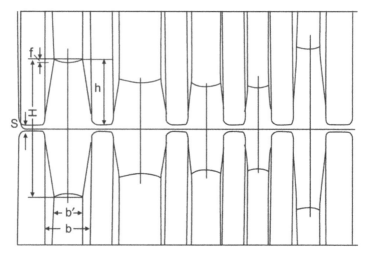

Fig. 7.3. Rolls with Edging Passes.

In designing the stepped rolls, the heights of the steps are determined on the basis of the possibility of rolling with various draught schedules. The width of each step depends on the maximum width of bar to be rolled.

2 Selection of Rolling Sequence for Rolling Flats

The depth of the grooves $(H-S)/2$, (Fig. 7.3), in the edging passes depends on the depth of the hardened layer of the cast-iron rolls. The height of the edging pass is set to the required width of the flat by raising or lowering the top roll as required.

The width of the edging pass at the bottom of the grooves depends upon the thickness of the flat h_{fl} and the draught Δh_{fl} in the finishing pass:

$$b = h_{fl} + \Delta h_{fl}$$

The amount of relief δ for an edging pass is from 10 to 15 per cent. The width of the pass at the break in the rolls equals

$$b' = b + \frac{\delta(H-S)}{2}$$

The bottom of the edging groove is made convex if square edged flats are to be rolled. This convexity is taken with a height f,

$$f = 0.5 \text{ to } 1.0 \text{ mm}.$$

A draught of 5 to 15 per cent is accomplished in the edging passes depending on the width of the flat.

Maximum reductions are permitted in the first few passings itself. The roll passes for flats are designed with due regard for power parameter s of the rolling mill and the strength of rolling stand.

The **second procedure** i.e., close design (tongue and groove) is applied in the rolling of wide and thin strip (Fig. 7.4).

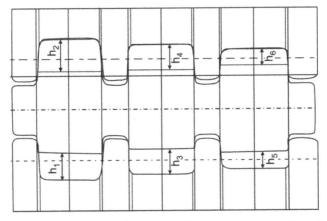

Fig. 7.4 Arrangement of tongue and groove passes on the rolls.

Rolling is performed in such passes with large draughts and restricted spread which is usually assigned from 55 to 90 per cent of the free spread for the given case. The relief is taken from 3 to 5 per cent for the roughing passes and 1.5 to 2 per cent in the strand and leader passes. Rounded fillets are provided at the corners of the pass. They have a radius equal to

$$r = (0.1 \text{ to } 0.2)\, h$$

The tongue and groove passes are designed in such a way, so that strips of various thicknesses, but of the same width can be rolled in the same pass. The difference in thickness of bars rolled in one pass may sometimes reach 30 or 35 mm.

As spread is restricted in this design of pass, it is necessary to turn the bar 180° after each pass to prevent overfills. This is accomplished by a corresponding arrangement of the passes on the rolls, as shown in Fig. 7.4 above.

The following are the disadvantages of using tongue and groove pass design for flat rolling:

1. It is necessary to use special devices to remove the scale from the bar.
2. Restricted spread leads to rapid roll wear.
3. The rolls can be only redressed to the amount of the width tolerance therefore more rolls will be required.
4. A separate set of rolls is required for each width of strip to be rolled.

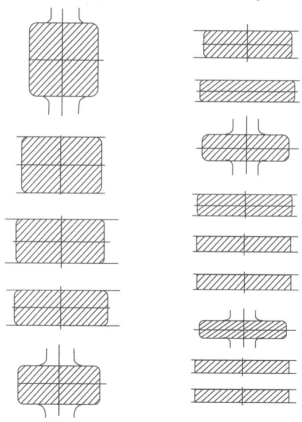

Fig. 7.5 Rolling of Flats in a specialized continuous mill (A, B, C,& D are vertical stands and stand 1–10 are horizontal stands.

2 Selection of Rolling Sequence for Rolling Flats

In modern section mills, having only one pass in each stand of the mill, strip and, flats are rolled in a sequence in which plain-barrelled rolls are placed alternate with edging passes. The number of edging passes and their order in the sequence, depends upon the definite conditions involved in each case of rolling. The smaller sizes of flats and strip can be rolled from square billets which are reduced to the required flat section in a minimum number of passes.

The most efficient method of rolling flats and strip is with alternate horizontal and vertical roll stands in specialized continuous mills. The rolling sequence for this type of mill is illustrated in Fig. 7.5.

Rolling in above mentioned sequence (Fig. 7.5) is continuous and after a definite number of passes, the strip is edged between vertical rolls. Advantages of this method is the complete removal of scale, well-worked edges of the strip, possibility of applying heavy draughts and the possibility of rolling a wide range of sizes in same rolls.

Higher draught is to be applied in initial passings of rolling of flats and strips, which can be only limited by bite conditions of rolls. It is necessary to reduce the draught in the succeeding passes due to rapid cooling of the thin strip. Flats upto 125 mm in width are rolled from square billets. Denoting one side of the billet as "a", the thickness of the flat as "h" and its width as "b", the following equation can be arrived (Fig. 7.6):

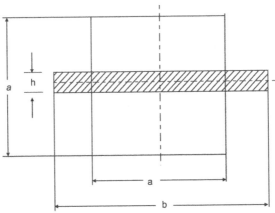

Fig. 7.6 Determining the side of a square billet in the rolling of flats.

$$b - a = k(a - h)$$

or

$$a = \frac{b + kh}{1 + k}$$

Where: k—the spread factor with a value from 0.3 to 0.5; the larger value being used in rolling thicker and narrower flats.

264   Chapter 7   Rolling of Flats and Squares

The total draught is equal to the ratio of the side of the square billet to the thickness of the finished flat.

After distributing the draught among passes, the thickness of the bar can be ascertained for each pass. Next step is to determine the draught and spread in each pass. After this, the draught values is to be checked to determine, whether it is well within the limit of mill parameters, subject to the bite conditions, power requirement and against the roll strength.

3 ROLL PASS DESIGN FOR FLAT 50 × 12 – 16 MM

First of all, Roll pass designer has to ascertain about the customer's requirement regarding type of product to be rolled and details of specifications under which it will be supplied. Its chemical and mechanical properties, tolerances on size, desired surface finish and above all, should know the detailed specification of the furnace, mill and finishing facilities, under which product is to be rolled.

3.1 Selection of Mill, in which section is to be rolled

Before the roll pass design of product, the roll pass designer should know the following:

- What are the strength and weakness of the mill, in which this product is planned to roll?
- The type and layout of the mill, production capacity of furnace, mill proper and finishing section of the mill.
- Size of billet available to roll this section.
- Temperature of input billet and finished product temperature.
- Type of layout of mill *i.e.*, whether it is an open, semi continuous or continuous mill.
- The details of mill facilities viz., mill configuration *i.e.,* number of rolling stands, nos of vertical stand available in each group. The distance between stands and in between groups, stand size, type of stand to be used for rolling *i.e.,* whether it is a housing-less, open or close type of housings which is to determine mill spring and ease of roll changing, type of roll used, specified rolling load and stand motor capacity, number of tillers in the mill and its spacing from preceding and succeeding stand. Shifting facilities to transfer the bar from one line to another and its length decides the maximum length of billet can be taken up for rolling.

All above mentioned parameters decide the type of roll pass design schedule, optimum numbers of passes are to be used in the sequence and also to determine the type of reduction pattern to be adopted.

3.0 Roll Pass Design for Flat 50 × 12 – 16 mm

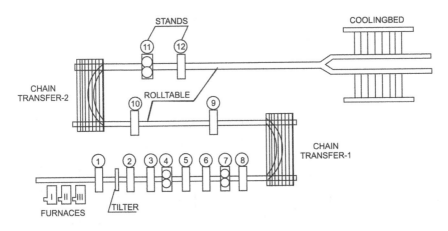

Fig. 7.7 Line Diagram of the Rolling Mill.

Here, the reference mill is a continuous cross country mill having 3 furnaces capacity of 60T/hr. Two furnaces are operated at a time and the third furnace is to be kept under repair. Mill is having 12 stands, out of which three stands, 1 (4, 7 and 11) are vertical stands. Mill is sub divided into three groups *i.e.*, roughing, intermediate and finishing group of stands. Tilter is placed between stand 1 and 2 and chain transfer are used to shift the metal from one line to another. Chain transfer (CT-1), *i.e.*, after stand 8 and CT-2, *i.e.*, after stand 10 are of 44 and 77 meters respectively.

Facility is provided for taking on –line sample with the help of a hot saw.

Cooling bed is of the 90 M length and with the half of bifurcates material can be transferred to either left or right side of cooling bed. Straightening machines and cold shears are placed on both the sides after the cooling bed.

Finishing section is having the facilities for stacking, making compact bundling and inspection and rail-road shipping facilities

3.2 Customer's Requirement *wrt* Specification, Tolerance etc.

Technical delivery conditions (TDC), given by the customer to decides the design of finishing pass in particular and roll pass design in general, *i.e.*, up to what extend finishing pass is to be designed on negative tolerance. The chemistry and size of input billet, alongwith selection of temperature of rolling at finishing stand are also important.

If BIS specification has to follow, then following are the details of tolerances of flats - 50 mm

Width tolerance ±1.0 mm

Thickness ±0.5 mm

Weight ± 2.5%

3.3 Rolling Scheme for Flat 50 x 12 & 50 x 16

Size of Flat- 50×12 Flat 50×16

Billet Size – $100 \times 100 \times 5$ m   Billet Size – $100 \times 100 \times 6$ m

Table 7.1

Std	h	b	t	D_k	n	h	b	t	D_k	n	Dc
1	74	70	6	456	300	75	70	6	456	300	515
	TILT90°					TILT 90°					
2	70	74	10	463	333	70	74	10	463	330	523
3	45	90	45	500	333	45	90	45	500	332	500
4(v)	55	55	13	385	300	55	55	13	385	300	425
5	35	65	35	500	368	40	63	40	500	332	500
	TILT90°					TILT 90°					
6	22	73	22	420	300	34	66	34	420	300	420
7(v)	49	22	17	416	460	46	36	14	416	424	448
8	18	51	18	420	367	25	51	25	420	368	420
9	16	52	16	420	300	20	54	20	420	300	420
10	14	53	14	420	300	17.5	56	17.5	420	300	420
11(v)	49	14	5	327	340	49	17.5	5	327	334	370
12	12	50	12	370	300	16	50	16	370	300	370

3.4 Pass Design of Flat 50 × 12 & 50 × 16 mm

Rolling scheme for rolling 50×12 and 50×16 mm flats are shown in Table 7.1. above. The billet size for 50×12 and 16 mm are taken as $100\times100\times5$ m and $100 \times100\times6$ m respectively to give a finished length of 78 m and 72 m.

The finishing pass is a barrel pass, while the edger pass is placed as a pre-finishing pass to give correct size of flats, because control of the width of bar is very much required to get the correct size of finished flat. The dimensions of final finished flat are based upon the roll pass design of this edger pre-finishing pass only. The number of edging passes and their order in the sequence depends upon the definite conditions involved in each case of rolling. The smaller sizes of flats and strip can be rolled from square billets which are reduced to the required flat section in a minimum number of passes.

The pass design of flat 50×12 and 16 is shown in Fig. 7.8. Other passes are kept as barrel, while vertical stand 4 and 7 are having either edger or box passes, as shown in Fig. 7.9 and 7.10 respectively.

3 Roll Pass Design for Flat 50 × 12 – 16 mm

3.5 Design of Pre-finishing Pass (Pass-11-Vertical Stand-Edging Pass)

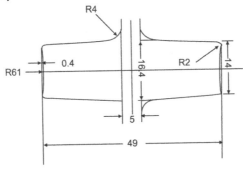

Flat-50 × 12

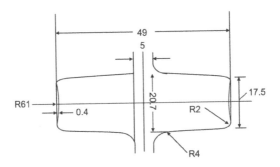

Flat-50 × 16

Fig. 7.8 Pre-finishingPass (Stand-11).

3.6 Pass-7 (Vertical Stand-Intermediate Group)

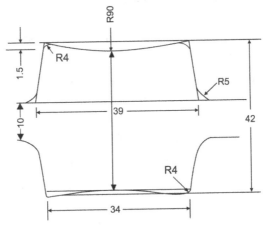

Fig. 7.9 Vertical Stand- Intermediate Group Stand-7.

Pass-IV (Vertical Stand - Roughing group)

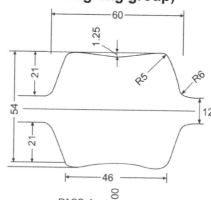

Fig. 7.10 Vertical Stand - Roughing group stand-4.

4 MILL SETTING AND SECTION CONTROL

When setting of Mill for Flats, special attention should be paid to avoid the distortion of metal in the edging passes. Twisting of metal around its axis, general bending of metal along the entire length and bending of front and back ends from all directions should be strictly avoided. This distortion of section in the edging and box passes may be due to the following reasons :

4.1. Under filling of the pass due to insufficient thickness of the bar fed. As a result of this, the flats will be twisted and may get cobble in other stands. Due to this, we will get finished section with diagonal difference.

4.2. Overfilling of the pass on account of excessive thickness of the section fed. This may happen due to excess reduction in the edging pass or feeding of more metal from previous pass. In this case, twisted and distorted section will come out from next stand.

4.3. Non-coincidence of the pass due to axial deviation of rolls, will also produce metal in twisted and distorted condition.

   (*i*) The thickness of the bar fed into the box pass should be such that it should be equal to the root of the pass with the deviation of ±2 mm.

   (*ii*) Overfilling of the edging pass should not be allowed.

   (*iii*) In case of twisting of section on the plain barrel on account of shifting of the rolls in the previous edger. The previous stand roll should be shifted in the axial direction towards the twisting of metal as observed after plain barrel.

5 Inspection and Checking of Dimensions of Flat Sections

When rolling thin flats, entry and delivery rolling tackles with worn out working surfaces should not be used.

4.4. While rolling thin and wide flats, care should be taken to see that roll tackles and the roll passes of intermediate, pre-finishing and finishing stands should not be excessively worn out. One of the signs of excessive worn out pass is that the delivery of flat is not straight, and it will have periodical looping in the vertical plane.

4.5. When thin flats are rolled, the gaps at the ends of roll barrels (collar gaps) should be set with minimum difference. If rolling tackles are fixed correctly (both entry and delivery), bending in different directions can be rectified by pressing the top roll on the side towards which the piece got bent or by pressing the previous stand top roll from the side opposite to bending. Pressing of previous stand is allowed only in case the dimensions of thickness of the bent flats which are uniform along the entire width.

4.6. When thick and narrow flats are rolled from same roll set, special attention should be paid to the twisting of bars around their axis and difference of diagonals and roughness of the surface of the edger stand. Twisting from vertical stand or roll cross in vertical stand gives rise to diagonal difference.

This twisting due to under-filling of the edging pass, because of insufficient thickness of the incoming bar or due to roll cross causes under filling of the opposite diagonals of flats.

4.7. Trapezoidal flats (the top of flat is wider than the bottom or *vice-versa*) may occur due to the following two reasons:

(*i*) Due to considerable wear out of the roll with plain barrel (upto finishing stand). The spread will be then unequal and will cause production of trapezoid flats.

(*ii*) Due to excessive reduction in the edging pass, the piece will get bend up or down.

5 INSPECTION AND CHECKING OF DIMENSIONS OF FLAT SECTIONS

5.1. When inspecting thin flats on the cooling bed, the width of flats should be checked atleast at three places *i.e.*, on the front, tail and also at the middle of the workpiece.

In case, variation of dimensions from the end to the middle is observed, then it is necessary to remove the tension of metal in the continuous group of stands.s

5.2. Quality of the edges of flats should be inspected next. Periodic dents on the edges of flats indicate the sticking of metal in the edging passes.

5.3. Scabs on the surface located uniformly also indicate defects of the same nature. In this case rolling should be immediately stopped and necessary steps should be taken.

6 DEFECTS AND ITS RECTIFICATIONS

6.1 Round Edges (convex)

Fig. 7.11 Convex sides.

This is due to less reduction in edging pass and too much reduction in finishing pass and barrel passes.

6.2. Concave Edges

Fig. 7.12 Concave sides.

This is due to insufficient draught in finishing pass and too much draught in edging pass and sometimes due to use of wrong edging pass also.

6.3 Dished

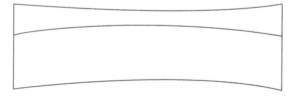

Fig. 7.13 Dished.

This may also due to insufficient draughting in the final edging and finishing pass.

6 Defects and its Rectifications

6.4 Dished and Convexed

Fig. 7.14 Dished and Convexed.

This may also be due to too much draught in final edging pass and not adhering to the draughting practices in other stand as per schedule given by roll pass designer.

6.5 Wedge

Fig. 7.15 Wedge.

Wedge is caused by either tension in the mill

- If edging pass too wide
- Entry guide not lined up with the edging pass,
- Not using a roller guide infront side of edging pass,
- Loop height is too high infront of edging stand.

6.6 Opposite Diagonals Rounded

This is due to roll cross of vertical stand, twisting of bar from vertical stand.

6.7 Rhomboid

Fig. 7.16 Rhomboid.

Rhomboid is due to if the edging passes are not square or setting of loose roller entry guide on edging pass or there is too much tension in the mill.

6.8 Sticker mark – due to Sticking of Metal on the Edger

6.9 Wear out Mark on Top and Bottom Edges–due to worn out flat pass or edger.

6.10 Lap– due to Fin from edger.

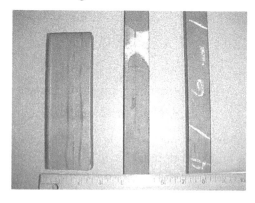

Fig. 7.17 Lap.

6.11 Diagonal Difference

The main reasons for the diagonal difference is either due to twisting from vertical stand or may be due to roll cross of vertical stand when twisting is not able to be detected.

7 ROLLING OF SPRING STEEL FLATS

Spring steel is a harder quality of steel. It will be difficult to get the desired shape for harder quality of steel, as harder will be the steel, more difficult it will be to get the desired shape. In addition, rolling of spring steel also causes the increase in load on the stand due to more spread and quicker wear out of the passes. As spring flats are used for special purpose only, the defects are undesirable. Defects like seams, scabs, marks, roughness, deformity of the section should be avoided.

During heating of this quality of steel, proper care should be taken so that temperature of rolling should not go high, otherwise, metal will easily get over burn and surface defect comes on the final section.

The bending of flat should be avoided as far as possible. Bending can take place on cooling bed and shear approach roll table, if metal is not transferred properly. It may also get bent on rope transfer, if dogs are not properly aligned or roll table plates are set uneven or may be due to careless operation of the rope transfer mechanism.

The most pertinent point in flat rolling is the sticker mark on the edges. This is due to sticking of metal in the edger passes. The most probable reason for sticking is the wedging action in the edger. The wedging action is more, if the thickness of the piece coming from the previous pass is higher. Another reason is the use of scabby metal. Efforts should be made for the avoidance of scabby billet to charge into the mill. Metal should be rejected from charging grate itself.

Sticker on the finished profile makes the product to be rejected. The vigil and timely detection of sticker will help in saving the metal from getting it rejected. So proper care should be taken right from the beginning to avoid sticker mark.

8 GUIDES USED IN FLAT ROLLING

Guiding and controlling the flat during the rolling process is of paramount importance.

8.1 Entry Guide

For maximum control, the entry guide should be kept as close as possible to the rolls with clearance ≥ 3 mm (Fig. 7.18). It must be checked periodically to ensure that it is "Inline" with the delivery guide.

To control and guide the flat, the entry guide should be set in such a way, that there should be a clearance of 3 to 5 mm over the entering bar width.

Fig. 7.18 Entry static entry box with inserts for rolling flats.

8.2 Delivery Guide

The bottom cast stripper guides are set in the guide box and the rest bar should be so adjusted, until it set to about 3 mm below the bottom roll. It should be in line with the entry guide, so as to prevent front end to get hooks and wave.

It is also necessary to check whether the guide is "flat" with a level and whether tip of guide is in contact with the bottom roll.

Accurate setting and adjustment of the entry and delivery guides, adjustment of required clearances for inter stand equipments and conveyors will prevent occurring of defects in flats.

Poor guiding and control may cause a number of defects, explained below:

1. Horizontal Wave

Fig. 7.19 Horizontal wave.

To prevent horizontal wave, the static entry guide should be set so that there should not be a clearance of ≥3-5 mm over the entering bar width. The exit guide on the finishing stand can be set to the finished width; this would allow the clearance on the guide to equal the spread on the finishing pass.

Fig. 7.20 Below shows that too much clearance in the entry guide allows the bar to deflect on side-ways create the horizontal wave

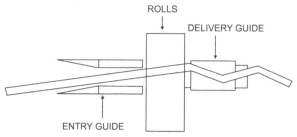

Fig. 7.20 Wrong Setting of Guides causing Horizontal wave.

2. Vertical Wave

Fig. 7.21 Vertical wave.

3. Hooked Front Ends

Fig. 7.22 Hooked front ends

4. Wave and Hooked Front Ends

Fig. 7.23, above shows the entry and delivery guide misaligned. This is a common cause of front end hooking, horizontal wave and in some cases sweep.

Other common causes of hooking and horizontal wave are:

1. The entry guide is not aligned with the shaped edging pass.
2. Worn out condition of drive spindles or couplings.

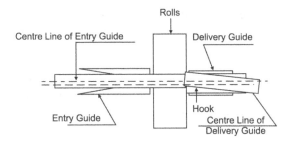

Fig. 7.23 Wrong setting of Guides causing wave and hoked front Ends.

3. Rolls not set parallell.
4. Entry piece is not square, which is entering to "flat" pass.

9 ROLLING OF SQUARES

There are two types of squares rolled:

(a) Sharp cornered square   (b) Round or radius cornered square

Fig. 7.24 Type of Square Section.

Rolled squares are used in virtually every manufacturing industry *e.g.*,

  (*i*) Automotive ring gear for engines shafts, tow bar frames.
  (*ii*) Ornamental Security screens fences and gates. Hand rail supports for ladders.
  (*iii*) They are also supplied to the cold drawing industries, when end users demand a close tolerance and clean surface finish.

10 SELECTION OF ROLLING SEQUENCE FOR ROLLING SQUARE

The most extensively used sequence for rolling squares is the one shown in Fig. 7.25. This sequence consists of a finishing square, leader diamond and strand square passes.

The bar obtained in the breakdown passes will enter into the strand square pass. Then the bar is turned 90° and entered into the leader diamond pass.

Upon delivery, it is again turned to 90° and entered into the finishing square pass. In the last pass, the major diagonal of the diamond will be vertically positioned.

It is evident that a diamond bar in this position will have a tendency to turn down, *i.e.*, to roll over with its major diagonal in a horizontal position. This tendency of turn down can be prevented by using guides to hold up the bar in entering the square pass.

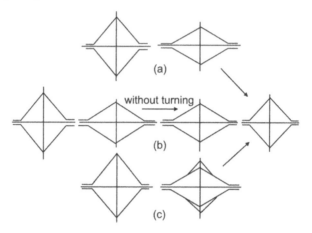

Fig. 7.25 Pass Sequences for Rolling of Squares.

The diamond-diamond and diamond-square breakdown sequences are usually used in the roughing stands. In rolling squares of the smallest sizes, an oval-square roughing sequence can be applied. The sequence shown in Fig. 7.25(a) and (c) are used in rolling squares of small and medium sizes.

Large steel squares are rolled with a different sequence, shown in Fig. 7.25(b), which has a strand diamond pass in addition to the leader diamond and a square pass precedes the strand diamond.

The bar delivered from the strand diamond pass enters into the leader diamond without turning. This measure will produce a leader diamond bar with properly filled corners, especially in rolling heavy bars. The dimensions of the finishing square pass can be determined on the basis of the shrinkage in cooling and provision of minus tolerance as per specification to be used. If a side of the finished cold square is denoted by a, the minus tolerance by Δa and the co-efficient of shrinkage is taken as 1.012 to 1.015 (finishing temperature of 850-1000°C), then the side of the finishing square pass a_{sq} will be equal to

$$a_{sq} = (1.012 \text{ to } 1.015)(a - \Delta a)$$

10 Selection of Rolling Sequence for Rolling Square

In rolling squares with sides upto 25 mm, the finishing pass is designed as a true square without rounding the corners. For larger squares, the finishing pass is designed such that its horizontal diagonal will be 1.42a and vertical diagonal will be 1.41a, where a is the size of square. In this case, the angle at the apex of the pass equals to 90°30'. The clearance between the rolls depends upon their diameters.

The design of the leader diamond pass is based on the condition that the co-efficient of elongation in the finishing pass should be within 1.10 to 1.15.

Then the area of the leader diamond pass will equal to:

$$F_d = (1.10 \text{ to } 1.15) F_{sq}$$

Where: F_{sq}—the area of the finishing square pass.

The size of the obtuse angle β in the leader diamond pass depends on the size of square being rolled. The smaller the square, the larger angle β is taken and *vice-versa*.

The height of the leader diamond pass depends on the spread (Δb) of the bar in the finishing pass. This spread is taken according to the amount of draught *i.e.*:

$$\frac{\Delta b}{\Delta h_m} = 0.3 \text{ to } 0.6$$

Where: Δh_m the mean draught.

$$b = h \tan \frac{\beta}{2}$$

The dimensions of the diamond pass are sometimes determined from its area. Thus, if the area (F_d) and height (h) are known, the width (b) of the pass can be determined from the formula:

$$b = \frac{2F_d}{h}$$

The dimensions of the strand square pass can easily be determined if the co-efficient of elongation is assigned for rolling the bar in the leader pass. The co-efficient of elongation in the diamond pass is usually taken from 1.20 to 1.25. Then the area and side of the strand square pass will equal

$$F_{a \cdot sq} = (1.20 \text{ to } 1.25) F_d$$

and
$$a = \sqrt{F_{s \cdot sq}}$$

The dimensions of the strand square pass are frequently determined on the basis of the spread when the square bar is rolled in the leader diamond pass. If this spread is known, the vertical diagonal of the square pass can be determined.

The spread in the leader diamond pass may be determined by the formula:

$$\frac{\Delta b}{\Delta h_m} = 0.25 \text{ to } 0.50$$

In rolling squares, the main problem is the difficulty in obtaining properly filled corners, particularly, those at the breaks of the rolls presenting the most difficulties in getting properly filled corners. This problem can be solved by the design of diamond passes with special care *i.e.*, the computation of the amount of spread, when it enters the square pass

11 ROLL PASS DESIGN OF SQUARE- 50 MM

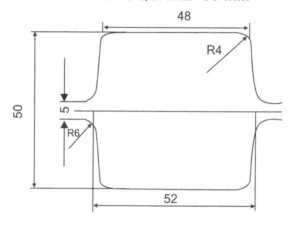

Fig. 7.26 Finishing Pass of Square 50 x 50 mm (Stand-12).

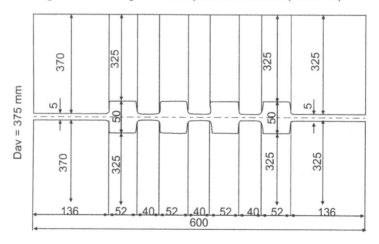

Fig. 7.27 Finishing Stand of Square 50 (Stand-12).

11 Roll Pass Design of Square- 50 mm

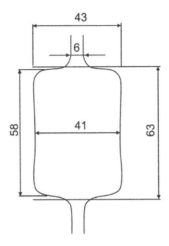

Fig. 7.28 Pre-finishing Pass (Stand-11).

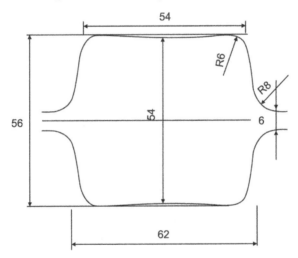

Fig. 7.29 Strand Pass (Stand-10).

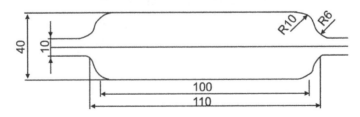

Fig. 7.30 Barrel Pass (Stand-7 and 9).

280 Chapter 7 Rolling of Flats and Squares

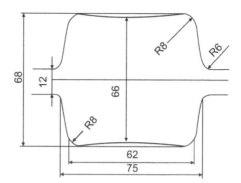

Fig. 7.31 Box Pass (Stand-8).

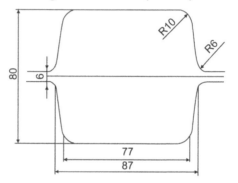

Fig. 7.32 Box Pass (Stand-6).

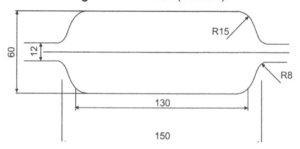

Fig. 7.33 Barrel Pass (Stand- 3&5).

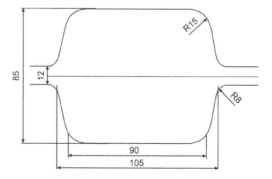

Fig. 7.34 Box Pass (Stand-4).

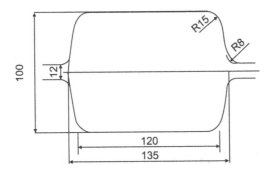

Fig. 7.35 Box Pass (Stand-2).

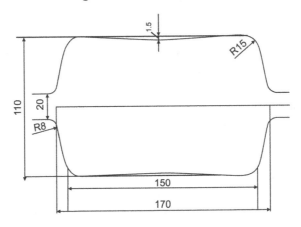

Fig. 7.36 Box Pass (Stand-1).

12 ROLLING SCHEME FOR SQUARE

Billet to be used – 150 × 150 × 6 m

Table 7.2

S.No	Stand	Shape of Pass	H	B	t	Dk	n	Dc
1.	(H)	Box Pass	120	156	30	425	300	515
		TILT 90°						
2.	(H)	Box Pass	121	128	33	435	340	523
3.	(H)	Barrel Pass	90	140	42	475	340	523
4.	(V)	Box Pass	105	100	32	352	300	425
5.	(H)	Barrel Pass	75	115	27	452	360	500
		TILT 90°						
6.	(H)	Box Pass	90	85	16	345	311	419
7.	(V)	Barrel Pass	62	100	32	395	345	425
8.	(H)	Box Pass	72	72	18	365	300	419

Contd...

9.	(H)	Barrel Pass	56	76	26	370	300	400
10.	(H)	Box Pass	60	60	10	369	300	419
11.	(V)	Box Pass	48	64	13	336	300	370
12.	(H)	Box Pass	50	50	5	325	331	370

(H)–Horizontal Stand, (V)–Vertical Stand
H– Height of Pass, B–Width of Pass, t–Roll Gap
D_k–Rolling Diameter, n–Motor RPM, DC–Average Diameter

13 DEFECTS AND ITS RECTIFICATIONS

13.1 Overfill

Normally caused by :

(*i*) Too much stock comes from leader pass, or the finishing square is too small. (If the sample is taken from tail end of the stock, then take a sample from middle portion to check for tension).

(*ii*) Entry guide is not set in line with the pass.

(*iii*) Diamond is leaning on entry to the square due to looseness of the entry guide.

13.2 Under Fill

Normally caused by

(*i*) Top and bottom corners of the pre-finishing diamond remain under filled.

(*ii*) Too little stock comes from the pre-finishing diamond pass.

(*iii*) Entry guides loose causing the diamond to lean.

(*iv*) The stock entering the diamond is too small and or under filled.

13.3 Lap

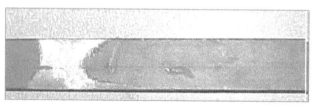

Fig. 7.37 Lap in Flat Rolling.

Overfill on the diamond and or earlier process pass. (The shallower the angle of the lap the earlier in the process it occurred).

13.4 Offsquare

Normally caused by either the finishing rolls are crossed and or the entry diamond is leaning.

❏❏❏

8

Rolling of Angles

1 INTRODUCTION

Generally Equal angles of size of 20×20 mm to 200×200 mm and unequal angles of size 30×20 mm to 200×120 mm are rolled in a rolling mill.

Fig. 8.1 An Angle Section.

2 TYPES OF SYSTEM USED FOR ANGLE ROLLING

In the pass sequence for rolling of angles (Fig. 8.2.I), the flanges are gradually reduced and bent in closed passes. The flanges of angles, rolled in this sequence are remain straight throughout the rolling and the central angle is gradually decreased from 125° or even from 145° in the first roughing pass to 90° in the finishing pass. The central angle may be decreased uniformly in each pass or make proportional to the draught of the flanges in each given pass.

The main disadvantage of this sequence is deep groove cut in the rolls, which causes large difference in the working diameter of top and bottom rolls. The latter factor leads to increased roll wear, higher power requirement and

284   Chapter 8   Rolling of Angles

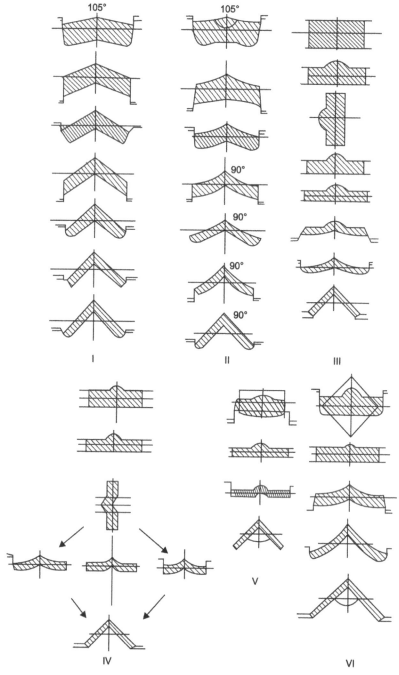

Fig. 8.2 Rolling Sequence of Angle Rolling.

poor surface finish of the final product. Keeping in view of above disadvantages associated with the above mentioned sequence, "Butterfly Design" is developed. (Fig. 8.2.II). The butterfly system has the central angle of 90° in

3 Selection of Rolling System

last three passes *i.e.*, finishing, pre-finishing and leader passes and in the first roughing pass, the central angle ranges from 125° to 145°, which is gradually decreases and made at 90° in the finishing pass. The flanges can be subjected to considerable draught; in addition legs are getting bent in the course of rolling.

A universal pass sequence (Fig. 8.2 III) is employed in modern high production mills to roll angle sections. The main advantage of this sequence is the bar can be rolled without spread restriction in roughing group. This facilitates the use of same sets of rolls for all sizes of angles. The edging pass in this sequence works the flange toes and control the length of the flanges. This sequence also advocates the use of less number of rolls to accommodate the rolling of wider size range of angles, in turn increases the mill output considerably and reduces the cost of production.

In addition to the roughing edging pass, the sequence shown in (Fig. 8.2. IV), has a second edging pass that usually precedes the leader pass. Here also, larger ranges of sizes can be rolled from one set of rolls. The sequence can be applied in a continuous mill, where edging passes are in the vertical stands of the mill.

Sometimes, angle sections are rolled by bending the flanges, as shown in (Fig. 8.2. V). The square or rectangular billet is first rolled into a flat bar, with the protrusion in the Centre of the top face. Further passes bend the flanges to give the required Centre angle. It is observed that certain difficulties are faced in the bending the flange in the finishing passes. This sequence is proved suitable for the rolling of thin- flanged angles.

There is one more sequence, in which a square billet enters into the first pass with one diagonal position vertically. (Fig. 8.2. VI). It gives the rapid increase in the width of section. This method is generally used for small sizes of angles and angles can be rolled in as few as minimum as in three passes only.

3 SELECTION OF ROLLING SYSTEM

Popular pass system for rolling angles are divided into mainly two categories:

(a) Rolling in closed passes with "Tight Widening"

The edges of grooves are made in same ways as rolling of strips in closed passes. The inclination and rounding of the side walls gives a flexibility to change the degree of tightness for spread or widening. With very tight widening, it is easier to achieve consistency of the width in every pass, and thus to attain correct size of the legs.

This method of rolling are not finding much popularity for high productive mill, as flexibility of operation is limited. It is used only for special steel production for specific requirement.

(b) Rolling in Open Passes with "Free Widening"

Rolling in open passes with "Free Widening" is widely used for angle production.

The advantages of open system are as follows:

- Considerable increase in co-efficient of deformation due to a smaller cut in the roll, higher reduction can be given in a particular pass, which makes it possible to reduce the number of passes; in turn increase the production of the mill.

- Partial or in some cases, it makes complete elimination of roll changing, while changing over to another size of angle. In order to change over to angles of intermediate sizes, rolling can be possible with the only change of pre-finishing pass.

- Rolling in open pass system, makes it possible of rolling in same roughing groove, angles as well as strips, without any roll changing. It helps in reducing the roll inventory.

- Further, simple pass design and absence of collars in the open system helps in reduction in the cost of rolls.

- Absence of any danger of scratches or laps in the finished product.

The main drawback of this system is the difficulty in getting the correct size of the leg due to free widening. On the contrary consistency of thickness of legs can be easily maintained in the close system with tight widening. That's why, pre-finishing or leader pass is made closed to prevent any variation in size of finished section. If more number of grooves with free widening follows the closed pre-finishing pass, then it will be more difficult to control the width of section.

Selection of the any sequence for rolling angles is based upon the following factors:

- Whether the design is based upon the principle of "One Pass One Stand"?
- Whether same groove will be used for rolling different sizes?
- Whether technological process is automatized and mechanized?

  (Automation assists in achievement of constant temperature of rolling and thus establishing the exact amount of free widening at all passes).

4 SALIENT FEATURES OF ROLLING ANGLE

4.1 Selection of Mill Size

A relationship with the mill size to the angle size is shown below:

4 Salient Features of Rolling Angle

Table 8.1

Roll dia	250	300	350	500	750-800
Angle size	20-50	40-75	60-90	80-140	120-200

The relation is not always definite. It depends upon the condition of different elements of mill *i.e.*, furnace condition, type of stand, motor capacity, finishing facilities etc.

4.2 Design of Finishing Pass

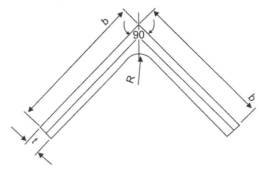

Fig. 8.3 Design of Finishing Pass.

The finishing pass is designed either with free or tight widening. It is proved that finishing pass with free widening will have following advantages:

- The possibility to roll angles of different sizes in one roll groove or pass.
- The possibility of rolling different thickness of angle of one size.
- The possibility of formation of fin is eliminated.

As amount of widening in finishing pass is very negligible, the edges obtained from rolling with free widening are of good quality. Considering all above mentioned facts, it is can be concluded that the design of finishing pass with free widening is the best.

4.3 Computation of the Size of Billet

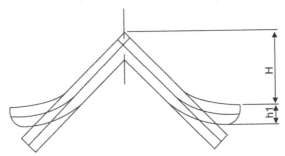

Fig. 8.4 Computation of size of Billet.

(a) Height of Billet

(*i*) Rolling in closed type of pass:

$$h = h_1 + 2H$$

Where h_1 is the thickness of the leg and H is height of apex above bend as shown in Fig. 8.4.

(*ii*) Rolling in open pass with free widening

$$h = h_1 + H$$

Where, h_1 is the thickness of flange and, H is the height of the apex, above bend.

(b) Width of Billet

$$b = b_{\text{pre-finishing}} - \Delta b_{\text{total}}$$

$b_{\text{pre-finishing}}$ is width of flared pre-finishing pass and Δb_{total} is the commulative spread from roughing to pre-finishing stand

4.4 Co-efficient of Deformation

Maximum deformation with free widening in case of small and medium angle, is generally recommended as 2.0 or more, while for bigger angle, it varies from 1.6-1.7. Higher deformation is not recommended in finishing group of modern mills, as it leads to faster wear out of the passes.

5 MODERN METHOD OF ROLLING ANGLES

In a modern mill, if suppose total 11 passes are used for producing angle 55×55 mm, then first four passes will be used as breakdown passes to facilitate rolling with the higher input size, which in turn will facilitate higher productivity with longer length of finished product.

Pass-5 is the first shaped pass. It is necessary that bar should be made free from fin before entering into the first shaped pass. It will help in getting the good surface finish of the finished toe of the angle. The design of the apex of pass-5 is to be very shallow, varies from 125° to 145°, to make the roll stronger and filling of apex in the pass is ensured.

Passes of roughing stands are used to produce the entire range of size of angle by merely raising or lowering of roll. It is possible because these passes are flat and raising of roll increases the length of leg very little, not enough to have any substantial effect on the overall width of the last roughing pass. In case of intermediate passes, raising of the roll increases leg length not only because of the lift, but with lifting of rolls, the legs are thickened proportionally more at ends, than at apex. With increased thickness at the

5 Modern Method of Rolling Angles

ends of the leg, the subsequent reduction in leader and finishing passes causes increased spread, resulting in more leg length. The reduction in thickness in pre leader passes is more important to ensure prevention of under-filling or overfilling in subsequent finishing pass.

To get a sharp apex is one of the problems in the rolling of angles. The rolling of heavier gauges presents maximum trouble in maintaining a sharp apex on the finished bar. To overcome this problem, it is proposed to roll angles with heavier gauges from a thicker slab. But, with thicker slab, there will be difficulty in maintaining the right balance of reduction between leg and apex. The increased elongation and spread of the legs stretches the metal from the apex. To overcome this problem of blunt apex, a new pass design is evolved with 90° apex design, not only for finishing and pre-finishing passes, but for strand and intermediate passes also. The roll pass design with fixed 90° apex design is explained below.

5.1 Roll Pass Design of Equal Angle

First of all, Roll pass designer has to ascertain the customer's requirement regarding tolerance on size, mechanical and chemical requirements and specifications under which product is to be certified, *i.e.*, its chemical and mechanical properties, tolerances on size, weight tolerance, desired surface finish. In addition, details specification and layout of the mill is to be studied thoroughly, under which, the section is to be rolled.

Different users of angle will be having different priorities on specification *i.e.*, manufacturers of transmission towers will demand for stringent tolerance on size and length, while construction contractors will ask for negative tolerance on weight to facilitate more length per tonne of weight.

5.2 Selection of Mill

Before the roll pass design of product, the roll pass designer should know the following:

- What are the strength and weakness of the mill, in which this product is planned to roll?
- The type and layout of the mill, production capacity of furnace, mill proper and finishing section of the mill.
- Size of billet available to roll this section
- Temperature of input billet and finished product temperature
- Type of layout of mill *i.e.*, whether it is an open, semi continuous or continuous mill.

- The details of mill facilities *viz.*, mill configuration *i.e.*, numbers of rolling stands, numbers of vertical stands available in each group, the distance between stands and in between each group, stand size, type of stand to be used for rolling *i.e.*, whether it is a housing-less, open or close type of stands, which is to be determine the amount of mill spring and ease of roll changing, type of roll used, specified rolling load and stand motor capacity, numbers of tillers in the mill and its spacing from preceding and succeeding stand. Shifting facilities to transfer the bar from one line to another and its length decides the maximum length of billets can be taken up for rolling.

All above mentioned parameters decide the type of roll pass design schedule, optimum numbers of passes to be used in the sequence and also to determine the type of reduction pattern to be adopted.

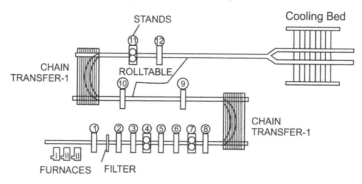

Fig. 8.5 Line Diagram of the Rolling Mill.

Here, the reference mill is a continuous cross country mill having 3 furnaces of capacity 60T/hr each. Two furnaces are generally operated at a time and the third furnace is kept under repair. Mill is having total 12 stands, out of which three stands (4, 7 and 11) are vertical stands. Mill is sub divided into three groups *i.e.*, roughing (5 stands), intermediate (4 stands) and finishing group (3 stands). Tilter is placed between stand 1 and 2. Metal get shifted from one line to another, with the help of chain transfers. Chain transfer (CT-1), *i.e.*, after stand 8 and CT-2 *i.e.*, after stand 10 are of length of 44 and 77 meters respectively.

Facility is provided for taking on line sample with the help of a hot saw.

Cooling bed is of the length of 90 M and with the help of bifurcates, material can be transferred to either left or right side of cooling bed. Straightening machines and cold shears are placed on both the sides of rolling line after the cooling bed.

Finishing section is having the facilities for stacking, making compact bundling and inspection and rail-road shipping facilities.

5.3 Customer's Requirement *wrt* Specification, Tolerance etc.

Technical delivery conditions (TDC), submitted by the customer decides the design of finishing pass in particular and roll pass design in general, *i.e.*, which specification has to be followed while designing the section and upto what extend one can go with negative tolerance, without affecting finished product properties.

If BIS specification is to be followed, then following are the details of tolerances for angle 55×55×6 mm:

(*a*) Leg Length — ± 2 mm *i.e.*, 53 mm to 57 mm.
(*b*) Out of Square — The leg of angles should be perpendicular to each other with-in a tolerance of ± 1°.
(*c*) The difference between length should not be more than 3 mm.
(*d*) Weight — + 5% on plus side and – 3% on negative side of the sectional weight is allowed.

Technical delivery conditions (TDC) of transmission tower manufacturer are generally very rigid *i.e.*, they generally demand tolerance on leg length only upto 1 mm.

6 PASS DESIGN OF ANGLE 55 × 55 × 6 MM

6.1 Design of Finishing Pass

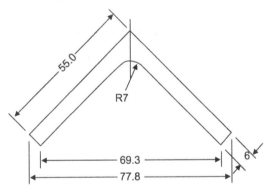

Fig. 8.6 Cold Section of Angle 55 × 55 × 6 mm.

Area of Angle = $t \times (2L - t) + ax$, where t and L are the thickness, length of angle and ax is the area of curvature.

$$l_{12} = 55 - 3 = 52 \text{ mm}$$
$$F_{12} = 644 \text{ mm}^2$$
$$\text{Weight} = \frac{644 \times 7.85}{1000}, \text{ where density of steel is 7.85 gm/cc.}$$
$$= 5.06 \text{ kg/m}$$

If the length of bar is taken as 85 m (length of cooling bed is 90 m)

Then total bar weight = 5.06 × 85

= 430.10 kg.

The input billet length will be 100 × 100 mm, its sectional weight is 76.2 kg/m

Then the length of billet will be = $\frac{430.10}{76.20}$ = 5.64 m

It is recommended that the length of billet is to be taken as 5.0 m to have the standardization of the billet size in billet yard .Billet length of 5 m will give total length of bar of about 76 m at cooling bed, which is within limit.

Finishing temperature after the finished stand is to be ensured around 850°C to have better quality and properties of the finished angle.

Flange Thickness (Hot) = Flange thickness (cold) × Co-efficient of elongation

= 55 × 1.012 = 55.56, say 55.6 mm

At the temperature of 850°C, the co-efficient of elongation will be around 1.0102.

But the leg length in this case is to be kept to the cold section size *i.e.*, 55 mm, to facilitate to rolling in negative size and to give more roll life even after the wear out of pass.

6.2 Computation of Thickness of other Passes of Angle Rolling

A thickness graph is to be prepared for co-efficient of reduction *vs.* pass number is shown below in the Fig. 8.7.

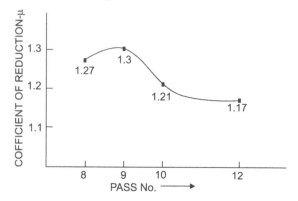

Fig. 8.7 Graph for computation of Co-efficient of Reduction for Closed Passes for Angle 55 × 55 × 6 mm.

It is seen from the graph that co-efficient of reduction is to be kept as minimum as possible in finishing passes, as to give more pass life due to less

6 Pass Design of Angle 55 × 55 × 6 mmt

wear out of the finishing pass. Reduction then gradually increases in subsequent passes *i.e.*, from pre-finishing passes going towards initial passings, whereas, it limited by angle of bite and motor capacity only.

The finishing pass should be designed with free widening. In this case, the finishing pass will so design, so that not only, it will facilitate to roll angles of size 50-65 mm size from the same pass, but also to roll thickness from 5 mm to 10 mm of same size from the same pass.

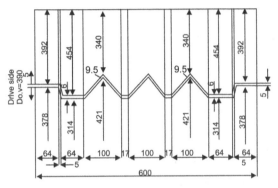

Fig. 8.8 Roll Diagram of Finishing Stand (Stand-12).

6.3 Pre-finishing Pass (Stand -10)

The computation of the pass and roll design of pre-finishing stand-10 is explained below and as shown in Fig. 8.9 and 8.10 below.

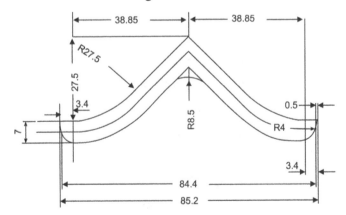

Fig. 8.9 Pass Diagram for Prefinishing Stand (Stand-10).

Thickness of flange is taken as 7 mm.

$$\mu_{12} = t_{10} / t_{12}$$
$$\mu_{12} = 7/6 = 1.17$$
$$\Delta h_{12} = 7 - 6 = 1 \text{ mm}$$

Spread is generally assumed as 30% for angle design

$$\Delta b_{12} = 0.3 \times 1 = 0.3 \text{ mm}$$

$$l_{10} = l_{12} - \Delta b_{12}$$

$$= 52 - 0.3$$

$$= 51.7 \text{ mm}$$

Length, $\quad l_a = 27.5 - 7/2$

$$= 24 \text{ mm}$$

Length of chord, $l_b = 3.14 \times 45/180 \times (27.5 + 3.5)$

$$= 24.3 \text{ mm}$$

Length, $\quad l_c = l_{10} - (l_a + l_b)$

$$= 51.7 - (24 + 24.3)$$

$$= 51.7 - 48.3$$

$$= 3.4 \text{ mm}$$

Assuming an 6% taper on each side

Then, $\quad a = 0.06 \times 7$

$$= 0.42 \text{ mm}$$

$$B = 2 \times (38.85 \text{ say } 39) + 3.4 = 42.4 \text{ mm}$$

$$= 2 \times 42.4$$

$$= 84.8 \text{ mm}$$

$$B_{max.} = B + a$$

$$= 84.8 + .42$$

$$= 85.22 \text{ mm or } 85.2 \text{ mm}$$

$$B_{min.} = 84.8 - 0.42$$

$$= 84.38 \text{ mm or say } 84.4 \text{ mm}$$

$$Y_{10} = Y_{12} \times l_{12}$$

$$= 1.165 \times 7$$

$$= 8.2 \text{ or say } 8.5 \text{ mm}$$

Rolling Diameter, $D_k = D_{av} - \dfrac{F_k}{B_{max}}$

$$= 425 - \frac{745}{85.2}$$

$$= 416.2 \text{ mm}$$

As finishing and pre-finishing stands (stand 10 and 12) are not the part of a continuous group, the N_{12} and N_{10} are assumed as 300 RPM.

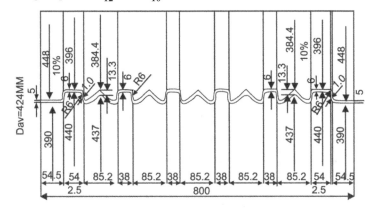

Fig. 8.10 Roll Diagram Pre-finishing Stand (Stand-10).

6.4 Strand Pass (Stand-09)

With the help of thickness graph, computation of pass and roll design of stand 9 has been made and explained and shown in the Fig. 8.11 and 8.12.

Thickness of flange is taken as 8.5 mm, as per graph Fig. 8.7.

$$\mu_{10} = t_9/t_{10}$$
$$= 8.5/7 = 1.21$$
$$\Delta H_{10} = 8.5 - 7 = 1.5 \text{ mm}$$

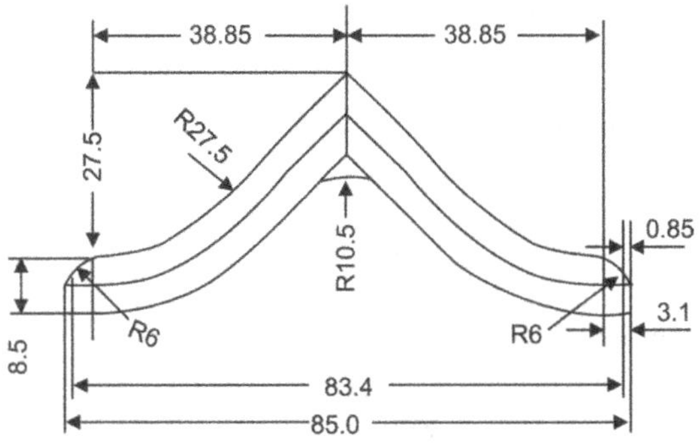

Fig. 8.11 Pass Diagram Stand-9.

Spread for angle design is taken as 30%

$$\Delta b_{10} = 0.3 \times 1.5 = 0.45 \text{ mm}$$
$$l_9 = l_{10} - \Delta_{b10}$$
$$= 51.7 - 0.45$$
$$= 51.25 \text{ mm}$$

Length, $\quad l_a = 27.5 - 8.5/2$

$$= 23.25 \text{ mm}$$

Length of chord, $l_b = 3.14 \times 45/180 \times (27.5 + 4.25)$

$$= 24.92 \text{ say } 24.9 \text{ mm}$$

Length, $\quad l_c = l_9 - (l_a + l_b)$

$$= 51.25 - (23.25 + 24.90)$$

$$= 51.25 - 48.15$$

$$= 3.10 \text{ mm}$$

Assuming a taper of 10% on each side

Then, $\quad a = 10/100 \times 8.5$

$$= 0.85 \text{ mm}$$

$$B = 2 \times (38.85, \text{ say } 39 + 3.10)$$

$$= 2 \times 42.10$$

$$= 84.20 \text{ mm}$$

$$\textbf{\textit{B}}_{\textbf{max.}} = \textbf{\textit{B}} + \textbf{\textit{a}}$$

$$= 84.20 + 0.85$$

$$= 85.05 \text{ mm or say } 85 \text{ mm}$$

$$\textbf{\textit{B}}_{\textbf{min.}} = \textbf{\textit{B}} - \textbf{\textit{a}}$$

$$= 84.2 - .85$$

$$= 83.35 \text{ mm or say } 83.4 \text{ mm}$$

$$\textbf{\textit{Y}}_9 = \textbf{\textit{Y}}_{10} \times \boldsymbol{\mu}_{10}$$

$$= 1.21 \times 8.5$$

$$= 10.28 \text{ or say } 10.5 \text{ mm}$$

Rolling Diameter, $D_k = D_{av} - \dfrac{F_k}{B_{max}}$

$$= 424 - \dfrac{895}{85}$$

$$= 413.5 \text{ mm}$$

Motor RPM, $\quad N = \dfrac{C \times i}{F_k \times D_k}$

$$= \dfrac{99.14 \times 10^6 \times 1.88}{895 \times 413.5}$$

$$= 503 \text{ rpm say } 510 \text{ rpm}$$

6 Pass Design of Angle 55 × 55 × 6 mm

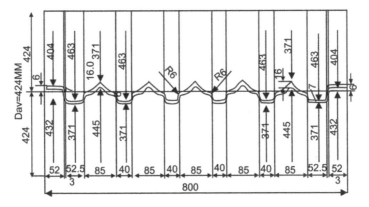

Fig. 8.12 Roll Diagram stand-9.

6.5 Intermediate Group

6.5.1 Stand-08

With the help of thickness graph, computation of the pass and roll design of stand 8, has been made and explained and shown in the Fig. 8.13 and 8.14, below.

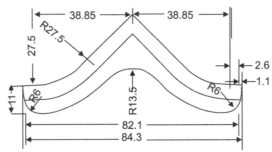

Fig. 8.13 Pass Diagram Stand-8.

Thickness of flange is taken as 11.0 mm, as per thickness graph Fig. 11.8.

$$\mu_9 = t_{10}/t_9$$
$$\mu_9 = 11.0/8.5 = 1.29$$
$$\Delta h_9 = 11 - 8.5 = 2.5 \text{ mm}$$

Spread for angle design is taken as 30%

$$\Delta b_9 = 0.3 \times 2.5 = 0.75 \text{ mm}$$
$$l_8 = l_9 - \Delta b_9$$
$$= 51.25 - 0.75 = 50.50 \text{ mm}$$

Length, $l_a = 27.5 - 11/2$
$$= 22.0 \text{ mm}$$

Length of chord, l_b = 3.14 × 45/180 × (27.5 + 5.5)

$$= 25.9 \text{ mm}$$

Length, $\quad l_c = l_8 - (l_a + l_b)$

$$= 50.50 - (22 + 25.9)$$

$$= 50.50 - 47.9$$

$$= 2.60 \text{ mm}$$

Assuming an 10% taper on each side

Then, $\quad a = 0.1 × 11$

$$= 1.10 \text{ mm}$$

$B_8 = 2 × (38.85, \text{ say } 39 + 2.6)$

$$= 2 × 41.6$$

$$= 83.20 \text{ mm}$$

$B_{\text{max.}} = B + a$

$$= 83.2 + 1.1$$

$$= 84.3 \text{ mm}$$

$B_{\text{min.}} = B - a$

$B_{\text{min.}} = 83.2 - 1.1$

$$= 82.1 \text{ mm}$$

$y_8 = y_9 × \mu_9$

$$= 1.29 × 10.5$$

$$= 13.50 \text{ mm}$$

Rolling Diameter, $D_k = D_{\text{av}} - \dfrac{F_k}{B_{\text{max}}}$

$$= 424 - \frac{1170}{84.3}$$

$$= 410.2 \text{ mm}$$

Motor RPM, $\quad N = \dfrac{C × i}{F_k × D_k}$

$$= \frac{99.14 × 10^6 × 2.026}{1170 × 410.2}$$

$$= \text{say 425 rpm}$$

6 Pass Design of Angle 55 × 55 × 6 mm

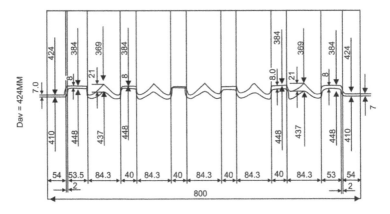

Fig. 8.14 Roll Diagram Stand-8.

6.5.2 (Stand-6 & 7)

The stand 07 is a vertical stand and its purpose is to facilitate the rolling of different section/thickness from same scheme/pass, by reducing the width of work-piece with the help of vertical roll and to bring the width of stock fed to stand 8 to a correct calibrated size and to eliminate the variation in the input billet size

(a) Pass-6

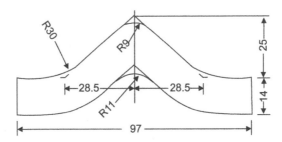

Fig. 8.15 Pass Diagram Stand-6.

The thickness of flange is taken as 14 mm with help of thickness graph (Fig. 11.8)

$$\mu_8 = t_8 / t_6$$
$$= 14/11 = 1.27$$
$$\Delta H_8 = 14 - 11 = 3.0 \text{ mm}$$

Spread for angle design is taken as 30%

$$\Delta b_8 = 0.3 \times 3 = 0.9 \text{ mm}$$
$$l_6 = B_8 - \Delta b_8$$

$$= 83.2 - 0.9$$
$$= 82.3 \text{ mm say } 82 \text{ mm}$$
$$B_6 = B_7 + \Delta h_7, \text{ assume } \Delta h_7 \text{ as } 15 \text{ mm},$$
$$= 82 + 15$$
$$= 97 \text{ mm}.$$

It is assumed that this pass will be used for the rolling angle sizes from 50 × 50 to 65 × 65 mm

$d = 28.5$ mm assumed, based upon the sizes of angles to be rolled, average size comes to 51 mm

Here, it is assumed as 98°
$$\tan 49° = \frac{28.5}{x}$$
$$x = \frac{28.5}{1.15}$$
$$= 24.8 \text{ or say } 25 \text{ mm}.$$

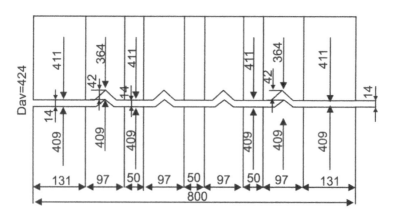

Fig. 8.16 Roll Diagram Stand-6.

Rolling diameter, $D_k = D_{av} - \dfrac{F_k}{B_{max}}$

$$= 424 - \frac{2400}{97} = 399.3 \text{ mm}$$

Rolling Constant $= F \times D_k \times \dfrac{N}{i}$

Where, F is the area of metal, D_k is rolling diameter,

6 Pass Design of Angle 55 × 55 × 6 mm

N_6 is the Speed of Motor, which is assumed as 300 rpm and i is the reducer ratio.

Then constant for intermediate group works out to

$$C = 2400 \times 399.3 \times \frac{300}{2.9} = 99.14 \times 10^6$$

(b) Pass-7

As explained earlier, the function of pass-7 is to press the flanges and buckle them to make to the stock to enter easily in pass-8, which is the first forming pass, designed with fixed angle i.e., 90°.

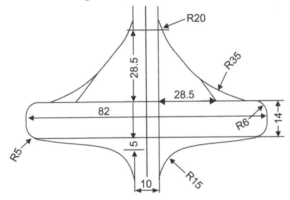

Fig. 8.17 Pass Diagram Stand-7.

The angle is considered between 95 to 105 degree and it is generally taken same as of pass 6 i.e., 98°.

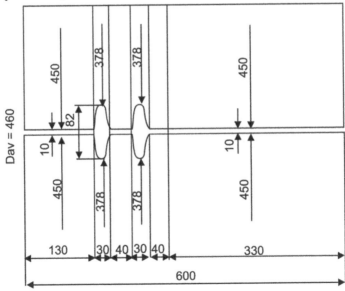

Fig. 8.18 Roll Diagram Stand-7.

Rolling Diameter, $D_k = D_{av} - \dfrac{F_k}{B_{max}}$

$= 460 - \dfrac{1710}{82} = 439.2$ mm

$N_7 = C \times i / F_k \times D_k$

$= 99.14 \times 10^6 \times 2.95 / 1710 \times 439.2$

$= 389$ rpm say 390 rpm.

6.6 Roughing Group

6.6.1 Stand -05

Pass design of stand 5, has been done as shown in the Fig. 8.19, shown below:

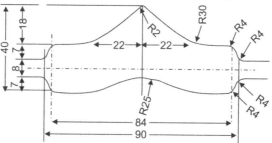

Fig. 8.19 Pass Diagram Stand-5.

ΔH_6 is assumed as 26 mm.

$H_5 = H_6 + \Delta H_6$

$H_5 = 14 + 26 = 40$ mm

Here, height of pass is so selected, so that it can cover the total height of pass 6.

And the spread in the pass is assumed about 25%, as there is restricted spread from side of the pass. Then spread will be

$\Delta b = 7$ mm

$B_5 = B_6 - \Delta b_6$

$= 97 - 7 = 90$ mm

$\mu_5 = \dfrac{4600}{3220} = 1.43$

Rolling Diameter, $D_k = D_{av} - \dfrac{F_k}{B_{max}}$

$= 424 - \dfrac{3220}{90} = 388.2$ mm

6 Pass Design of Angle 55 × 55 × 6 mm

6.6.2 Stand 1 to 4

Pass design of stand 1 to 4, has been done with square –rectangular system as explained below :

The roll pass design of square –rectangular is the simplest form of roll pass design system. Here, stand 1 to 4 will have square- rectangular design system.

First of all, it is to be decide first, which stand will have box pass and which stand will have barrel pass,

It is preferred to have box pass in first stand for following reasons:
- To take care of sectional variation of incoming billets.
- To shed of scale from the hot billet. It also acts as scale shedding pass.
- To avoid slipping of metal and better stability of work-piece.
- Take care of the variation of temperature, to avoid bending of piece, which will facilitate easy tilting by 90° before the stock enters in to stand-2.

Likewise box pass is also recommended in Stand-4, which has to provide the feeding of correct metal to the first cutting passes, stand-5 of angle design.

Other left over stands of roughing stands *i.e.*, 2 and 3, can have barrel passes, as these barrel passes will help in avoiding the tedious roll changing of stands of roughing group. Which takes a lot of time in roll changing, while changing from one section to another. Only disadvantageous is that there is no control over the section, as there will be free widening in barrel passes.

(a) Roughing Group Pass-4 (Vertical Stand)

The design of box pass of stand-4 is shown below (Fig. 8.20). The root of pass should be equal or within ±2 mm of the incoming width of the stock to facilitate the smooth entry of bar and to avoid faster wear out of pass.

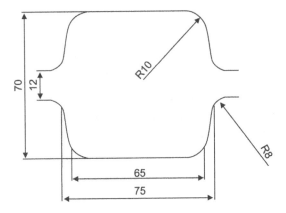

Fig. 8.20 Pass Diagram Stand-4.

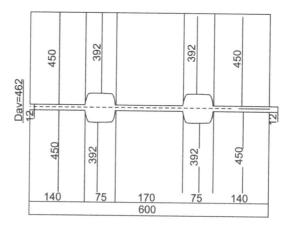

Fig. 8.21 Roll Diagram of Stand-4.

(b) Roughing Group Pass-1 (Horizontal Stand)

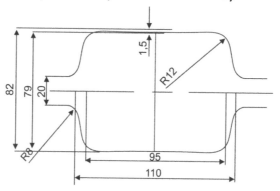

Fig. 8.22 Pass 1 (Box Pass).

(c) Computation of Rolling Diameter for Roughing Group

Here, D_k for roughing group (Stand 1 to 5), will be worked out as follows:

(i) **Pass 2 and 3**: These two passes are barrel passes, D_k for this two passes will be the Dc *i.e.*, collar diameter. D_k for stand 2 and 3 will be 523 and 510 mm respectively, which will be also Dc of these two stands.

(ii) **Pass 1 and 4**: These two stands have box passes. For Box passes *i.e.*, for stand 1 and 4, D_k will be the pass diameter, which comes to 476 and 392 mm for stand 1 and 4 respectively.

(iii) **Pass 5**: As stand 5 is the first cutting pass of angle design, D_k of sand 5, works out in same manner as shown for other closed passes of angle design.

$$\text{Rolling Diameter, } D_k = D_{av} - \frac{F_k}{B_{max}} = 424 - 3220/90 = 388.2 \text{ mm}$$

Planning of Cement Project

(d) Computation of Motor Speed N

$$\text{Rolling Constant} = F \times D_k \times \frac{N}{i}$$

Where, F is the area of metal,

D_K is rolling diameter,

i is the reducer ratio

(*i*) **Stand-1:** Here, it is recommended to use box pass (Fig. 8.22). The advantage of using box pass at stand-1 are numerous like stability of work-piece is more in box pass compare to the barrel pass where chances of twisting is less. It also acts as scale shedding pass, which in turn will help to improve the quality of product. The amount of spread is also less compare to barrel pass.

It is not part of the continuous group. Here, speed is kept such that to facilitate easier bite and should not give any chance of slipping of the metal, which effects adversely the mill equipments. Here, speed is assumed as 300 rpm.

Sometimes, ragging of rolls, use of sand and forced biting is given in stand-1 to facilitate smooth bite in stand-1.

Diameter of stand-1 plays a vital role in the biting of billets. The use of minimum diameter rolls should be avoided.

(*ii*) **Stand-2**

N_2 is the Speed of Motor, Which is assumed as 300 rpm

Then Constant for roughing group works out to

$$C = \frac{7250 \times 523 \times 300}{8.12}$$

$$= 140.1 \times 10^6$$

(*iii*) **Stand-3**

$$N_3 = \frac{C \times i}{F_k \times D_k}$$

$$= \frac{140.1 \times 10^6 \times 6.96}{5800 \times 510}$$

$$= 329 \text{ rpm say } 330 \text{ rpm}$$

(*iv*) **Stand-4**

$$N_4 = \frac{C \times i}{F_k \times D_k}$$

$$= \frac{140.1 \times 10^6 \times 3.81}{4600 \times 392}$$

$$= 296 \text{ rpm say } 300 \text{ rpm}$$

(*v*) **Stand-5**

$$N_5 = \frac{C \times i}{F_k \times D_k}$$

$$= \frac{140.1 \times 10^6 \times 4.96}{3220 \times 388.2}$$

$$= \text{say } 550 \text{ rpm}$$

7 ROLLING COMPUTATION SHEET FOR ANGLE 55 × 55 × 6 MM

Table 8.2 Input Billet – 100 × 100 × 5 mm

Stand	Shape of Pass	Height H Mm	Width B Mm	Redn H Mm	Spread ΔB	Gap T Mm	Area MM²	Coeff. of Elongatio	Length if Profile L Mts	Rolling Dia Dk Mm	Reduce Date Radio I	Speed of Motor	Dc	Dav
I	BOX	80	105	20	5	18	8150	1.19	5.95	476	9.58	300	515	535
		Tilter 90°												
II	Barrel	85	87	20	7	–	7250	1.13		523	8.12	300	523	535
III	Barrel	65	95	20	8	–	5800	1.25		510	6.96	330	510	530
IV	BOX	70	70	25	5	12	4600	1.26		392	3.81	300	450	462
V	Open	42	90	28	20	–	3420	1.48	14	488.2	4.69	550	500	424
VI	Open	14	97	27	7	14	2400	1.34		399.3	2.9	300	411/409	424
VII	Closed	82	14	15	–	10	1710	1.40		439.2	2.95	390	450	390
VIII	Closed	11	84.3	3.0	.9	7	1170	1.46	40	410.2	2.026	425	424/410	424
IX	Closed	8.5	85	2.5	.75	6	895	1.31	50.5	413.5	1.88	510	404/432	424
X	Closed	7	85.2	1.5	.45	5	745	1.20	64.5	416.2	1.59	300	448/390	424
XII	Angle 50-65	6	55×55	1.0	.3	5	644	1.15		381.8	1.15	300	392/378	390

8 ROLL PASS DESIGN OF UNEQUAL ANGLES

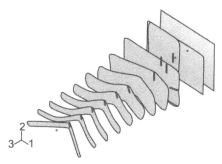

Fig. 8.23 Sequence of rolling Unequal Angles.

8 Roll Pass Design of Unequal Angles

The same pass sequence and similar draught calculation for equal angles explained, are applied to the rolling of unequal angle sections. The difference lies in the positioning of passes in the rolls.

8.1 Design of Finishing Pass

The pass setting is not simple for unequal angles. There are two main methods of arrangements of grooves for designing of passes for unequal method. In the first method, the bi-section is arranged vertically and in the second method, the bi-section is inclined, so that the projection of both flanges to the vertical axis is equal.

(*a*) **Method-1 (Bisection is Arranged Vertically)**

In the first method, the bi-section is arranged vertically (Fig. 8.24). The bi-sector of the central angle is positioned square to the roll axis.

Fig. 8.24 Method-1 (bisection is arranged vertically).

This arrangement has vertical projections of the flanges of different length, which causes a large end thrust and to consequential inevitable axial shift of the rolls, so that the longer flange will get thicker than the shorter flange.

It can be seen that rolls are subjected to the considerable axial forces, because force P_1, representing the resultant of the pressure of right flange of longer flange of the bottom rolls. It is more than the P_2 i.e., of shorter side and likewise P_3, the resultant of the pressure of right flange of longer side of the top rolls is more than the P_4 of shorter side as shown in Fig. 8.25 below.

A natural imbalance is observed in the rolling of unequal leg angles by this method. The long leg which represents a larger area, give greater resistance to deformation. It means, larger leg tries to push the roll apart. Due to this, the top roll tend to get displaced to the right and bottom rolls to the left, this displacement gives rise to flanges of different thickness. Also, the arrangement of passes in rolls will become very difficult with this design of rolling. A

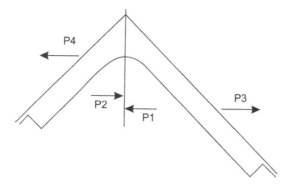

Fig. 8.25 Axial Forces in Method-1.

balance can be made by making the shorter leg bit heavier than the long leg and special measures should be for arresting the tendency of end thrust by assignment of draught in such a way to equalize the pressure acting on short and long flanges. This is accomplished by assigning less draught for long flange and more for short flange. If in spite of having all above measures, end thrusting condition still persists, then, twisted angle will be come out of the rolling in spite of accurate guide settings,.

The first method is limited to only when the small quantities of same size are to be rolled *i.e.*, less than what is required for to complete wear out of the pass.

(*b*) **Method-2 (Vertical Projections of Flanges are Equal)**

In the second method of positioning passes for unequal angles (Fig. 8.26), the vertical projections of flanges are equal *i.e.*, the toes of the two flanges are at same diameter. This position excludes horizontal shift of rolls and the above mentioned factors explained in method-1 get nullified.

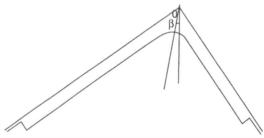

Fig. 8.26 Method-2(vertical projections of flanges are equal).

However, the same pass cannot be employed for rolling other angles of the same size, but with heavier or lighter thickness, as changing the distance between rolls. It will not change the thickness of the two flanges uniformly.

8 Roll Pass Design of Unequal Angles

In other word, the change of elements between rolls causes due to unequal change of thickness of the left and right flanges.

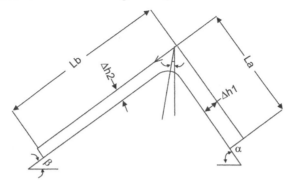

Fig. 8.27 Design of Finishing Pass (unequal angle).

The above mentioned method, will cause vertical axis is to be arranged at some angle ($\gamma°$) to the vertical line. However, the change of clearance between rolls causes unequal change of thickness, of left and right flange

$$\Delta h = \Delta h_1 - \Delta h_2$$
$$\Delta h_1 = \Delta h \cos \beta$$
$$\Delta h_2 = \Delta h \cos \alpha$$
$$\tan \alpha = L_b / L_a,$$
$$\beta = 90 - \alpha$$
$$\gamma = \beta - \alpha / 2$$

To decide what widening, whether tight or free is to be provided in the finishing pass is of great importance. It is proved that finishing pass design with free widening has big advantages because of.

- Possibility of formation of fin is absent
- Provide to roll in one groove, angle of different thickness, as with tight widening, the length of profile changes simultaneously with thickness and separate pass is to cut for angle of same size with different thickness.

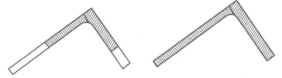

Fig. 8.28 Angle of different size rolled from same pass.

- As shown above in Fig. 8.28, it is possible to roll in one finishing pass, angle of some different sizes.

The value of widening in finishing pass is generally not much and cannot go beyond the tolerance limit, hence there is no doubt of getting the correct size of flange. Edges obtained from free widening are of good quality.

Considering all above facts, it may be concluded that finishing pass with free widening is always recommended and all other passes will of closed type.

8.2 Determination of Sizes of other Passes

The calculation of other passes is done against the direction of rolling. The finishing pass is only having the straight flange, while all other passes are having flared flanges. According to data available from several designs, it is recommended:

(*a*) While rolling in closed pass

$$H = (0.35 - 0.45) \times L$$

(*b*) While rolling in open pass

$$H = (0.25 - 0.35) \times L$$

Rolling of angles in open passes is done with lesser no of passing (shaped) and from a flat billet *i.e.*, from a rectangular shape. Rolling in open shape makes the rolling more stable.

8.3 Computation of Pre-finishing Pass

Dimension of other elements of pre-finishing pass are determined with following formula:

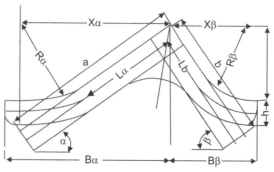

Fig. 8.29. Computation of Pre-finishing pass.

$$X_\alpha = R_\alpha \times \frac{1 - \cos \alpha}{\sin \alpha} + H \times \frac{\cos \alpha}{\sin \alpha}$$

$$X_\beta = R_\beta \times \frac{1 - \cos \beta}{\sin \beta} + H \times \frac{\cos \beta}{\sin \beta}$$

9 Mill Setting of Angle Rolling

For long flanges

$$B_\alpha = L_a - 0.5h\,(1 - 1.414\sin y) + (2R_\alpha + h)\,(\frac{1 - \cos\alpha}{\sin\alpha} - 0.00872\,\alpha)$$

$$- [H + 0.5\,h\,(1 - 1.414\cos\alpha)]\,(\frac{1 - \cos\alpha}{\sin\alpha})$$

For short flanges

$$B_\beta = L_b - 0.5\,h\,(1 - 1.414\sin\gamma) + (2R_\beta + h)\,(\frac{1 - \cos\beta}{\sin\beta} - 0.00872\,\beta)$$

$$- [H + 0.5\,h\,(1 - 1.414\cos\beta)]\,(\frac{1 - \cos\beta}{\sin\beta})$$

The above formula for determining X_α, X_β, B_α, B_β are quite complicated, their use is simplified depending on the relation L_a/L_b.

For simplification of these calculation, these values, X_α, X_β, B_α, B_β can be determined from the chart below in Table 8.3.

Table 8.3

a/b	α	β	γ	B_α	B_β	X_α	X_β
1.00	45°	45°	0	L_a-0.329h +0.044R_α- 0.0414 H	L_b-0.392 h +0.044R_β -0.414 H	0.414R_α + H	0.414R_β + H
1.333	36°52'	53°08'	8°08'	L_a-0.32 h +0.023 R_α- 0.333 H	Lb-0.464 h +0.072 R_β -0.5 H	0.333R_α + 1.333H	0.50 R_β + 0.75 H
1.444	34°42'	55°18'	10°18'	L_a-0.302h +0.020 R_α+ 0.313 H	Lb-0.483 h +0.082 R_β -0.524 H	0.313R_α + 1.444H	0.524R_β + 0.693H
1.50	33°41'	56°19'	11°19'	L_a-0.294h +0.018 R_α-0.303 H	Lb-0.491 h + 0.088 R_β -0.535 H	0.303R_α + 1.5H	0.535R_β + .667H
1.666	30°58'	59°02'	4°02'	L_a-0.272 h +0.015 R_α + 0.278 H	L_b-0.513 h + 0.103 R_β -0.567 H	0.278R_α + 1.666H	0.567R_β + 0.60 H

After calculating the sizes of pre-finishing pass, as per above calculation, width of following pass are determined with the computation of the spread, as explained above with free widening or with tight widening in close pass.

9 MILL SETTING OF ANGLE ROLLING

9.1 Setting of Roughing and Intermediate Stands

Before setting mill for angle rolling, the roller should ensure that the

1. Proper passes should be placed at all stands as per the rolling scheme.

312   Chapter 8   Rolling of Angles

2. It should also be ensured that either thinner gauges passes(6-8 mm) or the thicker gauges (10-12 mm) are used in the rolling at a time. Generally, passes are designed separately for the thinner and the thicker sizes for intermediate and pre-finishing by the designer to facilitate the correctness of the finished dimension as per the specification. These thick angle passes are deeper and narrower than the thinner angle passes. Correctness of the width of metal from the first cutting pass is to be ensured as spread will be different for thinner and thicker angles. Inserts are also be different in pre-finishing and finishing stands for thicker or thinner pass and if by mistake thicker gauge pass is taken in the pre-finishing stand, then while rolling thinner section and then load in finishing stand will shoot up very high and the leg will become short. If thicker pass is taken in the intermediate group, then to get the correct thickness, we have to press more, causing less collar gap and formation of fin because of more spread.

3. Pass burning is to be done in stands, wherever reduction is very high, like the first pass of the intermediate group and also in the first cutting pass, it is a must. Atleast 5 mm extra gap should be kept for pass burning.

4. While operating screw down of the vertical rolls, care should be taken while pressing or closing of stand for the guard gap, otherwise nose of guard will get bend or broken.

5. The level of the roller table w.r.t. roughing stand should be checked, especially in first stand otherwise piece will not get grip by the stand and get held up in the roller table, if reduction given in stand-1 is less.

6. Cooling water to the bottom roll of stand-1 should be set in such a manner, so that it should not come in contact with the front end of piece, otherwise metal will get cold, which may lead to more spread of metal and may also be responsible for the cobble in the roughing group.

7. The correct feeding and the alignment of the centre of the gap of entry guide should coincide exactly with the apex of the first cutting pass, which is required to get equal and correct leg of the profile. The pass filling is also very much important for getting the sharpness of the apex. It can be ensured by rejecting the piece after the first cutting pass to check the size of width, difference in the thickness and the length of the leg and the sharpness of the apex of the piece.

8. The vertical stands of roughing and intermediate group are used to control the legs of the finished angle by opening or closing of rolls of these vertical stands. It should be tried to get the correct width of the work piece after these stands.

9. Apex of passes of the intermediate group stands is to be slightly rounded off for good flow of metal and to avoid breakage of roll from sharp apex.

9 Mill Setting of Angle Rolling 313

10. It is to avoid giving any loose packing in the guard of the closed pass, which may get disturbed during the course of rolling. Height can be reduced by bringing down the rest bar. Also check the line of rolling and the entry rest bar, if it is down, then the entry rest bar is to be taken up.

11. The metal entry into box pass should not be more or less than ± 2 mm of the root of the box pass, otherwise twisting will takes place. Suppose the root of the box pass is 68 mm, then the metal entering to the box pass, should have the metal size of 68 ± 2 mm.

12. It is to understand that angle rolling is more or less of conversion of a flat into the angle. If we concentrate at width of metal which enters to the close pass, two contradicting things get happen :

    (*a*) Metal has a tendency to spread under given reduction.

    (*b*) The height of metal is rising in butter fly design, as the metal get approaches to finishing stands *i.e.*, the apex angle is decreasing from 140° of first cutting pass to 90° of finishing pass during the course of rolling in case of butterfly design. This will try to bring down the width. Roller should be vigilant to ensure the correctness of width.

13. In the closed passes, overfilling or under filling are not allowed. Overfilling will cause the fin formation; while under filling will cause the variation in the leg width of the finished profile.

    Following parameters are checked during the mill-setting, of the rejected bars:

    (*a*) Proper filling of the apex.

    (*b*) Flange thickness of both sides.

    (*c*) Width of the bar with the width of pass.

    (*d*) Approx. leg length from the apex for both the sides and its variation.

14. It is to ensure that the straight delivery of piece from the stands with closed passes. The bending of bars towards right or left can be rectified by shifting the delivery roll tackles towards the side opposite to the bending.

15. Twisting of bar in closed pass (if not due to rolling tackles) can be rectified by shifting the top roll towards the side opposite to twisting.

16. Bending of front end in the upward or downward direction is caused either by guards which are either worn out or by new ones, if not set properly. Less guard gap and less gap between top and bottom exit guides also cause upward movement of the bar.

17. Knocking of metals in the first cutting pass causing more kick load is due to lesser gap in the entry guide, which should be strictly avoided.

18. Twisting of angle may be due to too much pressing of an angle by one side exit guide or because of less gap between exit guides.

19. Cutting from stands is because of more width of metal from previous stand, more reduction in stand or bad setting of entry tackles.

When rolling thin angles, wearing out of taper in closed passes should be checked or otherwise collaring of rolls may takes place. One of the features of collaring of rolls in the process of rolling will be periodical bending of pieces in the vertical directions. The following steps are to be taken:

(*a*) The load of the stand is to be reduced,
(*b*) The timely changing of passes.

20. Mostly collaring of rolls takes place due to wearing out of the close pass, while rolling either high grade steel or when feeding of bar in the close pass at lower temperature.

21. Roller should ensure the proper filling of closed passes in the intermediate group to avoid variation in the leg length of the finished profile.

22. Cutting of metal from the stands is because of :

(*a*) More metal in width from previous stand.
(*b*) More reduction in the stand.
(*c*) Bad setting of the entry guide.

23. Fin in the tail end of the bar coming out of the pre finishing stand causes cobble in the entry guide of finishing stand. This may be due to:

(*a*) One of the entry guides of the pre finishing stand *i.e.*, stand 10, got cut or open more.
(*b*) Entry box level is too high.
(*c*) The length of guides is short i.e. not going up to the pass.
(*d*) Entry box level is too high.
(*e*) More packing has been given at the front end of the entry box.

24. Roller should also ensure that the approach guide roll table should be in sufficiently raised condition, so that bar should not hit the bottom of entry box.

25. Approach guide gap should be such that to avoid the bar to feed wrongly or otherwise it will cut the entry guides and will cause bending of bars from pre-finishing stand.

26. Roller should check rollers of roll table after the pre-finishing stand, if it is not working, metal may get bend more.

27. Dogs alignment of the chain transfer should be checked to avoid poor drag of the piece.

9 Mill Setting of Angle Rolling

9.2 Setting of Finishing Stands and Profiles for Angle Rolling

1. Entry inserts or rollers used in the entry guide of the finishing stand should not be worn out and they should be as per the pass width of pre finishing stand.

2. If inserts are used in the finishing stand, then the gap between top and bottom inserts should be sufficient in respect to the thickness of the profile rolled. If it is more, it will cause metal play in the inserts, which in turn leads to the dropping of the apex. If it is less, then metal from pre-finishing stand will not enter the finishing stand.

3. Entry box should be set in such a way, so that apex of the insert and that of roll should exactly coincide.

4. Entry box insert should not touch the top or bottom roll, otherwise there will be difficulty in reducing the thickness of the profile as roller will not able to press the top roll and it will also hamper the movement required for the axial shifting of the top roll. It will also result in the loosening of the box due constant jerking, because of continuous contact with roll, which will result in the apex fall.

5. Gap of exit box should be adjusted with the help of the sample piece of finished product.

6. Though the thickness of leg is not a criterion for angle rolling, but weight of the section is always adjusted by either increasing or reducing the thickness of the section.

7. The roller should examine the entire length of the bar and should see the apex sharpness, which should be good at every point of the work piece. Dimension should also be taken atleast not less than three places, including the middle section of the bar, because the tension in the group will cause shortening of either both or one leg. The sample are taken at sample saw. Back end of the piece should be on the higher side of the nominal leg length by atleast + 0.5 mm.

8. The wear pattern in the finishing pass is as follows:
   (a) The reduction is maximum in thickness at apex.
   (b) The tip area of bottom roll presses the metal upward with more thrust.
   (c) As there is more mass of metal at apex, it requires more cooling. In the absence of adequate cooling wear out increases due to rubbing action, in addition, it will generate more thrust.

9. The function of the finishing pass is also to straighten the piece coming out of pre-finishing stand hence top roll got worn out.
   - The roller should see the quality of the apex, of the bar, rolled out of the finishing stand by means of the pressing wood packing and

checking of piece at the sample saw. Roller should also bear in the mind that as wear out of apex increases, as profile goes to higher side of weight tolerance, hence care should be taken by roller, to press less in the beginning of rolling and more at the end of the shift.

10 DEFECTS AND ITS RECTIFICATION

1. Wear out of apex

Wear out at the apex of finishing stand is due to:

- The reduction at the apex is at maximum.
- More thrust at the apex due to tip area of bottom roll presses the metal upward.
- More wear out of top roll due to the straightening of metal took place in the finishing pass.
- More mass and thrust at the apex causes rubbing.
- In the absence of effective cooling at the apex.

2. Blunt Apex

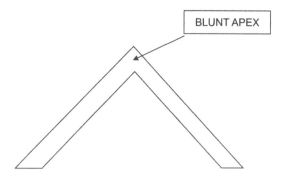

Fig. 8.30 Blunt Apex.

Bad apex from previous stands causes the formation of blunt apex. If previous passes are excessively wear out, then those passes are to be changed.

- If both sides are blunt, then the entry guides are to be raised upward by lifting entry side rest bar.
- Both sides apex blunt can also be made ok by opening the vertical stand of roughing group and the pressing of the vertical stand of the intermediate group, to lift the metal up in the apex.
- If blunt apex is found only at one side, it may be mainly due to bad alignment of the entry rollers/inserts. They are to be taken towards the blunt apex side, only if apex is dropping towards shorter side. In case apex

10 Defects and its Rectification 317

is falling towards the longer side, then before the shifting of the rollers/ guides of finishing stands, both legs of the metal of the pre-finishing pass should be checked and adjusted, if any discrepancies is noted.

- The tail end of the bar get bent due to more play in entry inserts of finishing stand and which may cause blunt apex in the tail end of the bar.
- If the apex is blunt, sometimes left or sometimes right, throughout the length of bar. It is due to play, the entry box may be loose or insert/ roller are loose. Entry rest bar should also be tightened and may be taken up.
- Blunt apex also comes, when rolls got turned with larger apex radius than design. In this case, pass should be immediately changed.
- One of the cause for blunt apex may be due to incorrect setting of stand.

3. Both legs short or long

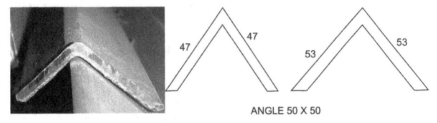

Fig. 8.31 Both legs Short or long.

- Side control basically is made from vertical stands in the sequence for major difference.
- Otherwise for the minor difference, the control is to be made from the entry guides of pre-finishing and finishing stands.
- Whenever there is a change from smaller thickness to higher thickness, then precautions is to be taken to increase the metal from the roughing group, otherwise there leg may be shorter.

4. Leg width difference

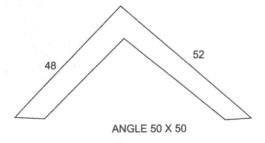

Fig. 8.32 Leg Width Difference.

- The entry guides of the pre-finishing stand are to be shifted towards the shorter leg, if the leg difference is less than 0.5 mm. For difference of 0.5 mm to 1 mm, previous two stands and if it is more than 1 mm, then guides of further stands to be shifted towards shorter leg.
- If one side is longer and thinner and other side is thicker and shorter, then the top roll of finishing stand is to be taken opposite to the thicker leg.
- If one side is longer as well as thicker, then before adjusting entry guides, top rolls of pre-finishing stands are taken towards the thinner leg, thus making legs equal and then only other adjustments are made.

5. **Weight of angle section:** Weight of angle section is adjusted by first reducing or increasing the thickness of flange first.
6. **Cutting of metal:** It may be due to:
   - Entry guides are either more open or due to use of wear out guides.
   - Mis-alignment of rolls
   - Reduction in particular pass is more than the required.
   - More metal from vertical stand.
7. **Fish tail formation from pre finishing stand:** It may be due to:
   - Wrong setting of entry guide, it may taper towards one side of the pass. It should be set at centre of the pass.
   - Reduction in the pre-finishing pass should not be more.
   - Entry guides should be closer to closer to pass, especially lower portion of the guides should be set nearer to the bottom roll as possible.
8. **Lap in the upper portion of the leg**

Fig. 8.33 Lap in the Upper Portion of the Leg.

Continuous line defect on inner or outer legs, results of overfilling of stands, where parting is at the top of the pass.

10 Defects and its Rectification

While for lap at the bottom portion of legs results from passes, where parting is at bottom.

Remedy is to Identify stand/pass producing the fin and then

Reduce stock in identified pass by lowering preceding stand.

9. **Fire cracks:** When rolling angles, the bottom side of the angle should be checked thoroughly for the fire cracks caused by less cooling of these portions of the bar.
10. **Guard marks**
11. **Laps on legs at equal intervals:** May be due to chipping of pass.
12. **Bottom of apex is not rounded due to bad turning.**

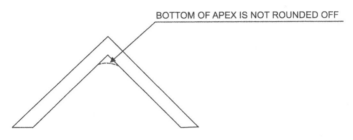

Fig. 11.35 Bottom of Apex is not Rounded.

13. **Under filling**

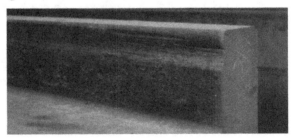

Fig. 11.36 Under Filling.

Under filling is due to Lack of material in later parts of the rolling process.

Possible causes

- Earlier stands wider than later part of finishing train.
- Rolls get turned larger than design.
- Insufficient stock entering later mill passes.

Remedies

- Confirm widths and dimensions of rolls used.
- Check and balancing the load throughout the mill and try to give more work at the later passes .

14. Shelly Toes

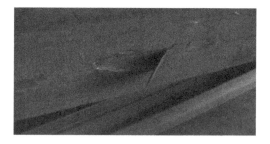

Fig. 11.37 Shelly Toes.

This defect is mainly found in toes of angle sections

Possible causes are:
- Furnace discharge rollers.
- Tilter fingers causing corners to come up.
- Under filling of passes.

Remedies are:
- Adjust furnace rollers.
- Modify pass settings at finishing and reversing mills to remove defect.

15. Slivers / Shelling

Fig. 11.38 Slivers/Shelling.

These defects are observed at outer flanges / legs

Possible causes are:
- Pass wear causes indentations/grooves, leading to tearing/sharp edges.
- Sharp edges get pick-up material rolled in on the following bloom and roll back in on the next pass. The slivers develop into shelling as deterioration occurs.

Remedies are:
- Immediate change of pass
- Grind the pass as soon as possible to remove sharp edges.

❏❏❏

9

Rolling of Channels

1 INTRODUCTION

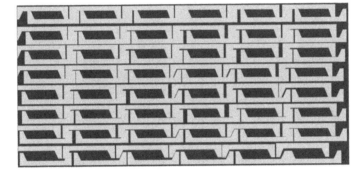

Fig. 9.1 Channel Section.

In channel rolling several close passes follow one after another, unlike beam rolling, where final profile is obtained by with alternate combination of open and close flange passes. Pass designer has to device some way to get the required finished channel profile with minimum consumption of energy, minimum wear-out of rolls and rolling within the shortest time to give higher production and to avoid roll breakage due to temperature fall in flanges. The determination of pass schedule is also be subjected to the availability of mill facilities *viz.*, roll specification and size, power of motors of roll stands and quality of the rolled metal desired.

Lesser slope (6° *i.e.*, 10%) and greater width of channel, further adds additional difficulties in rolling. The insignificant slope on the flanges of channel (10%) compares to the beam profile (14%) forces the designer to reduce the amount of side reduction on flanges. As the amount of reduction of flange is always linked with their slope. In addition, deeper cut on rolls due to lesser slope in closed pass, affects the strength of rolls, which in turn also force the designer to assign lesser reduction in channel profile, compares to the beam rolling.

2 METHODS OF PASS DESIGN

2.1 Conventional Method of Rolling Channel

There are three conventional methods for channel roll pass design:
- Pass design as per beam method
- Pass design with increased outlet
- Pass design according to bending method
- Each of these methods has their own advantage and disadvantage and use of each method is justified by the definite condition of the mill.

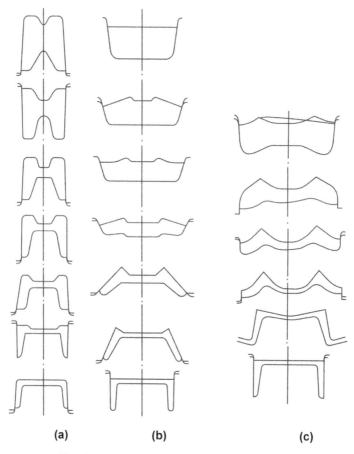

(a)　　　　(b)　　　　(c)

Fig. 9.2 Pass Sequences for Rolling of Channels.

2 Methods of Pass Design

(a) Beam Method or Counter Flange Method

The most common and popular method of rolling channel is beam or counter flange method. This method of channel design follows closely the method and principle used in beam method, as shown in [Fig. 9.2 (a)].

The unique feature of Beam method is the use of common roughing passes for rolling both for channel and beam. It means, the web-cutting pass and subsequent one or two passes are commonly used for rolling of both channel and beam of the approximately same sizes.

The main advantage of this method is the use of less numbers of roll and also will require fewer changes of rolls, while changing over rolling from channel to beam and *vice-versa*. However, the necessity of high reduction for false flange in roughing passes makes this method irrational from the point of view of the consumption of energy. The major drawback of this method is the low outlet of pass (< 3%), which makes the use of beam method is to be limited and it also decreases the total numbers of redressing of rolls.

Serious consideration must be given to the cutting-in action of first few passes by rolling channel with the use of Beam method. The tongue or "V" of the roll should compress the pass in such a way, so as to form the flange in same and subsequent passes.

The tongue of the "V" pass can be compared with the edge of an axe. The narrower and thinner will be the blade or tongue, more it get sink deepen to the bloom with ease. There is a practical limit to this thinning of blade also, keeping in view of roll strength of roll.

The more blunt of the tongue or "V' will be, the greater will be the force required to sink it into the hot bloom and in the absence of that action, the flange will get pull down. In other words, it means the metal will run into elongation towards web, rather than to flow upto fill the flanges. Speed of rolling also affects the flow of metal towards the flange. Higher will be the speed of rolling, the greater will be the tendency of metal to flow towards the elongation.

It is also to note that the width of bloom or billets entering to the first cutting-in pass should be tight fit to the width of the pass. The advantages of tight-fit entry of bloom are as follows:

1. The closer the fit, the greater will be the pulling in effect of the rotating collars of "V" pass to bite the metal into the pass. In other word, it can be said that with the improved biting, the need of ragging of the roll will be reduced, which will otherwise necessary to make the roll to bite.

2. If there is too much clearance between the bloom and to the width of "V" pass, then metal which otherwise will require to flow towards of the

flange, now will go for spread towards the sides, this will result in the robbing of the flanges.

3. With too much clearance, there is also a chance for bloom to be get shifted towards one side of the pass, which will otherwise results in uneven height of flange of finished profile.

It is also not proper to design the bloom to the exact width of the pass, which will cause rapid wear out of the pass. It is advised to give a reasonable spread in all succeeding passes. *i.e.,* to make the succeeding pass should be made slightly wider than the preceding pass. This procedure helps in preventing the excessive wear out of the pass.

The use of beam blank method also is advantageous from the point of view of yield. With the use of beam blank, a good balance can be managed in between flange and web reduction, which will produce a finished bar that will not have any extended tongue at web due to excessive elongation. With the minimum extended tongue, the end crops will be reduced, thereby increasing the yield of the mill.

(*b*) Design with Increase Outlet

It is evident From [Fig. 9.2(b)] that in design with increase outlet, all passes except the finishing pass are turned with quite a significant outlet. Due to large outlet, wear out of pass gets minimized and there will be a possibility to restore the pass with the minimum loss of diameter, thus giving good life of the roll.

Besides this, the increased outlet also makes it possible to apply higher reduction in all passes, which will certainly reduce the numbers of passes require for rolling. In this design, sufficient bend is required to be given in pre-finishing pass, to avoid sharp bend of flange in the finishing pass, which otherwise, will lead to the appearance of harmful stresses in corners and the formation of wrinkles on the inner surface of profile *i.e.,* at the joint of the flange and the neck.

Major drawbacks of increased outlet design are:

- The difficulty in the entry of the work-piece in the finishing pass, due to significant difference in width of pre-finishing pass with the finishing passes (less taper in the finishing pass).

- With the increase in width of each pass, roll will accommodate less numbers of passes with the increased pass outlet design, though this disadvantage will get neutralized by the requirement of less number of passes for rolling channel, as higher reduction can be given with increase outlet design.

2 Methods of Pass Design

(c) Pass design according to bending method

This design is the further development of the increased out-let design explained above. [Fig. 9.2(c)] shows pass design of channel by this method. It is sometimes also termed as angle design method, as it is similar to the angle design. In this method, the maximum reduction is to be given in initial passings and the bending is to be performed in final passing. The uniformity of roll diameter in this design will facilitate to provide a higher rate of reduction, as strength of rolls will be improved with higher diameter rolls. The shallow cut in rolls of roughing group passes of angle design in comparison to other designs will facilitate to render higher amount of reduction in roughing group.

Main drawback of angle design is the problem associated with the entry of work piece in subsequent passes of the sequence. Significant increase of width in roughing passes reduces the dimensions of collars as well affects the stability of the work piece by the fluctuation in the dimension of flanges of the profile.

2.2 Modern Method of Rolling Channel
Universal Method of Rolling Channel

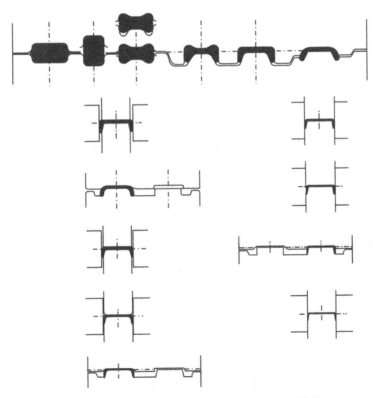

Fig. 9.3 Rolling of Channel in a Universal Mill.

326   Chapter 9 Rolling of Channels

Channels can be rolled in either reversing universal mills or in continuous mills:

Advantages of universal rolling of channel:

- Ability to produce tailor sizes to suit application.
- Web and flange thicknesses can be adjusted independently.
- Capacity to increase/decrease flange length.
- Reduced set up times and costs.
- Quicker section change.
- Better surface quality.
- Better mechanical properties.
- Channels are easier to straighten.
- Roll wear is less, so longer campaign can be planned.
- Reduced roll costs.
- There will be greater flexibility in rolling different gauges.

3 SALIENT FEATURE OF CHANNEL PASS DESIGN

3.1 Draughting

Balanced draughting for web and flange is an important aspect of the channel design. But, it is not possible to maintain the balanced draught throughout the entire reduction schedule. To do so, it requires that every channel rolled should be made from such a bloom, which is roughly shaped to the final channel dimension. But as rectangular shaped blooms are generally used for channel rolling, because roll pass designer has no other option, but to give unequal draught. It is always preferred to give unequal draught at the start of the rolling *i.e.*, in roughing passes for the following two reasons:

Firstly, because steel is hot and plastic at this stage of rolling and will have less resistance to the deformation due to higher temperature, to make it possible to have higher reduction in initial stage of rolling to facilitate the rapid formation of flange.

Secondly, also it gives an opportunity to remove any irregularities caused by non-uniform draught before it reaches to the finishing pass.

Channel pass design is to start by establishing first the number of passes and then to determine the co-efficients of elongation separately for the web. Flange and counter flange in each pass and once the total co-efficient of elongation get established for the web, flange and counter flange. It becomes necessary to divide the total flange area into its component parts *i.e.* areas of the flanges and of the counter flanges.

3 Salient Feature of Channel Pass Design

3.2 Work Feature

The feature of work in the design of beam passes also applies to the channel design. The drawing or shortening of the flanges by work and principle of web buckling are same for both profiles. If too high reduction applied on thin web and flanges receive only lighter reduction in finishing pass, then it makes web to buckle or wave. The buckling of web is due to firstly because the web is prevented from spreading and secondly, it is not sufficiently strong to push the flanges along. Some designer reduces the flanges slightly more than the web in order to setup slight tension in web. This is based on the idea that slight tension reduces the contraction of the flanges during cooling.

3.3 Fillets

Another part of channel section must be given a careful consideration is the design of fillet. Proper selection of fillet is very essential. If the fillet at the junction of web and flange (R_1) is too large, then the flange will tend to be drawn down or "Robbing".

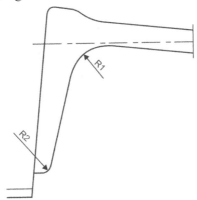

Fig. 9.4 Fillets for Channel.

Then again the fillet must not be too small or a gall (a wiping or pinching on the surface of the steel), which will result in producing "Fold". Such a defect may cause rupturing of the flange and the web.

In the beam method, the fillets at the outer edges at the bottom of flange serve as a buffer against fins. If the fillet is too small, the metal worked out in the open hole, will produce a fin, which will be further rolled over to form a lap. This defect will further be carried forward to the finished bar, regardless of the amount of subsequent reduction.

However, if fillet is too large, then there is chance of pass will get under filled and if it be remain under-filled in subsequent following passes, then it will create a round back on the surface of the finished channel, because of insufficient metal available at that portion, due to large fillet. The fillet on the outside toe (R_2) of the flanges also helps in for elimination of fins.

3.4 False or Counter Flange

While rolling a channel profile, false or counter flanges are necessary for better production of edges of flange and to protect them from cooling. Their absence may also lead to creation of residual stresses in the finished product. Presence of false flange in pass design will cause uneven and additional deformation of the channel profile, which results in extra consumption of energy for rolling and excessive wear out of rolls. Hence, it is necessary to design channel profile with minimum dimension of false or counter flanges.

If designer uses common roughing passes for channel and beam, then it is not possible to control the dimension of false flanges, which means losing of energy for rolling. Different authors give different dimensions and contradicting advices regarding choice of the size of false flanges. The main reason is the difficulty to create a quite basic theory or calculation for such determination.

Considering the necessity of simplifying the calculation to determine the height and thickness of false flanges, following empirical formulae has been worked out for computing different elements of false flange.

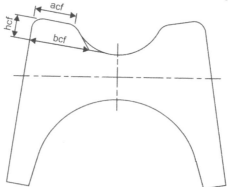

Fig. 9.5 Elements of counter-flange of channel Design.

Ht. of false or counter flange = $h_{cf} = (0.03-0.05) h \times (n-1)$

h = height of actual flange

n = Pass no. counting back from finishing pass

Lower side of co-efficient i.e., 0.03 may be used for passes which follow the closed one.

The thickness of counter flange at the root depends on the thickness of the flange. In the web cutting pass, the thicknesses are equal at the root, while in last pass having counter/false flange, the thickness at the root is taken 1.5 times that of flange. In intermediate passes, it varies from 1 to 1.5 times to the flange.

The tip thickness of false or counter flange can be determined from the formula

$$a_{cf} = (0.4 \text{ to } 0.5) b_{cf}$$

Where b_{cf} – counter flange thickness at the roof.

The height of counter flange, h_{cf} is found by formula

$$h_{cf} = \frac{2F_{cf}}{a_{cf} + b_{cf}}$$

F_{cf} – counter flange area
a_{cf} – the thickness at lip
b_{cf} – the thickness at roof

3.5 Profile Radii

The profile radii in the channel passes are computed as follows (Fig. 9.6).

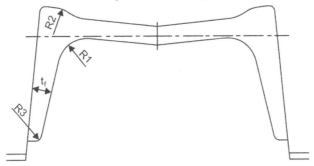

Fig. 9.6 Computation of Different Radii of a Flange Profile.

Where, $R_1 = t_f$,
$R_2 = (0.4 \text{ to } 0.5) R_1$,
$R_3 = 0.5 R_1$

As design of pass starts from finishing pass, the formation of false flange should start from pre-finishing pass itself. The above formula may be used only when there is specific design for rolling channel. In case of using common roughing pass for beam and channel, the height of false flange is calculated in such a way that within the given number of passings, it should be brought gradually to the dimension of actual flange. In order to avoid formation of fins, reduction in the height of false flange should not be made too high and it may be chosen approximately proportional to the co-efficient of deformation in those particular pass.

330 Chapter 9 Rolling of Channels

An analysis of actual roll pass design of channel shows that the co-efficient of elongation of flange and counter flange, if it is taken together, then it will be equal to the co-efficient of elongation of the web. A deviation from this rule is found in the finishing pass where the co-efficient of elongation for the web is less than that for the flanges to ensure that the latter are properly formed.

Table 9.1: Counter flange Area in Percent of the Total Area of Flanges and Counter flanges for the Size Range of Channels

Channel	Counter flange area to total flange area ratio in %								
	From first *and following passes having counter flanges								
Size in mm	1	2	3	4	5	6	7	8	9
180–300	6	9	12	18	24	30	36	42	48
120–160	6	12	18	24	30	36	42	48	–
50–100	8	16	24	32	40	48	56	–	–

* Pass Computation is from Finishing Pass

Using the data of table 9.1, the areas of the flange and counter flange are determined for all passes upto the one where the total flange area equals to that of the flanges. In preceding passes (in the order of rolling), the flanges and counter flanges are almost symmetrical and almost equal in area. Such passes can be also used in the rolling of beams.

3.6 Spread in Channel Rolling

Spread in channel passes is determined by the same empirical formula used for rolling of beams:

$$B = 0.01 \, B_1 + (n - 1)$$

where: B_1 – width of the finishing pass;

n – The pass number counting back from the finishing pass.

The following data may also be used to assign the spread value in individual passes of channel:

Table 9.2 Determination of Spread Value.

Channel	In the finishing pass, mm	In the other passes, mm
50–80	0.5	0.5 + (n–1)
100–180	1.0	1.0 + (n–1)
200–225	2.0	2.0 + (n–1)
300–400	3.0	3.0 + (n–1)

While designing channel, the rigid alteration of closed and open flanges is absent. In channel design, several open passes follow one another, which help in the better and quicker formation of flanges of the profile.

3 Salient Feature of Channel Pass Design

The increase in flange height in live hole is from 0.5 to 1.5 mm (larger value for larger sizes).

Height of flange can only be obtained in closed passes in which they undergo reduction along the height. The height of the flanges in open passes is always indefinite, because it is impossible to access the exact increase in height. Therefore, to regulate and to control the height of flange, it is necessary to have semi close or close passes, as a control pass in channel design (Fig. 9.7).

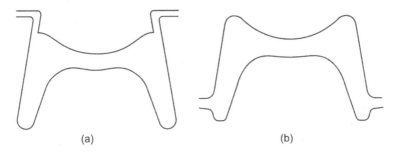

Fig. 9.7 Closed(a) and Semi closed(b) Pass for Height control.

Usually one or two control passes are kept in the rolling sequence of channel. However, more numbers of closed passes can also be used, to meet stringent tolerance requirement by the customers.

Ideally, it would be proper to make finishing pass of channel as the control pass. However, by making the finishing pass as control pass. It will get wrinkle, with the slightest overfilling of the pass, especially when the reduction is insufficient.

Due to above mentioned reason, designer always recommend to make pre-finishing pass as control pass, which is one before the finishing pass. Here, one may not afraid of getting exact height of flange or correct (sharp) edges, as reduction in the finishing pass is usually insignificant.

It is preferable to have the second control pass at the first or the second pass (along the direction of rolling) after the first forming pass. It eliminates the inexactness in height of flanges which is due to fluctuation in dimensions of blooms. While designing channel, it should be kept in mind to limit the number of control passes because their use is connected with decrease in deformation co-efficient and consequently causes increase in number of passings un-necessarily. It also requires the increase of the height of bloom/billet. While rolling channel in 3-high stands, the presence of additional closed pass makes the impossibility of getting open flange being cut in roll without the formation of high double collars. In other words, the effective utilization of roll length will get reduced.

The semi closed pass differs from closed passes with regard to the character of deformation, as it can be seen from [Fig. 9.7(b)], in semi closed pass, the middle and top part of flange are shaped by the working of top and bottom rolls, which facilitates the possibility of reduction of thickness of flange of semi close pass by the relative movement of top and bottom rolls. The bottom most part of semi close pass is worked as a dead hole, which is in one roll only. Due to closed nature of bottom part, reduction of thickness of flange at bottom part is impossible due to squeezing action and only reduction in height can takes place. One may not be afraid of fins at the gap because the form of the flange obstructs displacement of metal towards its ends. Deformation at other part of pass is similar to the deformation in open pass

Other advantages of semi-closed passes are:

1. Semi closed pass has an inclination of the flange in same direction as in the open pass, due to this reason, when strip enters to the semi closed pass, then there will be no harmful bend of the flange as can takes place in close pass. Bend of the flanges causes difficulty of entry of the strip in next pass and in addition, it also requires additional energy to make the bend. Fig. 9.8 shows strand pass in dotted line, entering into pre-finishing pass in dark line of channel design. Because of semi-close nature of pass, having the same 10% taper, there is no bending observed, while stock enters into pre-finishing pass.

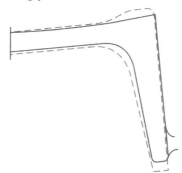

Fig. 9.8 The Strand Pass enters into Semi-close pass of Same Taper.

2. Absence of flange bending in semi closed pass make it possible to increase the outlet of adjacent passes which is favourable for the life of rolls and increases the numbers of their reuse.

3. In semi close pass, the difference of diameter between top and bottom roll is less, compare to the closed control pass, thus avoiding backlash of top and bottom roll.

3 Salient Feature of Channel Pass Design

3.7 Pitch Line Location

Proper pitch line placement in channel rolling is very important. If it is located incorrectly then, it may cause heavy backlash or clashing of top and bottom rolls. Placing the pass out of line results in greater speed of one roll over the mating roll; larger diameter drags the smaller along. The bar when going through the rolls act as a transmitting agent. However, upon leaving the rolls, there is no such agent and a clash occurs, which is called as BACKLASH.

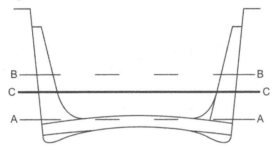

Fig. 9.9 Positioning of Pitch line in channel Rolling.

If the pass is placed according to line *A-A*, the bottom roll works on the channel at higher speed than does the top Roll and therefore drag the top roll along.

If the pass is placed to line *B-B*, then top roll works on the channel at higher speed than does the bottom roll and therefore, drags the bottom roll along. In either case, the result will be the clashing of rolls, when bar either enters or leave the roll.

That's why ,pass is to be placed *C-C* at suitable pitch line in between *A-A* and *B-B* to balanced speed difference between the top and bottom roll and to avoid the "Backlash" of top and bottom roll.

The smaller the average diameter in relation to the height of pass than greater will be the impact. It is therefore, for this reason only, the selection of larger diameter roll is better for section rolling mill, as a large diameter roll for any given depth of passes produces less difference in linear velocity. If the pass is correctly placed on pitch line the wear due to frictional grinding will be at minimum because the difference in roll diameters will get reduced. It is the frictional grinding that produces the "curly marks" on the flange of a channel. If it is possible to have all points along the flange rolling at the same speed, then there is no chance of formation of these marks.

3.8 Condition of Entry of Strip Into Pass

Different cases of entry of strip in to succeeding passes are shown; Entry of strip in to the pass when the inclination of flange and of the strip and that of the succeeding pass is same as shown in Fig. 12.10, in this case entry is performed very smoothly and perfectly.

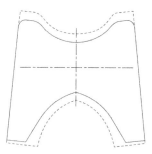

Fig. 9.10 Entry of Strip in a Pass, where Inclination of Strip and Flange are Same.

In case of over bending of the flange of pre-finishing pass in the finishing pass (Fig. 9.11), the condition of entry of the strip into succeeding finishing pass does not differ from these of closed flanges of the beam. Here bending of flange takes place, causing fast wear out of the finishing flange.

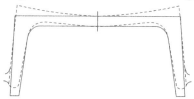

Fig. 9.11 Bending of Strip (Pre-fishing Strip Enters in Finishing Pass).

3.9 Guides Used in Channel Rolling

3.9.1 The purpose of guides in channel rolling to facilitates the following:

- To get the desired section with quality,
- Section should be free from surface defects,
- Minimum on-line down time attributable to change the "tackle",
- Easy Maintainability,
- Ease of setup—on or off the mill,
- Repeatability of its use.

3.9.2 Types of Guides used in Channel Rolling

(*a*) **Web Guide**

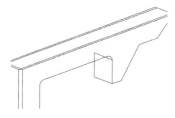

Fig. 9.12 Web Guides Bottom Roll.

(b) Flange Guides Top Roll

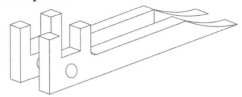

Fig. 9.13 Flange Guides Top Roll.

4 ROLL PASS DESIGN FOR CHANNEL 100 × 50 MM

4.1 Selection of Mill

Roll pass designer should know the following, before the start of channel:

- What is the strength and weakness of the mill, in which this product is planned to roll?
- The type of mill, production capacity of furnace, mill proper and finishing section of the mill.
- Size of billet available to roll this section.
- Temperature of input billet and finished product.
- Type of layout of the mill *i.e.*, whether it is open, semi continuous or continuous mill.
- The details of mill facilities *viz.*, mill configuration *i.e.*, number. of rolling stands and its configuration, nos of vertical stands, the distance between stands and in between each group of stands, stand size, type of stand to be used for rolling *i.e.*, whether it is housing-less, open or close type of housings to determine mill spring and ease of roll changing, type of roll used, specified rolling load and stand motor capacity, numbers of tillers in the mill and its spacing from preceding and succeeding stand. Shifting facilities to transfer the bar from one line to another and its length decides the maximum length of billets can be taken up for rolling.

All above mentioned parameters decide the type of roll pass design schedule, optimum numbers of passes to be used in the sequence and also to determine the type of reduction pattern to be adopted.

Here, the reference mill is a continuous cross country mill having 3 furnaces of capacity 60T/hr. Two furnaces are operated at a time and the third furnace is kept under repair. Mill is having 12 stands, out of which three stands (4, 7 and 11) are vertical stands. Mill is sub divided into three groups *i.e.*, roughing, intermediate and finishing group of stands. Tilter is placed between stand 1 and 2 and with the help of chain transfer metal is get shifted from one line to another, it is carried out with the help of chain transfers. Chain transfer

(CT-1), *i.e.*, after stand 8 and CT-2 *i.e.*, after stand 10 are of 44 and 77 meters respectively.

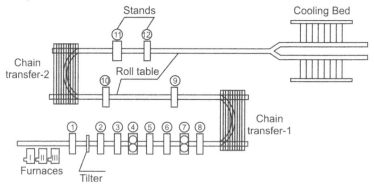

Fig. 9.14 Line Diagram of the Rolling Mill.

Facility is provided for taking on–line sample with the help of a hot saw.

Cooling bed is of the length 90 M and with the help of bifurcates, material can be delivered to either left or right side of cooling bed. Straightening machines and cold shears are placed at the both sides after the cooling bed.

Finishing section is having the facilities for stacking, making compact bundling and inspection and rail-road shipping facilities.

4.2 Customer's Requirement wrt Specification, Tolerance etc.

Technical delivery conditions (TDC), submitted by the customer decides the design of finishing pass in particular and roll pass design in general, *i.e.*, which specification has to be followed while designing the section and what are tolerances and upto what extend one can go for negative tolerance, without affecting its properties.

If BIS specification has to follow, then following are the details of channel 100 × 50 as per BIS.

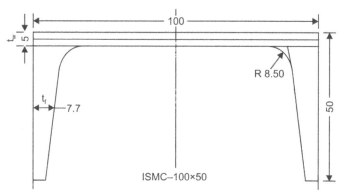

Fig. 9.15 ISMC100 as per BIS Specification.

4.2.1 BIS ISMC –Details of Specification

Table 12.3

ISMB 100 × 50	Sec Wt. kg/m	Area cm²	Depth D, mm	Breadth B, mm	Thickness of Web t, mm	Flange Thickness D, mm
	9.56	12.18	100	50	5.0	7.70

Tolerances as per BIS on Channel 100 × 50

   (*i*) Weight tolerance: ± 2.5%
   (*ii*) Depth of Channel: ± 2 mm
   (*iii*) Breadth of Channel: (+) 3 mm
                                              (–) 2 mm
   (*iv*) Flanges out of square/out of parallel: upto 1 in 60 is allowed
   (*v*) Camber: The maximum permissible camber of channel is 0.20% of the length.

4.3 Selection of Billet

Channel can be rolled using beam blank, if the beam method of rolling is used. The use of beam blank is advantageous from the point of view of yield, as a good balance can be maintained between the reductions of web and flange. The use of beam blank, in place of rectangular billet for channel rolling avoids differential elongation, which produces tongue formation as shown above in Fig. 9.16. This tongue formation reduces the yield drastically.

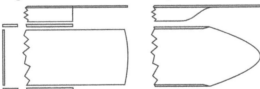

Fig. 9.16 Tongue formed due to differential Elongation.

The use of beam blank also assists in preventing fin formation in roughing stands; otherwise heavy web reduction in rectangular bloom/billet sequence will cause overfilling at the parting.

Generally, a square or rectangular billet/bloom is used for the channel rolling. One or two roughing passes are used to reduce billet/bloom into required cross section, needed to enter the first forming pass.

Selection of billet size and its length is determined by following:

- The length of stock after stand-1 as tilting of 90° is required after stand-1. Length of stock should be such that, it should get free from stand-1, before get tilted.

- Length of stock after stand-8, as chain transfer-1 can only accommodate the length upto 40 m.
- Length of stock after stand-10, as chain transfer-2 can only accommodate the length upto 70 m.
- Length of cooling bed-Stock should have length around 75-80 m as the max. length of cooling bed is 90 m.

A square billet with size 100 × 100 × 6 m is found suitable as it will give a bar length, which will get accommodated in all above mentioned area.

4.4 Determination of Co-efficient of Reduction for the Different Elements of Channel Design

Roll pass designer has to first establish the balance in co-efficient of reduction between different elements of channel

Following elements of channel are to be considered separately for designing a channel section in a rolling mill.

(a) μ_{Web} – Co-efficient of reduction of web,
(b) μ_{Ft} – Co-efficient of reduction of Flange tip,
(c) μ_{Fr} – Co-efficient of reduction of Flange root.

It is the skill of designer to prepare a balance approach in reduction pattern of above mentioned elements of a channel section, to achieve the desired dimensions of finished product.

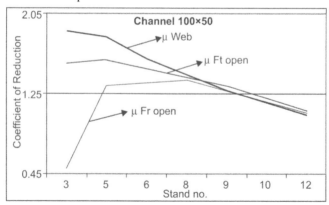

Fig. 9.17 Reduction Graph for Different Elements of Channel.

The total reduction in finishing pass (Pass no. 12) is to be kept as minimum as possible to reduce the fast wear out of pass. It will help in to avoid the frequent change of pass.

μ_{Ft} (co-efficient of reduction at the tip of live flange) in the finishing pass should be kept slightly higher than the μ_{Web} (co-efficient of reduction of web)

5 Computation of Elements of Pass Design

to get the correct dimension of flange and to avoid the pulling down of flange. Designer prefers to reduce the flange slightly more than the web in order to set up slight tension in web, which avoids the contraction of flange during cooling.

While rolling in pre-finishing and strand passes (Pass no. 10, 9), the thrust of designer should be on the correct formation of flange, especially to the height of flange, as there will be always a possibility of pulling down of flange, if web reduction becomes abnormally higher than the flange in these passes.

It is required to give maximum reduction at the tip part of flange so as to metal to flow to the tip part of live flange. It can be seen from the graph that, μ_{Ft} is at maximum, which is followed by the co-efficient of reduction of the flange root on live side $i.e.$, μ_{Fr}.

The co-efficient of reduction of flange at close or dead side is not of much importance. Close pass lies in one roll only and as it name implies, it is also called as dead hole. There will be no side reduction at dead flange. Here, only the height reduction will be affected. The incoming flange of preceding pass to the dead hole of succeeding pass will have more height and of slightly less or equal thickness.

In the early passes $i.e.$, first web cutting pass and succeeding one or two passes, the maximum thrust is for web formation. Knifing action of web cutting pass and high temperature of metal helps in cutting in action of the metal, which is only limited by the angle of bite and the motor capacity of mill.

5 COMPUTATION OF ELEMENTS OF PASS DESIGN

(*a*) Flange Dimensions

Table 9.4

Pass No	Flange	Flange height mm	Tip thickness mm	Root thickness mm	Pull down mm	Height increment mm
	Counter flange	0.40		15		
12	Close	46.3	4.75 (1.09)	9.5 (1.05)		0.8
10 (semi close pass)	Counter flange	2.0	–	20	3.0	
	Semi-close	45.5	5.2	10.0 (1.16)		
9	Counter flange	4.0	15			
	Close	48.5	4.8 (1.33)	11.6 (1.27)		1.5

Contd...

8	Counter flange	6.0	17			
	Close	47.0	6.4 (1.4)	14.8 (1.33)		1.0
6 (semi close pass)	Counter flange	9.0	19		1.0	
	Semi-close	49.0	9.0	19.8 (1.35)		
5	Counter flange	11.0	20			
	Close	50.0	9.0 (1.53)	26.8 (1.34)		6.0
3	Counter flange	15.0	21			
	Close	44.0	14.0 (1.64)	36		8.0
2	Counter flange	18.0	23.0			
	Close	36.0	21.0	36		

Fig. under brackets show the co-efficient of reduction of different elements

(b) Web and Pass Dimensions

Table 9.5

PASS No.	Width mm	Thickness mm	Total height mm	Draught mm	Spread mm	Pass area mm²	Co-efficient of elongation μ
12	99.8	4.6 (1.08)	51.3	0.4	1.0	1200	1.05
10 (semi closed pass	98.8	5.0 (1.16)	52.5	0.8	1.3	1260	1.22
9	97.5	5.8 (1.28)	58.3	1.6	1.5	1540	1.31
8	96.0	7.4 (1.44)	60.4	3.3	2.0	2020	1.46
6	94.0	10.7 (1.60)	68.7	6.4	2.0	2965	1.31
5	92.0	17.1 (1.81)	78.1	13.9	2.0	3889	1.49
3	90.0	31.0 (1.87)	90.0	27.0	0.0	5826	1.37
2	90.0	58.0	109			8027	

Fig. under brackets show the co-efficient of reduction of different elements

5.1 Design of Finishing Pass

Passes are to be designed on hot dimension, while the above mentioned BIS dimension is of cold dimension. Hot dimension is generally taken 1.010 -1.015 times more than of cold dimension at the temperature above 800°C.

5 Computation of Elements of Pass Design

But to match marketing strategies i.e., customer's demand to have roll product with negative tolerances and to have more length of finished product with the same weight for cost saving purpose. Roll pass designer always keeps hot dimension of different elements of pass are to either equal or even less than the cold dimension, to take advantages of negative side of weight tolerance (-2.5%). Following are the criteria for selecting different element of pass:

5.1.1 Web Thickness

Cold dimension of Web thickness (t) given in the design is 5 mm. Web thickness should be such, to take care of the Mill spring, which is directly proportional to the rigidity. Rigidity depends upon the selection of stands i.e., whether it is conventional or housing fewer stands.

Mill Spring of a conventional stand varies from 0.3-0.5 mm, then web thickness d, will be assumed as, $d = 5 - 0.4 = 4.6$ mm

In addition, designing with thinner web will also consider problems like faster wear out at rolling at lower temperature.

5.1.2 Flange Height

Flange Height i.e., Breadth of section is given as 50 mm.

Tolerance on flange height is ± 2 mm. Flange is always designed on positive side of tolerance, because roller always faces problem of flange shortness during rolling of channel due to pull down of flanges, because of various reasons explained earlier. Pass height in this case assumed as, $B = 50 + 1.3 = 51.3$ mm

Height of flange can also be slightly adjusted by pressing the roll.

5.1.3 Tip and Root Thickness of Flange

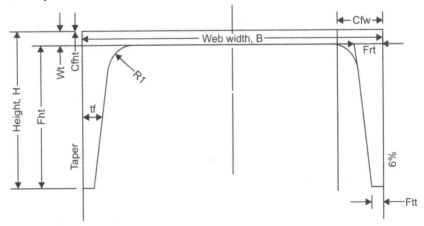

Fig. 9.18 Computation of Elements of Channel.

342   Chapter 9   Rolling of Channels

The average flange thickness, it is given as 7.7 mm.

In case of channel, the inner side of flange is having a taper of 6°, which comes to 10.5%. To compute the tip and root thickness of flange (Fig. 9.18).

$$\text{Flange height} = \text{Flange ht.} - (\text{web thickness} + \text{ht. of counter flange})$$

$$= (51.3 - (4.6 + 0.4) = 46.3 \text{ mm}$$

$$y = 46.3/2 \times 10.5/100 = 2.4 \text{ mm}$$

Tip thickness F_{tt} = 7.7 -2.4 = 5.3 mm

Root thickness F_{rt}= 7.7 + 2.4 = 10.1 mm

Tip thickness is always kept less than the computed value, due to wear out of rolls, that's why live side tip thickness (F_{tt}). It is generally assumed 0.3-0.5 mm less than the computed value, it works out to 4.75 mm.

Root thickness(F_{rt}) are also generally taken 0.4-0.7 mm less than the value to give more life to the pass and less off is to be taken during the restoration of pass. Here, it is taken as 9.5 mm.

5.1.4 Width or Depth of Pass

Width of a channel 100 is given as 100 mm as per the design. The allowable allowance is ±2 mm. As the restoration of the width of pass is very difficult; width of pass becomes the most critical item on point of view of design. That's why, the hot dimension B is even taken less than the cold dimension, but here, It should always be kept in mind that width or depth of pass should not go beyond the negative side, otherwise it will lead to rejection of material.

In this case B is taken as 99.8 mm $(D = 100-0.2)$ for hot section, which will give a cold section of 99 mm.

Height and width of counter flange: Height of counter flange is to at minimum or zero in the finishing pass, but generally it is preferred to be taken in the range of 0.2-0.5 mm to give sharp corners at finishing pass. Here it is assumed as 0.4 mm

Width of counter flange is generally taken 1-1.5 times to the root width of flange. Root width is 9.5 mm; counter flange width is taken as 15 mm.

5.1.5 Weight of Pass

Pass area is to be computed either mathematically or by auto cad. Area of pass comes to 1190 mm^2, which will give the sectional weight as 9.34 kg/m, (–2.3%) less than the normal sectional weight of 9.56 kg/which is within the weight tolerance of 2.5%.

One need not worry with the computed weight, even if, it is on negative side, it can be adjusted by raising the roll gap.

5.1.6 Taper of Pass

In designing the finishing pass for channels, it is advisable to keep the taper of pass as large as possible to permit easy delivery. Advantage should be taken from the fact that the flanges get contract during cooling. The angle of pass gets automatically reduced, while cooling in cooling bed. Here, it is taken as 6%, against 1.5% in case of Beam design.

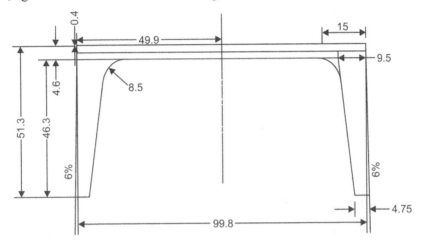

Fig. 9.19 Pass Diagram of Finishing Stand.

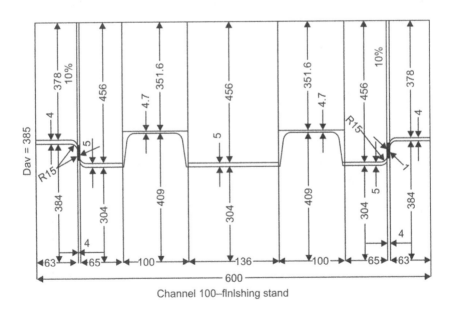

Channel 100–finishing stand

Fig. 9.20 Roll Diagram of Finishing Stand (Stand-12).

5.2 Pre-finishing Pass (Pass-10)

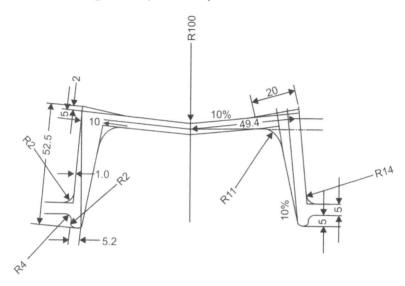

Fig. 9.21 Pass Diagram of Pre-finishing Stand.

Pre-finishing pass is to be of semi-close pass design to give a correct feed to the finishing pass, to avoid variation in flange length due to different changing elements of rolling, including temperature, will be taken care by semi-close pre-finishing pass in channel design.

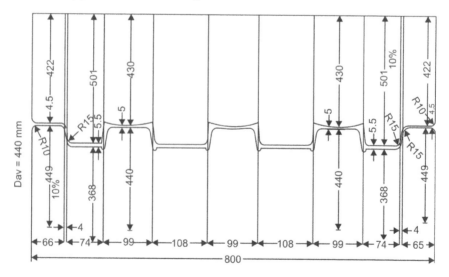

Fig. 9.22 Roll Diagram of Pre-finishing Stand (Stand-10)

5 Computation of Elements of Pass Design

Different elements of flange, counter flange and web are chosen based upon the graph for different elements of flange profile already shown in (Fig. 9.17).

5.3 Strand Pass (Pass-9)

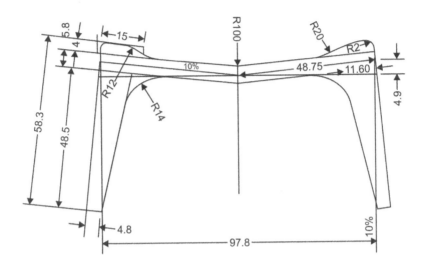

Fig. 9.23 Pass Diagram of Strand Stand (Pass-9).

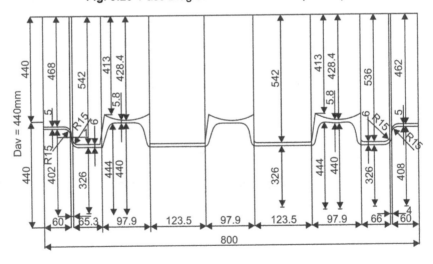

Fig. 9.24 Roll Diagram of Strand Stand (Pass-9).

5.4 Pass-8

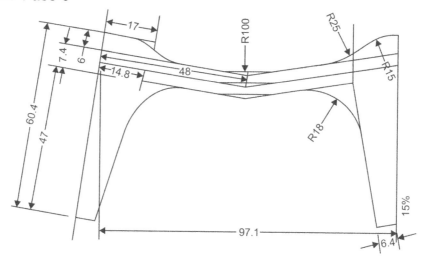

Fig. 9.25 Pass Diagram of Stand-8.

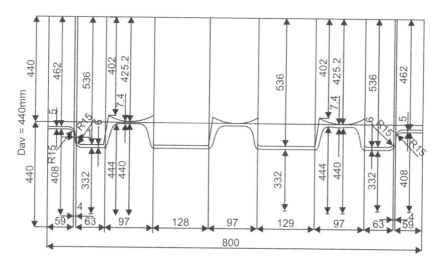

Fig. 9.26 Roll Diagram of Stand-8.

5 Computation of Elements of Pass Design

5.5 Pass-6

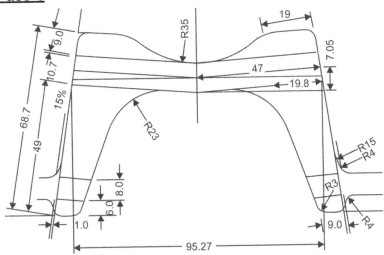

Fig. 9.27 Pass Diagram of Stand-6.

It is second control pass in the design of channel. It is desired as it eliminates the inexactness in height of flanges which is formed due to fluctuation in dimensions of blooms.

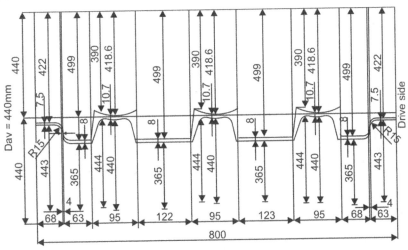

Fig. 9.28 Roll Diagram of Stand-6.

5.6 Pass-5

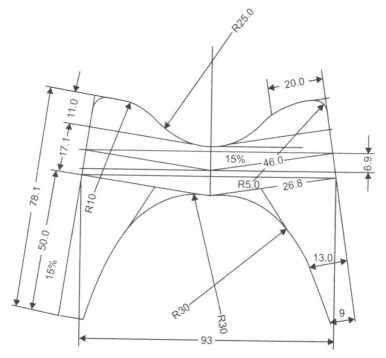

Fig. 9.29 Pass Diagram of Stand-5.

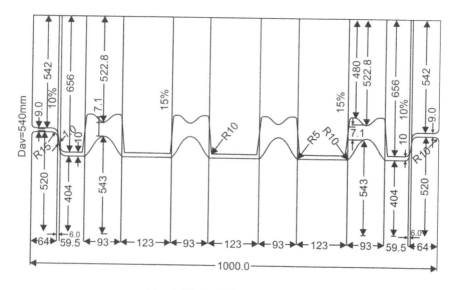

Fig. 9.30 Roll Diagram of Stand-5.

5.7 Pass-3

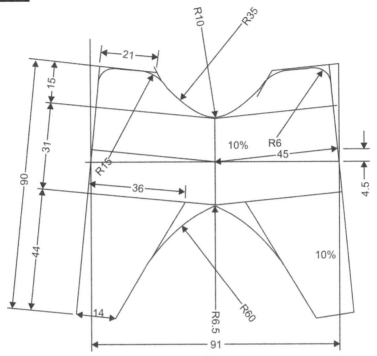

Fig. 9.31 Pass Diagram of Stand-3.

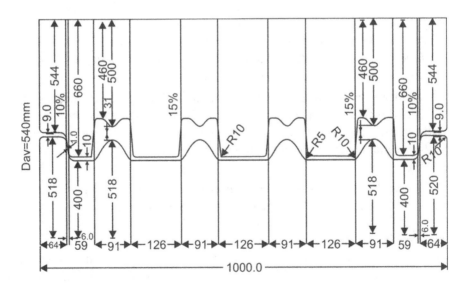

Fig. 9.32 Roll Diagram of Stand-3.

5.8 Pass-2

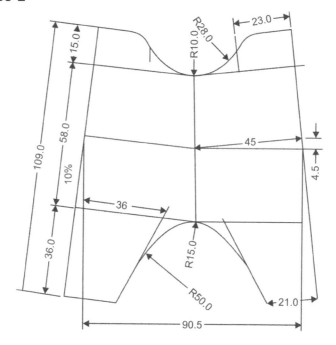

Fig. 9.33 Pass Diagram of Stand-2.

This is the first forming pass in the design of channel and is called as "V" or cutting-in pass. The tongue or "V" of the roll should be designed in such a manner, so that it compresses the pass in such a way, to form the flange. The tongue of the "V" pass should be like an edge of an axe. The narrower and thinner will be the blade or tongue, then more easily it get sunk deepen to the billet with ease, but there is a practical limit to this thinning of blade also, which shall be restricted by strength of roll.

Higher reduction is recommodated in initial passings, firstly, because steel is hot and plastic at that stage of rolling and due to higher temperature, there will be less resistance to the deformation, which make it facilitate to give higher reduction and help in the rapid formation of flange in initial stage of rolling.

Secondly, also it gives an opportunity to remove any irregularities caused by non-uniform draft before it reaches to the finishing pass.

5.9 Stand-1

Here, it is recommended to use box pass. The advantage of using box pass at stand -1 are numerous like stability of work-piece is more in box pass compare to the barrel pass, chances of twisting is less. It also acts as scale shedding pass, which in turn will help to improve the quality of product. The amount of spread is also less compare to barrel pass.

6 Mill Setting of Channel

It is not part of the continuous group. Here, speed is kept such that to facilitate easier bite and should not give any chance of slipping of the metal, which effects adversely the mill equipments. Here, speed is assumed as 300 rpm.

Sometimes, ragging of rolls, use of sand and forced biting is given in stand -1 to facilitate smooth bite in stand-1.

Diameter of stand-1 plays a vital role in the biting of billets. The use of minimum diameter rolls should be avoided.

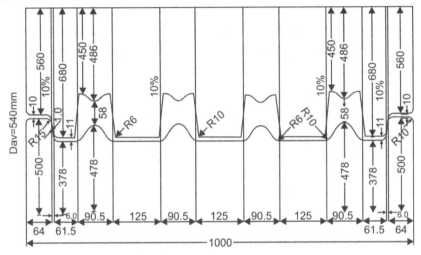

Fig. 9.34 Roll Diagram of Stand-2.

6 MILL SETTING OF CHANNEL

General rules of Mill setting for channel are:

1. Billets for rolling of channels should be free from scabs and cracks and should have smooth surfaced billets. The presence of scabs exceeding to 20 mm in billets should be rejected.
2. Billets for channels with bad, un-cropped front or tail should be rejected on the charging device itself.
3. Temperature of metal as recorded in roughing group stand should not be less than 1100°C. Before start of the rolling, billets which lie before the discharge door of furnace should be returned, as these billets are usually cold, may create problems in stability of rolling. Generally, 2-3 billets are to be rejected before the start of rolling.
4. Selection of pass for all stands of the mill for rolling should be made by the Mill (Opr.) I/c., and in shift by Shift Foreman (Opr.). Checking of passes should be made with the help of templates for the close part of the pass (groove). The width of the previous pass selected for rolling

352   Chapter 9   Rolling of Channels

should have width always be less than the width of the next pass or in certain cases, may be made equal to that. It is forbidden to use pass of the previous stand where the width of the pass is more than that of the next pass. The information about the width of passes obtained from the roll turning shop should be again be checked in the shop.

5. While fixing the gaps between rolls, special attention should be paid to the accuracy of the gap of the end of runners of rolls.

6. Final setting of the rolling tackles is allowed only after final setting of gaps between the rolls and checking the same with filter gauge.

7. Entry and delivery guides should be fitted according to the maximum width of the close pass of every stand.

8. In case if considerable changes has been made in gaps between rolls during the process of setting the mill, it is necessary to check the setting of guards again at the time of fresh setting of roll.

9. At the time of mill setting, dimensions of the bar should be checked after all every group *i.e.*, after all roughing, intermediate and finishing stands, as well as in both chain transfers, cooling bed and after tilter of roughing group. After tilter, measurements can be taken on the hot metal. After each group, stand metal to be rejected for checking the dimensions of metal. Pieces after pre-finishing stand may not be rejected as a sample can be taken from the saw keeping XII stand in idle running. During checking the following dimensions should be checked :

   (*a*) The web thickness,

   (*b*) The flange height,

   (*c*) Approx. flange thickness,

   (*d*) Checking for distortion should also be made.

   After final adjustment has been made over, slight changes in reduction can only be allowed after judging the loads on stand.

10. Final estimate of the section may be given after examination and checking of finished products on the cooling beds and also after examination and checking the dimensions of the sample at hot saw. It is necessary to blow away the water from the bar behind stands after first forming pass and subsequent passes of intermediate group, otherwise the web gets cooled easily.

Control of Section Weight

1. Weight on the piece depends of the:

   (*a*) Thickness of the web,

   (*b*) Thickness of flange,

7 Defects of Channels and its Rectification

(c) Filling up of inside corner,
(d) Taper of the leg,
(e) Leg lengths,
(f) Width of the piece.

Roller has very little control over (b), (c) (d) and (f). So to keep the weight within the tolerance limit. The roller has the only control over the web thickness and leg length. If web thickness is decreased by pressing finishing and pre-finishing stand, leg lengths also get slightly reduced and weight will become less and with the increase of web thickness weight will obviously become more.

Thickness of web and leg lengths can be decreased upto a certain limit, but when this limit is reached Roller has no way, but to change the pass.

Roller should see that passes should last long.

For this:

- Uniform load on all stands should be ensured.
- The temperature of the of piece should not be low, to avoid faster wear out of pass.
- The practice of rolling with matching of passes should be adhered.
- Uniformity of the profile from each and every stand should be ensured.
- Overfilling and under filling of passes should be avoided.
- Guards should be checked periodically to avoid marks in the passes. Pass, it should be grounded when a slight mark is noticed on the pass.
- Guards shouldn't be pressing too much on the pass due to heavy counter weight or use of heavy guards to avoid the mark on the pass.
- Guard should not be too short and setting of the guards should not made it to act as a cutting tool, the sitting of the guard on the pass should be uniform, the material of the guards should not be too hard.

7 DEFECTS OF CHANNELS AND ITS RECTIFICATION

1. Legs are short (both)

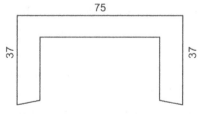

Fig. 9.35 Legs are short (both).

This is due to less metal is coming from roughing group. Too much hot metal and tension between stands should be avoided. Metal on the leg should be increased from roughing group. If the pass of pre-finishing stand wears out beyond the limit, then leg thickness-will more and legs become shorter. The passes should be changed.

2. Legs are longer (both)

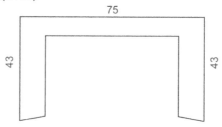

Fig. 9.36 Legs are longer (both).

This is due to just opposite action as mentioned in (1) above.

3. Legs are unequal

This is mostly due to the wrong setting of entry guides in first cutting-in pass ("*V*"). It requires guides should be set properly. Guides should be set more towards the shorter leg side.

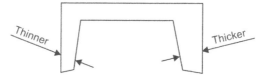

Fig. 9.37 Legs are Uequal.

If setting of entry guide is found correct, then top roll is to be taken towards longer side. When difference is less, then top roll of finishing and pre-finishing stand should only be shifted.

Entry guides of roughing groups and rolls of roughing groups are only be adjusted, if difference is more.

If necessary, one piece may be rejected after roughing group to check up the flange lengths and adjustment done accordingly. If the piece fed to first "V" stand twisted, then one leg may will become shorter, while other get longer.

4. One leg is thicker and other is thinner: The top roll of finishing stand is to be taken towards thinner side.

5. Both corners are blunt: This is due to not filling up of corners of pass of finishing stand.

This may be due to less metal is taken from pre-finishing stand. Top roll of pre-finishing stand should be raised.

If the finishing stand pass is not filled properly due to less reduction then top roll of finishing stand should be pressed.

If the turning of pre-finishing stand pass is not correct *i.e.*, if corners are turned less, in this case also the probability of production with blunt corners are more. To avoid this trouble, the pass of pre-finishing stand should be checked. If the pass of pre-finishing stand is too narrower than finishing pass, the rounded corners are expected. In this case, wider pass of pre-finishing stand to be taken or narrower pass of finishing to be taken for change. Rounded corners may be due to worn out pass of finishing stand, pass should be changed. Worn out pass of pre-finishing stand will help to fill up the corners.

6. **One side of corner is blunt:** This may also be due to above reasons and if the rolls of finishing or pre-finishing stand is cross. Top rolls should be moved towards blunt side in pre-finishing stand and opposite to blunt side in finishing stand.

7. **Bend web:** This is due to less reduction in finishing stand or more reduction in pre-finishing stand. Adjustment should be done accordingly. When the finishing stand pass is excessively worn out that may also cause bend web. Pass should be immediately changed.

8. **Inside corners are not filled up:** This is due to reasons explained in 5 and 6 above.

9. **Waviness of the web:** This is due to unequal reduction between flange and web. Sudden more reduction in web (pressing of top roll) will result in this.

Fig. 9.38 Waviness of Web.

10. **Under filling mark on the top:** This is due to reasons explained at 5 and 6 above.

11. **Worn out mark on the flange, web, fire crack mark on inside corners:** In such case if mark is more, pass should be immediately changed. Pass can be ground and used again, in case of less worn out mark.

It is to be noted that the bottom roll groove of all stand gets heated up very quickly as accumulation of heat takes place at bottom roll due to difficulty in fixing the proper cooling system.

Water cooling for bottom roll should be made more efficient.

12. **The impression on the web is nil or faint:** This is due to less reduction in finishing stand or more reduction in pre-finishing stand. Adjustment of reduction should be done accordingly.

Fig. 9.39 The Impression on the Web is Faint.

13. **Wear out of Branding:** In such case, pass should be changed immediately.
14. **Fins & Bevelled Corners on Channels**

Fig. 9.40 Fins & Bevelled Corners on Channels.

It is due to excess or lack of stock protruding from top of web/flange junction

Possible causes are:
- Rolls set incorrectly.
- Rolls turned with thick /thin flanges.
- Rolls turned to different fishing widths.
- Under/over work on 2-Hi/edging stands.

7 Defects of Channels and its Rectification

Remedies are:
- Reset rolls.
- Realign rolls.
- Check roll passes for flange thickness.
- Compare fishing widths of top and bottom rolls in same stand.
- Increase/decrease work on 2-hi/edging stands to improve corner profile.

15. Holes in Webs

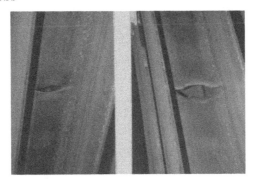

Fig. 9.41 Holes in Web.

It results of web being stretched due to too much flange work.

Possible causes are:
- Draughting ratios out of synchronisation between web and flanges.
- Buckle appears due to either too much work on the flange or insufficient web work (inverse of web buckle).

Remedies are:
- Adjust roughing stand to allow more stock into finishing stand.
- Adjust finishing mill to balance web / flange ratio.
- Check for possible inter stand tension in a continuous mill arrangement.

10

Rolling of Beam

1 INTRODUCTION

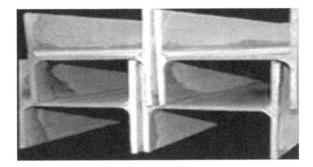

Fig. 10.1 Beam Profile.

Beams are the most typical flanged sections, if judged by the methods of rolling and roll pass designing. The rolling of beam design comprise two distinct stages of deformation;

(*a*) Obtaining the initial rough beam profile.

(*b*) Rolling the rough profile in to beam of the specified size.

The first stage is characterized by a sharp transformation from a rectangular cross section to rough beam form. The second stage consists in a gradual transformation from rough to the finished section.

2 METHODS OF BEAM DESIGN

2.1 The First Method

The first method Fig. 10.2 (i) is generally used for rolling of smaller beam sizes from 100 mm to 300 mm. It comprises only close beam passes. Non uniform deformation is given in the first forming passes *i.e.*, in a web cutting pass with *V* cutting element. The input material is generally rectangular in shape. In subsequent passes, under the influence of more or less uniform deformation of the separate elements of the profile, the strip gradually gets the size of the final profile. The open and close elements of flanges are changed alternatively between top and bottom rolls thus equalizing tensions. To ease the delivery of the strips from the rolls and for easy redressing, the wall of the groove is given a taper of 2-3% in the dead part and sometimes 5-6% is also recommended for the live side of the beam.

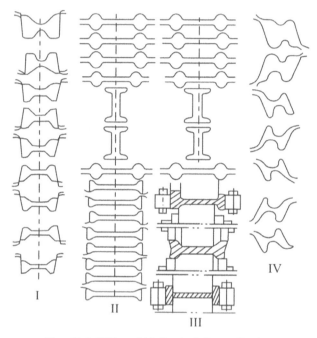

Fig. 10.2 Different Method of Beam Design

Widening or spread of profile is considerable in the roughing passes and it gradually decreases towards the end of the rolling. In the last pass, it is generally kept to 0.01 of the width of the whole profile.

2.2 Second Method

Medium and heavy beams, sizes from 300 mm to 700 mm are rolled by sequence shown in Fig. 10.2 (*ii*), The input *i.e.*, rectangular billet is first rolled in open beam passes of blooming mill and in rolls of two high reversing roughing stand of structural mill. Several passes are performed in each groove by changing the gap between rolls. Overfills, on sides are eliminated by turning the bar 90° and passing it through in barrel pass. Deformation in open pass will continue till web thickness is reduced to a value from three to five times the final thickness. Further rolling will be performed in close beam passes.

2.3 Third Method

Third Method Fig. 10.2 (*iii*) is used in rolling of wide flanged beam, including parallel flange beam. The same procedure and rolling schedule explained earlier, is used here also to get the rough section. The difference is that the rough beam section is further rolled in universal stands.

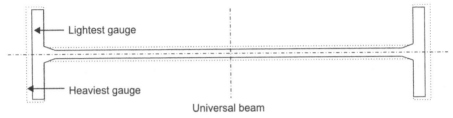

Fig. 10.3 Universal Beam Section.

Advantages of Universal Rolling;

- Ability to produce tailor sizes to suit application:
- Web and flange thicknesses can be adjusted independently.
- Capacity to increase/decrease flange length.
- Reduced roll costs.
- Reduced setup times and costs.
- Quicker section change
- Better surface quality.
- Better mechanical properties

2.4 Fourth Method

Fourth method Fig. 10.2 (*iv*), Diagonal method of rolling, uses close beam passes, inclined to the horizontal (diagonal) has found wide application in beam rolling due to following reasons:

- Due to the inclination of the open flange, the reduction in profile can be given high and it will lead to decrease in total number of passing.
- Off requires for the redressing of roll become reduced and due to this, life of roll get increased.
- It is also possible to reduce the profile without spread and even to give some reduction along width.
- With shallow cut in rolls, roll become more stronger and chances of roll breakage will be less and it also facilitates more reduction over work piece.

Disadvantages of using diagonal method of rolling are as follows:

- It makes the necessity of using both rolls with increased initial diameter, because close flanges are simultaneously in the top and bottom part of the groove.
- Appearance of the considerable axial forces on the rolls due to high inclination of roll.

Diagonal method is popular in those designs of wider flange beam, where height of flange does not permit to use the straight beam method. The co-efficient of reduction in the diagonal rolling will be always greater than the close flange beam rolling design.

3 MILL LAYOUT

3.1 Conventional Rail and Structural Mill

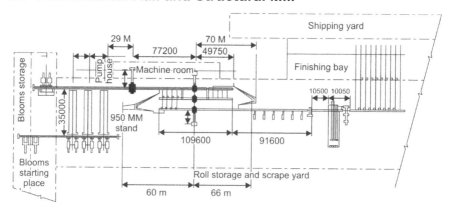

(a) Stand arrangements in two line (1 Roughing stand in one line and 2 intermediate and 1 Finishing stand In other line).

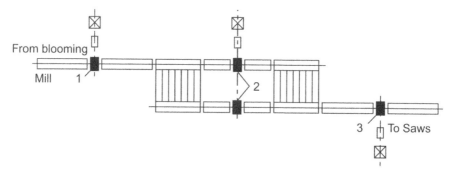

(b) Stand arrangements in three line (1 Roughing stand in one line, 2 intermediate stand in second line and 1 Finishing stand in third line)

Fig. 10.4 Layout of Conventional Heavy Structural Mill.

3.2 Universal Rail and Structural Mill

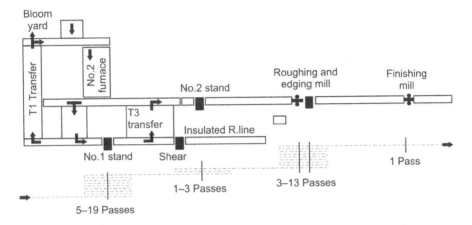

Fig. 10.5 Layout of Universal Rail and Structural Mill.

3.3 Continuous Combined Bar and Structural Mill

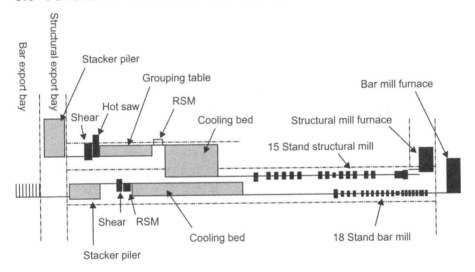

Fig. 10.6 Continuous combined Bar and Structural Mill.

4 SPECIAL FEATURES OF BEAM DESIGN

In design of rolling beam, the first consideration is to displace the metal from the web as rapidly as possible; of course enough flange metal is to be made available to permit proper proportioning of web and flange draught at later passes.

In early passes, designer should take advantage of the higher plasticity of metal, because of higher temperature of the stock. The metal should be displaced quickly from the center for the formation of the web and flanges. It is stated that the cutting-in action of the first shaping pass of a beam, the center is held back by less reduced outer pass and the outer parts are dragged along by center. An attempt to roll in a too wide groove will cause too great an elongation; with the result that height of the flange of groove will get reduced drastically. Another factor affecting the height is the clearance between the bloom and side of the pass, if the difference between the incoming stock and pass is too high, then the pressure of knifing action will force the metal to the side of the pass and metal which could be moved for upward movement for flange formation, has gone in for spread. The clearance between the pass and bloom is also equally important in developing flanges of equal height. First question arises in the design of the first shaping pass is "How deeply a groove can be cut in the roll or how much reduction can be given"? In actual practice, this reduction is usually limited by the ability of roll to bite.

Ragging is used to improve the ability of roll to bite. However, in some instances, particularly in conventional design, the strength of roll is also an important consideration. The strength of roll is most important, as flanges are cut so deeply into the roll, by that roll gets weakened. The fillet at the bottom of flange also plays an important role in control of the strength of roll. If fillet is small, then the stress concentration developed will make the roll weaker at such places The power of motor and strength of housing also plays a vital role in some mill.

In beam rolling, reduction takes place on account of two actions, *i.e.,* one is the slabbing action or by indirect draught in the live hole with the tongue of one roll and the collar of the mating roll. The other is the direct reduction *i.e.,* in the horizontal plane of the web. As there will be no slabbing action in the dead hole, here steel will be subjected to such an action, which reduces the height of metal rather than its thickness. Thus, flange thickness get reduce in live hole by indirect draught, while height will get reduce in dead part of beam design by direct draught. The thickness of flange of one side is always thicker than the other side in beam rolling, except in the finishing pass. This thickness variation of flanges will get decrease, as the finishing pass is approaching.

At first thought, it might appear that, if beam pass has all flanges designed with equal reduction, will all flanges will be filled equally? This is not true, as this design will produce heavy overfills in live flange and short flange in dead holes. The live flange has slabbing action and therefore, it offers the least resistance to the flow of steel.

5 IMPORTANT CONSIDERATIONS FOR ROLL PASS DESIGN OF BEAM

5.1 Selection of Bloom Size and Numbers of Passes

The selection of Bloom size and number of passes depends on the many factors-at one hand the Mill layout dictates, whether an odd or even number of passes are required and what length of stock can be handled in such layout. On the other hand, the availability of the power screw down on the roughing stand determines whether more than one pass can be taken in the initial passing's, hence it also decides the number of groove required. The size of roll diameter and barrel length affects roll strength and biting condition, which alongwith the power of the motor limits the maximum reduction, which can be under taken in a pass. Rolling temperature also plays a vital role in deciding numbers of passes. If flanges are thicker in early passes, then heat can be retained for a longer time, in turn, which may allow to increase no of passings. In view of above stated factors, it is always advised to use beam shaped bloom/billet for light and medium structural mill.

We may compute the size of rectangular bloom/billet from the following consideration:

Bloom height, $A = 2 \times$ height of flange $+ 20$ mm

Bloom width, $B =$ width of the Beam $- B_n$

Bloom width depends on the total amount of spread, *i.e.*, B_n allowed from first to last, will be deducted from the web width of the beam.

However, experience in rolling medium and heavy rolling shows that actual widening found out to be less than the computed from above mentioned formula.

Bloom Area $= 1.47 \times A \times B$

5.2 Advantage of using Beam Shaped Bloom/Billet

The advantage of using Beam shaped bloom/billet will assist in preventing the fin formation in roughing stands, otherwise heavy web reduction will be required for rectangular bloom/billet; to be taken as input material will cause overfilling at the parting. In rolling with the beam blank, the reduction of web and flanges are to be designed with a balance. The speed and temperature of rolling play the deciding role in the filling of the flanges. When steel is hot and speed of rolling is less, it makes the flow of steel into the flanges. Shaping pass operating at high speed and low temperature would result in the under-filling of the flanges of the shaped blank. Sometimes, designer design the "Cutting – in "pass with double V for following advantages:

- With two locations available for ragging with double V shape, a greater amount of reduction can be applied in the first cutting-in pass, as drastic ragging in subsequent passes may lead to rejection of the finished product, due to surface defects.

- By having sharper 'V' in double V shaped design, the metal will flow in to the flange more readily than comparing to the single V design of the first cutting-in pass. Blunt "cutting-in" pass will lead to metal to elongate *i.e.*, metal to flow into the web more readily and metal to withdraw from the flanges.

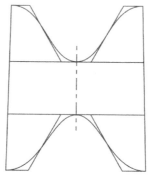

Fig. 10.7 Cutting-in Pass for Beam Rolling.

5.3 Elements of Beam Section

Design a beam profile is complicated in nature. The profile is divided into several simple elements; wherein degree of deformation can be controlled separately to achieve the ultimate dimension of the profile.

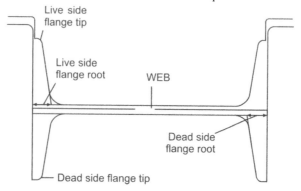

Fig. 10.8 Different Elements of Beam Section.

Web	Thickness, Length, Area
Live flange	Flange height, Tip thickness, Root thickness, Area
Dead flange	Flange height, Tip thickness, Root thickness, Area

5.4 Draughting Strategy for Beam Design

In designing passes for a beam profile, the first step is to displace the metal from the web, as quickly as possible, to take advantages of the higher rolling temperature and the consequent increase in the plasticity because to higher rolling temperature. But at the sametime, enough metal should be at the flange to facilitate further proportioning of draught of the web and flange in later passes. If no reduction is planned on the flanges at the later passes, then flange height will get automatically reduced with the further reduction of the web in succeeding passes, resulting in elongation. Furthermore, if the flange thickness is too thick from one pass to another pass, then it will be impossible to fit properly the live flange into the dead hole.

Designer should try to design live flange to fit the dead hole exactly and to facilitate heavy reduction on the live side of flange and to the web. It will result in for the work of these two parts of the section and drawing down the height of the dead hole, which sometimes, is also called as "wire drawing".

Entering the live flange into dead hole must not cause too much choking at the entry. The slight choke in the dead hole creates a resistance of the entry of live flange, to be driven down and holds up the live flange. With the result, the slight reduction of work on live flange and slight choke in the dead hole will produce a more uniformly flange beam shape.

Designers are sometimes uses double tapers on inside of the live flanges to eliminate some of the down push, resulting in achieving the height of the flange in live hole. It also helps to improve the entry conditions. Here, it is to be noted that these double taper produces flange, which is thicker at the base, than in the single taper design. This aspect is to be kept in mind, while designing next dead flange.

5.5 Spread

It is a design feature, which is of a vital importance. Bar is forced to spread against the sides of the grooved roll due to side work on the flange and the pressure exerted by the tongue on biting of the bloom in the course of the working of web. The bar is forced to spread against the sides of the grooved roll, causing wear. In addition, it helps in delivering the stock from the rolls, since it relieves some of the otherwise heavy side pressure of the steel against the side of pass will get generated, which would be there, if no provision is made for the allowable spread.

368 — Chapter 10 Rolling of Beam

Too much spread will cause a loss of height of the flange, since metal follows the path of least resistance, flows out in width of web. Generally, cutting-in passes are designed with the minimum or zero spread, to facilitate the metal to flow into the flange height. In pre-finishing and finishing pass of the beam design, the allowance should be such as to take care of ease of delivery and wear out factor of the pass, to avoid frequent pass changing.

Widening affects the pull-down of flanges in closed flanges and height increments in open flanges. If width of pass is taken more than to compensate the natural spread, then to compensate to this, shrinkage of flange takes place in both the flanges.

Total spread, $B_n = 0.01\, B + (n - 1)$.

Where n is the number of passing from direction of rolling and B is the width of that pass, the above formula may be used in those cases, where it is necessary to define the summarized widening, then

$$B = b_1 + b_2 + b_3 + \ldots\ldots + b_n$$
$$b_1 = 0.01 \times B_1 + (1 - 1)$$
$$b_2 = 0.01 \times B_2 + (2 - 1)$$
$$b_3 = 0.01 \times B_3 + (3 - 1)$$
$$b_4 = 0.01 \times B_n + (n - 1)$$

5.6 Flange Design

The consideration, which is vital for the production the sound finished beam is the amount of the work applied on the tip or toe of the flanges. If no work is applied, then there will be a tendency of the steel to get tear, causing cracks to form. That 'why alternate working on live and dead flange are given in beam design. As the finishing pass is approaches, reduction has to be reduced due to following reasons:

- To eliminate any unforeseen or unpredictable results, most likely to be caused by heavy reduction.
- With the decrease in the angle of flange, the higher reduction cannot be applied in later passings, as higher work on sides will cause the wear of flanges get faster and it will force to take higher off during restoration of the diameter or the change of pass.

It can also be explained by the fact that because of differential speed of different parts of the flange due to difference in diameter at various points of flange and web set up friction, resulting in wear (Fig. 10.9).

5 Important Considerations for Roll Pass Design of Beam

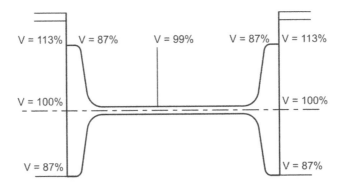

Fig. 10.9 Differential Speed at different Elements of Beam.

In addition, due to bar gets colder at later stage of rolling; it will further aggravate wear of pass. That's why, rolls used in the finishing stand are of hard cast iron rolls, having good wear out properties, but at the same time, these cast iron rolls are not sufficiently strong to provide higher reduction.

5.7 Buckling of Web

It has been stated that the rapid reduction should be made on the web, so that proper proportioning of work between web and flange can be attained as soon as possible in the reduction schedule. The need of such proportioning also causes the problem of "Buckling of the web". This condition is most prevalent on rolling of thin web and heavy flange draughts. Thin web is not able to pull along. Having a larger diameter at the web, the web has a tendency to throw. This results in the buckling of the web. On the contrary, if heavy reduction is applied on the web, the web cannot push the flange which is already very thin at that time and neither had it spread to any further extend as the spread allowance from pre-finishing to finishing pass is very small. In order to relieve itself, web develops small weaviness, which is called "Buckling of web".

Buckling can also be formed by flanges, got stick in the rolls and not gets released readily from the roll, with the result; web tends to run out into the elongation. This condition can be eliminated by giving very small draught on the web in later passes, so that having web pulled by the flanges, rather than flanges pulled by the web.

5.8 Fillets

It should be given special consideration during the design of beam. First there is a fillet at the junction of web and the flanges. If too large, the work placed on the fillet will result in pulling down or the robbing other flange height. If the fillet will be too small, galling may result, which could facilitates the rupture between flange and the web.

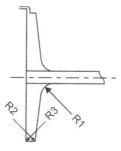

Fig. 10.10 Fillets at Beam Design.

Second, fillet will be at end or what is commonly termed as the toe of the flange. In case of dead hole, working into the live flange, outside fillet is used to eliminate the fin formation at the opening of the roll in the live flange. Fillet on the inside of a live flange is used to give a worked surface to the bar and particularly in the case the finisher stand, to produce bar with good surface finish.

5.9 Thrust

Unequal filling of the flanges in the straight method of rolling of beam produces an end thrust due to the unbalanced side work. In addition, if one flange becomes colder than the other, then also, it may cause end thrust. However, in these cases, the quantum of thrust is so minor that no provision is necessary to be made in the roll and thrust in this design is taken care by the thrust bearing of the roll chocks.

However, in case of the diagonal rolling, "Thrust" must be considered into account. Thrusting causes one roll to shift to one direction and other to another direction, resulting in increased thickness of one flange, while other gets reduced. If too thick flange will become wedged in the succeeding dead hole, it will result in the under filling in the succeeding dead hole and overfilling in the opposite live flange.

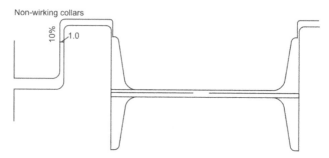

Fig. 10.11 Bearing Collars of Mating rolls to Prevent Thrust.

6 Roll Pass Design of Beam 125 × 70 mm

Due to this reason "Bearing collars" *i.e.,* the inclined collars are cut into the rolls, mates each other and prevent side shifting of the rolls. But, there are two drawbacks of bearing collars:

- They are wasteful from the point view of the roll usage, requiring part of roll body, which otherwise might be used for the working pass.
- They wear rapidly, because of the grinding action of one surface of roll against the opposing turning surface of the mating roll. It is for this reason, collars are always inclined, and this inclination reduces the working surface.
- At times, roll is required to be redressed because of the wearing of the thrust collars, though the condition of the working pass at that time remains good. It is necessary, because any wearing of the thrust collars will result in the change of the live flange thickness. Lubricants are generally used to reduce the wear of the bearing collars.

6 ROLL PASS DESIGN OF BEAM 125 × 70 MM

First of all, Roll pass designer has to ascertain the customer's requirement regarding tolerance on size, mechanical and chemical requirements and specifications under which product is to be certified, *i.e.,* its chemical and mechanical properties, tolerances on size, weight tolerance, desired surface finish. In addition, details specification and layout of the mill is to be studied thoroughly, under which, the section is to be rolled.

Different users of angle will be having different priorities on specification. Different users will have different specification for moment of inertia on x, y, z axis and radius of gyration etc.

6.1 Selection of Mill

Roll pass designer should know the following, before the start of beam:

- What is the strength and weakness of the mill, in which this product is planned to roll?
- The type of mill, production capacity of furnace, mill proper and finishing section of the mill.
- Size of billet available to roll this section.
- Temperature of input billet and finished product temperature.
- Type of layout of the mill *i.e.,* whether it is an open, semi continuous or continuous mill.
- The details of mill facilities *viz.* mill configuration *i.e.,* numbers of rolling stands and its configuration, numbers of vertical stands, the distance

between stands and in between each group, stand size, type of stand to be used for rolling *i.e.,* whether it is housing-less, open or close type of housings to determine mill spring and ease of roll changing, type of roll used, specified rolling load and stand motor capacity, numbers of tillers in the mill and its spacing from preceding and succeeding stand. Shifting facilities to transfer the bar from one line to another and its length decides the maximum length of billets can be taken up for rolling.

All above mentioned parameters decide the type of roll pass design schedule, optimum numbers of passes to be used in the sequence and also to determine the type of reduction pattern to be adopted.

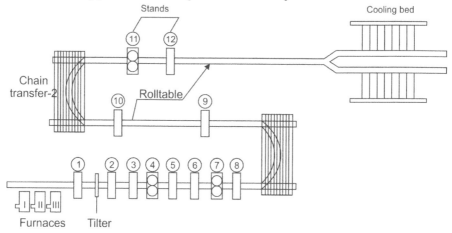

Fig. 10.12 Line Diagram of the Rolling Mill.

Here, the reference mill (Fig. 10.12) is a continuous cross country mill having 3 furnaces of capacity 60T/hr each. Two furnaces are generally operated at a time and the third furnace is kept under repair. Mill is having total 12 stands, out of which three stands (4, 7 and 11) are vertical stands. Mill is sub divided into three groups *i.e.,* roughing (5 stands), intermediate (4 stands) and finishing group (3 stands). Tilter is placed between stand 1 and 2. Metal get shifted from one line to another, with the help of chain transfers. Chain transfer (CT-1), *i.e.,* after stand 8 and CT-2 *i.e.,* after stand 10 are of length of 44 and 77 meters respectively.

Facility is provided for taking on –line sample with the help of a hot saw.

Cooling bed is of the length of 90 M and with the help of bifurcates, material can be transferred to either left or right side of cooling bed. Straightening machines and cold shears are placed on both the sides of rolling line after the cooling bed.

Finishing section is having the facilities for stacking, making compact bundling and inspection and rail-road shipping facilities

6.2 Customer's Requirement wrt Specification, Tolerance etc.

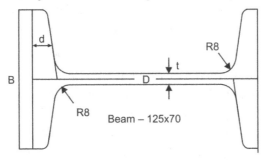

Fig. 10.13 ISMB125, its Element.

Technical delivery conditions (TDC), submitted by the customer decides the design of finishing pass in particular and roll pass design in general, *i.e.*, which specification has to be followed while designing the section and upto what extend one can go with negative tolerance, without affecting finished product properties. If BIS specification has to follow, then following are the details of Beam 125 × 70 as per BIS.

6.3 ISMB–Details of Specification as Per BIS

Table 10.1

ISMB 125 × 70	Sec Wt. kg/M	Area Cm²	Depth D, mm	Breadth B, mm	Thickness of Web t, mm	Flange Thickness d, mm
	13.30	17.0	125	70	5.0	8.0

Tolerances as per BIS on Beam 125 × 70

    (*i*) Weight tolerance: ± 2.5%

    (*ii*) Depth of beam: ± 2%

    (*iii*) Breadth of beam: (+) 3 mm

                                      (–) 2 mm

    (*iv*) Flanges out of square/

          Out of parallel: 1 in 60

    (*v*) Camber: The maximum permissible camber of beam is 0.20% of length.

6.4 Design of Finishing Pass

Passes are to be designed on hot dimension, while the above mentioned BIS dimension is of cold dimension. Hot dimension is generally taken 1.010 -1.015 times more than of cold dimension at the temperature above 800°C.

    But to match marketing strategies *i.e.*, customer's demand to have roll product with negative tolerances to have more length of finished product with the same weight for cost saving purpose. Roll pass designer always keeps hot dimension of different elements of pass are to either equal or even less than the cold dimension, to take advantages of negative side of weight tolerance (–2.5%).

Following are the criteria for selecting different element of pass:

(a) **Web thickness:** Cold dimension of Web thickness (t) given in the design is 5 mm. Web thickness should be such, to take care of the Mill spring, which is directly proportional to the rigidity. Rigidity depends upon the selection of stands i.e., whether it is conventional or housing fewer stands.

Mill Spring of a conventional stand varies from 0.3-0.5 mm, then web thickness d, will be assumed as, $d = 5 – 0.4 = 4.6$ mm.

In addition, designing with thinner web will also consider problems like faster wear out during rolling at lower temperature.

(b) **Flange Height:** Breadth i.e., flange height, B is given as 70 mm. Tolerance on flange height, as per BIS is ± 2mm. Hot dimension of flange is always taken more, firstly because roller always faces problem of flange shortness due to pull down due to various reasons explained earlier and secondly height can be easily reduced by pressing the web thickness, Thus, pass height is assumed as, $B = 70 + 2 = 72$ mm

(c) **Tip and Root thickness of flange**

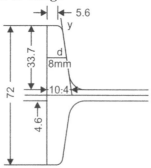

Fig. 10.14 Computation of Tip and Root Thickness.

The average flange thickness, df is given as 8 mm.

In case of beam, the inner side of flange is having a taper of 8°, which comes to 14%. To compute the tip and root thickness of flange (Fig. 13.14),

Flange height on one side = (Flange ht – web thickness)/2
$$= (72 – 4.6)/2 = 67.4/2 = 33.7 \text{ mm}$$
$$y = 33.7/2 \times 14/100 = 2.4 \text{ mm}$$

Tip thickness, $a = 8 - 2.4 = 5.6$ mm

Root thickness, $b = 8 + 2.4 = 10.4$ mm

Live side tip dimension are always kept less than the dead side, due to fast wear out due to relative motion of rolls. That's why, live side tip dimension is taken as 5.5 mm.

Dead side tip dimension has been kept more or less equal to live side tip dimension, because here is less wear out, because it is a dead part of roll that means, there is no relative movement of roll.

Here, it is assumed as 5.6 mm.

Root side dimensions are taken lesser than the computed one. It will give more life to the pass and less off to be taken during the restoration of pass. It is assumed as 10.1 mm and 10.2 mm on the live and dead side respectively.

(*d*) **Width or depth of Pass:** Width of a pass D is given 125 mm as per the design. The allowable allowance is +2 mm. As restoration of the width of pass is not feasible, it becomes the most critical item on point of view of design and roll consumption. That's why, the hot dimension D is taken even less than the cold dimension. But here, it should always be kept in mind that width or depth of pass should not go beyond the negative side, otherwise it will lead to rejection of material.

Here, in this case it is taken as $D = 125-1 = 124$ mm.

(*e*) **Weight of pass:** It is to be computed either mathematically or by auto cad. Area of pass comes to 1660 mm^2, which gives a weight of 13.03 kg/m, against the weight given by BIS specification *i.e.,* 13.30 kg/m.

One need not worry with the computed weight, even if, it is on negative side, it can be adjusted by raising the roll gap.

6.5 Selection of Billet

The use of beam blank is always advantageous from point of view of yield, as a good balance can be maintained between the reduction of web and flange, to avoid differential elongation, in case of using rectangular billet. This tongue formation reduces the yield drastically.

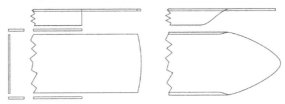

Fig. 10.15 Tongue formed due to differential Elongation.

It also assist in preventing fin formation in roughing stands, otherwise heavy web reduction to be given in rectangular bloom/billet will cause overfilling at the parting.

It is always advantageous of using Beam shaped Bloom/billet will assist in preventing fin formation in roughing stands, otherwise heavy web reduction requires in rectangular bloom/billet will cause overfilling at the parting.

A beam blank as shown in Fig. 10.16 is selected as a input material. It is of rectangular shaped 124 × 125 mm, having a shallow groove formation at the center. This shaped billet is rolled in the billet mill.

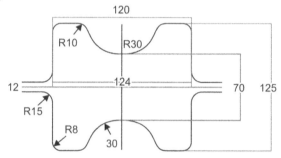

Fig. 10.16 Beam Blank Input for Beam 125 × 70 mm

6.6 Determination of Co-efficient of Reduction for the Different Elements of Beam Design

Roll pass designer has to first establish the balance in co-efficient of reduction between different elements of beam profile.

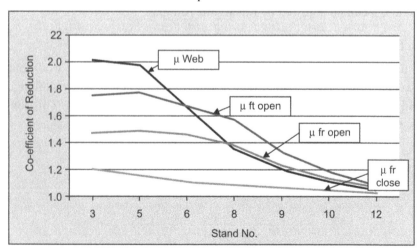

Fig. 10.17 Reduction of different Elements of Beam Design.

6 Roll Pass Design of Beam 125 × 70 mm

Following are the elements of beam design, which has to be computed separately for designing a beam section in a rolling mill.

(a) μ_{Web} – Co-efficient of Reduction of web.

(b) μ_{Ftopen} –Co-efficient of reduction of Flange tip open $i.e.$, on Live side.

(c) μ_{Fropen} – Co-efficient of reduction of Flange root open $i.e.$, on Live side.

(d) $\mu_{Frclose}$ – Co-efficient of reduction of Flange root close $i.e.$, on dead side.

It is the skill of designer to prepare a balance approach in reduction pattern of above mentioned elements of a beam section to achieve the desired dimensions of finished product.

The reduction in finishing pass (pass no.12) is to be kept as minimum as possible to reduce the fast wear out of pass, in turn will avoid the frequent changeover of pass.

μ_{Ftopen} (co-efficient of reduction at the tip of live flange) should be kept slightly more than the μ_{Web} (co-efficient of reduction of web) in the finishing pass to get the correct dimension of flange and to prevent the pulling down of live flange. Some designer prefers to reduce the flange, slightly more than the web in order to setup slight tension in web, which avoids the contraction of flange during cooling.

While rolling in pre-finishing and intermediate group passes (Pass no.10, 9, 8 & 6), the thrust of designer should be always on the correct formation of flange, especially to the height of flange, as there will be always a possibility of pulling down of flange, if web reduction becomes abnormally higher than the flange in these passes.

It is required to give maximum reduction at the tip part of flange so as to metal to flow to the tip part of live flange. It can be seen from the graph that, μ_{Ftopen} is at maximum (1.09), which is followed by the co-efficient of reduction of web (1.04) the flange root on live side $i.e.$ μ_{Fropen} (1.03)

The co-efficient of reduction of flange at close or dead side is not of much importance, as, dead part lies in one roll only, as the name implies. It is also called as dead hole. There will be no reduction at dead side of flange, only height reduction is affected. The incoming live flange enters into the dead hole will be of more height and of slightly less or equal thickness.

In the early forming passes $i.e.$, first web cutting pass and other roughing passes of roughing group (pass-2, 3, 5) of the mill, the maximum reduction is to be given for web formation of beam section. Knifing action of web cutting pass and high temperature of metal helps in cutting-in action of the metal, which is only limited by the angle of bite and the motor capacity of mill

378 Chapter 10 Rolling of Beam

7 COMPUTATION SHEET FOR BEAM 125 × 70 MM

7.1 Computation Sheet for Flange

Table 10.2

Pass No.		Flange Dimensions				Coff. of Elongation	Pull Down h mm	Height Increments In mm
		Flange Height hf mm	Root Thickness Rt mm	Tip Thickness tt mm	Flange Area ff mm^2			
12th	Dead	33.5	10.2 (1.03)	5.6	264	1.05	1.5	–
	Live	34.0	10.1 (1.02)	5.5 (1.09)	265	1.00	–	+1.5
10th	Dead	32.5	10.3 (1.065)	6.0	265	1.11	2.5	–
	Live	35.0	10.5 (1.105)	5.4 (1.165)	278	1.16	–	+1.0
9th	Dead	34.0	11.6 (1.05)	6.3	322	1.08	4.0	–
	Live	35.0	11.0 (1.215)	5.8 (1.275)	294	1.27	–	+1.0
8th	Dead	34.0	13.4 (1.06)	7.4	374	1.12	3.0	–
	Live	38.0	12.2 (1.26)	6.1 (1.39)	348	1.34	–	+1.0
6th	Dead	37.0	15.4 (1.13)	8.5	467	1.16	7.0	
	Live	39.0	14.2 (1.365)	7.1 (1.505)	416	1.43	–	+1.0
5th	Dead	38.0	19.4 (1.155)	10.7	595	1.17	6.0	–
	Live	43.0	17.4 (1.45)	8.3 (1.63)	544	1.49	–	+1.0
3rd	Dead	42	25.2 (1.21)	13.5	813	1.25	4.0	–
	Live	42	22.4 (1.48)	10.6 (1.70)	693	1.52	–	+1.0
2nd	Dead	41	33.2	18	1050	–	–	–
	Live	46	30.5	14.0	1011	–	–	0

Fig. under brackets show the co-efficient of reduction of different elements

7 Computation Sheet for Beam 125 × 70 mm

7.2 Computation for The Web

Table 10.3

Pass No.	Web Dimensions			Web Deformation		
	Width B mm	Thickness d mm	Web Area FW mm^2	Coff. of Elongation μ	Draught Δhw mm	Spread Δbw mm
12th	124 (1.04)	4.6	571	1.035	0.2	1.0
10th	123 (1.08)	4.8	591	1.07	0.4	1.5
9th	121.5 (1.15)	5.2	632	1.14	0.8	1.5
8th	120 (1.25)	6.0	720	1.23	1.5	2.0
6th	118 (1.45)	7.5	885	1.37	3.0	2.5
5th	115.5 (1.62)	10.5	1212	1.58	6.5	3.5
3rd	112 (1.88)	17	1904	1.88	15	4.0
2nd	108	32	3584	1.18	-	8.0

Fig. under brackets show the co-efficient of reduction of different elements

7.3 Computation for Pass height, Pass Area

Table 10.4

Pass No.	Full Height of Pass H mm	Total Height Pull Down	Pass Area F mm^2	Co-efficient of Elongation for Whole Pass m
12th	72.1	.2	1660	1.05
10th	72.3	1.9	1750	1.11
9th	74.2	3.8	1940	1.17
8th	78.0	5.5	2270	1.23
6th	83.5	7.0	2800	1.31
5th	91.5	9.5	3670	1.48
3rd	101	18	5430	1.51
2nd	119		8200	

8 PASS DESIGN OF BEAM 125 × 70

8.1 Pass Design of Finishing Stand (12th)

The pass design and roll design of stand-12 has been done, as shown in Fig. 10.18 and 10. 19 below:

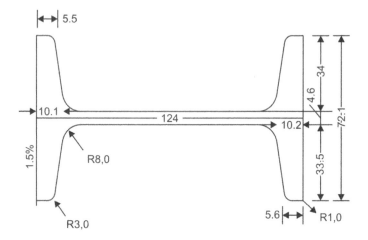

Fig. 10.18 Finishing Pass (Pass-12).

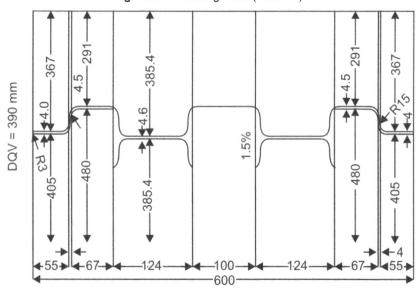

Fig. 10.19 Roll Diagram of Stand-12.

8.2 Design of Pre-finishing Stand (10th)

The pass design and roll design of stand-10 has been done, as shown in Fig. 10.20 and 10.21 below:

8 Pass Design of Beam 125 × 70

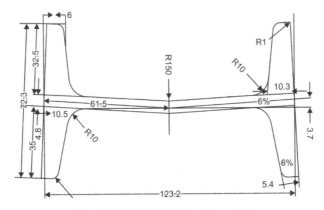

Fig. 10.20 Pre-finishing Pass-10.

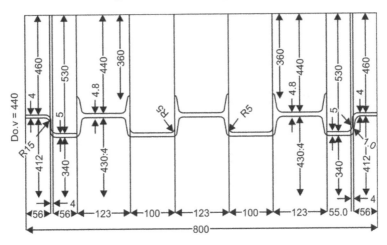

Fig. 10.21 Roll Diagram of Stand-10.

8.3 Design of Strand Pass (9th)

The pass design and roll design of stand-8 has been done, as shown in Fig. 10.22 and 10.23 below:

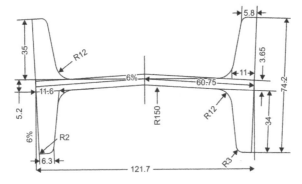

Fig. 10.22 Strand Pass-9.

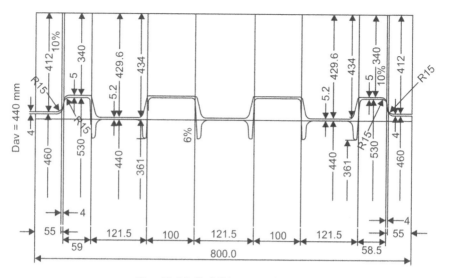

Fig. 10.23 Roll Diagram of Stand-09.

8.4 Intermediate Group (Stand-8)

The pass design and roll design of stand -8 has been done, as shown in Fig. 10.24 and 10.25 below:

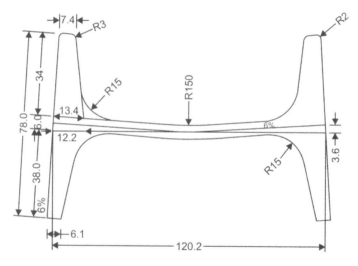

Fig. 10.24 Pass Diagram of Stand-8.

8 Pass Design of Beam 125 × 70

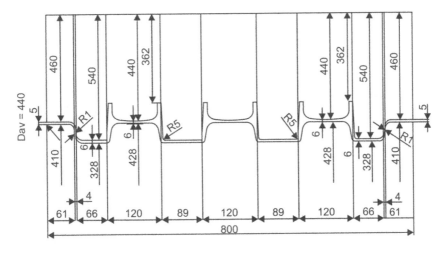

Fig. 10.25 Roll Diagram of Stand-08.

8.5 Intermediate Group (Stand-6)

The pass design and roll design of stand-6 has been done, as shown in Fig. 10.26 and 10.27 below:

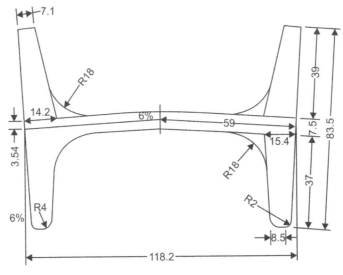

Fig. 10.26 Pass Diagram of Stand-6.

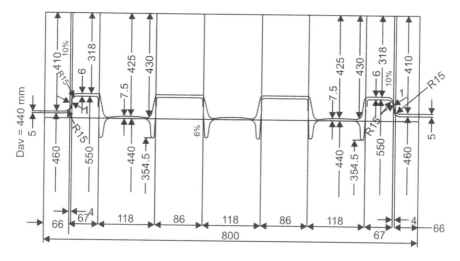

Fig. 10.27 Roll Diagram of Stand-06.

8.6 Design of Pass-5

The pass design and roll design of stand-5 has been done, as shown in Fig. 10.28 and 10.29 below:

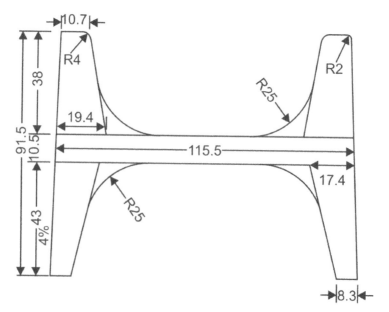

Fig. 10.28 Pass Diagram of Stand-5.

8 Pass Design of Beam 125 × 70

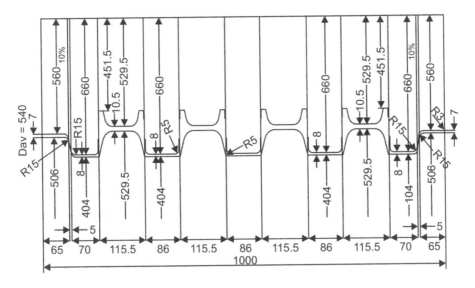

Fig. 10.29 Roll Diagram of Stand-05.

8.7 Design of Pass-3

The pass design and roll design of stand-3 has been done, as shown in Fig. 10.30 and 10.31 below:

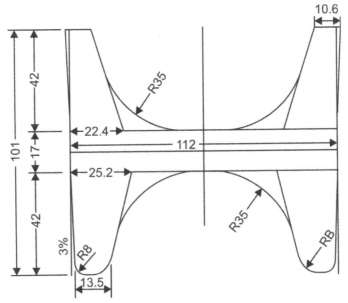

Fig. 10.30 Pass Diagram of Stand-3.

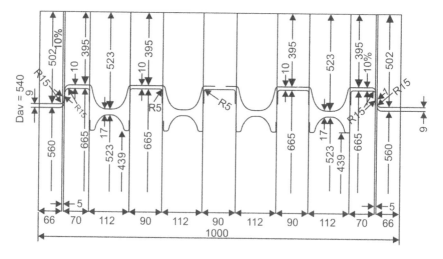

Fig. 10.31 Roll Diagram of Stand-03.

8.8 Design of Pass-2 ("V"Pass)

The pass design and roll design of stand -2 (**"V" Pass)** has been done, as shown in Fig. 10.32 and 10.33 below:

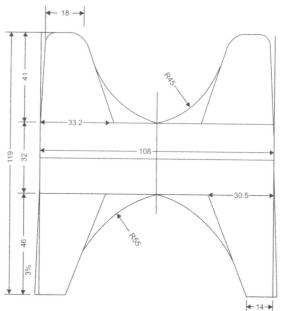

Fig. 10.32 Pass Diagram of Stand-2.

8 Pass Design of Beam 125 × 70

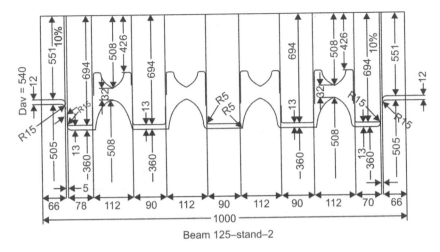

Fig. 10.33 Roll Diagram of Stand-2.

8.9 Design of Pass-1

The pass design and roll design of stand-1 (Box Pass) has been done, as shown in Fig. 10.34 below:

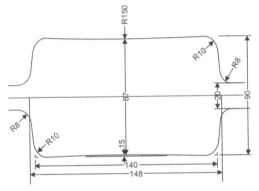

Fig. 10.34 Pass Diagram of Stand-1.

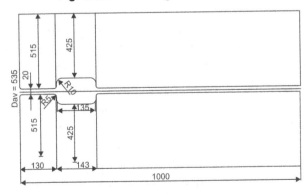

Fig. 10.35 Roll Diagram of Stand-1

9 ROLLING SCHEME FOR BEAM 125 × 70

Input Beam Blank Size- $122 \times 135 \times 80$ mm × 6 m length

Table 10.5

Stand	H	B	d	t	Dk	N	i	Dc
1	100	140		30	451	300	9.58	515/515
	Tilter 90°							
2	119	108	32	12	508	300	8.12	551/505
3	101	112	17	9	523	380	6.96	502/560
4	–	–	–	–	–	–	3.51	–
5	91.5	115.5	10.5	7	529.5	378	4.69	560/506
6	83.5	118.2	7.5	5	428	346	2.90	410/460
7	–	–	–	–	–	–		–
8	78.0	120.2	6.0	5	430	300	2.026	460/410
9	74.2	121.7	5.2	4	432	300	1.88	412/460
10	72.3	123.2	4.8	4	432	300	1.59	460/412
11	–	–	–	–	–	–		–
12	70	124	4.6	4	385	300	1.15	367/405

Note : H- Height of Stock, B –Width of Stock, d- Web thickness
t- Roll gap, D_K –Rolling Diameter, N- Motor Speed
D_C –collar Diameter, i –Reducer Ratio .

10 MILL SETTING OF BEAM

General rules of Mill setting for beam profile.

1. Billets/blooms for rolling beam should be free from scabs and cracks and should have smooth surface. If size of scabs exceeds 20 mm, then such billets/blooms should be rejected at storage yard

2. Billets/blooms with bad, uncroped front or tail end, should be rejected on the charging device itself.

3. Temperature of metal at roughing group should not be less than 1100°C. Billets, which lie near or before the discharge door of furnace (2-4 pieces) should atleast be returned, before the start of the rolling, as these billets are usually cold.

4. Selection of passes on stands is to be made by the Oprn. I/c and Shift Foreman (Opr.), in a shift, incase of emergency. Checking of the close part of the pass (groove) is to be made with the help of templates. The width of the previous pass should always be less than the width of the next pass or in certain cases like cutting-in pass, may be kept equal to that. It is forbidden to use passes of the previous stand with width more than of

11 Control of Weight of Beam Section 389

the next. The information about the width of passes received from the roll turning shop should again be counter checked in the shop.

5. While fixing the gap between rolls, special attention should be paid to the accuracy of the gap at the end runners of rolls.

6. Final setting of the rolling tackles is be done only after the final setting of gap between the rolls and checking the same with filler gauge.

7. Entry and delivery guides should be fitted according to the maximum width of the closed pass of every stand.

8. In case considerable changes of gap between rolls is required during the process of mill setting, then it will also be necessary to check the setting of guards again.

9. At the time of mill setting, dimensions of the bar should be checked at places, where measurements can be directly taken on the hot metal, otherwise, piece is to be rejected for checking of dimensions after each group *i.e.,* roughing, intermediate and finishing group of mill. Piece after pre-finishing stand may not be rejected, as in this case, sample can be taken from the saw, keeping finishing stand in idle running. During checking the following dimensions should be checked:

   (*a*) The web thickness,

   (*b*) The flange height,

   (*c*) Approx. flange thickness.

   After the final adjustment of mill, slight changes in reduction is only allowed be allowed after judging the loads on stand.

10. Final estimate of the section, in case of beam may be arrived after examination and checking of the dimensions of the sample and pieces at cooling beds. When rolling beams, it is necessary to stop the flow of water on the bar in stands of intermediate and finishing group, otherwise the web gets cooled easily.

11 CONTROL OF WEIGHT OF BEAM SECTION

1. Weight on the piece depends of the:

   (*a*) Thickness of the web

   (*b*) Thickness of flange

   (*c*) Filling up of inside corner

   (*d*) Taper of the leg

   (*e*) Leg lengths

   (*f*) Width of the piece.

Roller has very little control over (*b*), (*c*) (*d*) and (*f*). So to keep the weight within tolerance, Roller has to control the web thickness and leg length. If web thickness is decreased by pressing finishing and pre-finishing stand, then leg lengths also get slightly decreased and weight becomes less. On the contrary, with the increasing web thickness weight obviously becomes more.

Thickness of web and leg lengths can be decreased upto a certain limit, afterwards there is no other way, but to change pass.

Roller should see that passes should last long to give maximum possible life. Following parameters should be continuously monitored:

(*a*) Uniform load on the stand.

(*b*) The temperature of the piece.

(*c*) Matching of passes, especially for passes of pre-finishing and finishing stand.

(*d*) Uniformity of the profile from every stand of the mill.

(*e*) Overfilling and under filling of passes should be avoided.

(*f*) Guards should be checked periodically to avoid marks in the passes.

(*g*) Pass should be changed when a slight mark is noticed on the pass.

To avoid the mark on the pass, it should be seen that guards should not press too much on the pass, may be due to more counter weight or with the use of heavy guards. Guard should not be too short and setting of the guards should not be such that, it may act as a cutting tool. The sitting of the guard on the pass should be made uniform; the material of the guards should not be too hard.

In case of beam, roller has control over (*i*) web thickness and (*ii*) leg lengths but does not have much control over leg thickness.

To control the web thickness and leg length the methods of control has been explained earlier. Here, roller should be borne in mind that alternate passes are close at top and open at the bottom. To control the leg length, both the top and bottom flange should be taken into consideration separately. When both flanges are found short and are required to be increased, then metal should be increased on flanges from roughing group. When bottom one side flange is to be increased, the top close pass rolls to be adjusted towards opposite side. Similarly, in case of one flange adjustment rolls should be adjusted as stated above.

Filling of corners is important. The top roll of previous stand should be shifted opposite to the corner and the top roll of the stand from which under filling is coming to be shifted towards the corner. In case of bottom corner the entry guides may be adjusted accordingly.

12 SPECIAL PRECAUTION TO BE TAKEN DURING BEAM ROLLING

The sticker mark on the flange is noticed very often during rolling of beam profile. Sticker marks comes due to sticking of metal on the pass. This sticking is mostly due to wedging action in the pass. If the thickness of the flange is more from previous stand the wedging action will be more. If one side flange is having frequent sticker mark the probable reason is as explained above and that flange thickness should be reduced. If both flanges are found thicker than previous stand then pass should be changed immediately.

The second cause of sticking is due to loose portion of metal on the flange. This is probably due to scabby metal. If scabby metal is seen while it is passing through roughing and intermediate group it should be rejected.

13 DEFECTS OF BEAM AND ITS RECTIFICATION

Fig. 10.36 Defects of Beam.

1. **Legs are short (both):** This is due to less metal, coming out from roughing group. Too much hot metal and tension should be avoided. Metal on the leg should be increased, from the roughing group. If the pass of pre-finishing stand wears out beyond the limit, then also leg thickness will be more and legs will become shorter. The pass should be immediately changed.

2. **Legs are longer (both):** This is due to just opposite action as mentioned in (1) above.

3. **Legs are unequal:** This is due to mostly wrong setting of entry guides in first cutting pass, so guide should be set properly. Guides should move towards the shorter leg side. If setting of entry guide is correct, then the top roll is to be taken towards longer side.

   When difference is less, the top rolls of finishing and pre-finishing stand should be shifted. If difference is more, then entry guides and rolls of roughing groups are to be adjusted. If necessary, one piece may be rejected after the roughing group to check the flange lengths and adjustment are to be done accordingly. If the piece fed twisted to the stand after the tilter, then also, one leg may be become shorter and other gets longer.

4. **One leg is thicker and other is thinner:** The top roll of finishing stand is to be taken towards thinner side.

5. **Both corners are blunt.**

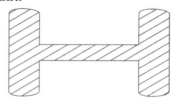

Fig. 13.37 Both corners are blunt.

This is due to not filling up of corners of the pass of finishing stand.

This may be due to less metal from pre-finishing stand. In this case, the top roll of pre-finishing stand should be raised. If the finishing stand pass is not filled properly due to less reduction, then top roll of finishing stand should be pressed. If the turning of pre-finishing pass is not correct *i.e.*, if corners are turned less, then the chances of blunt corner will enhance. To avoid this trouble, the pass of pre-finishing stand should be checked. If the pass of pre-finishing stand is too narrower than the finishing pass, the rounded corners are expected. In this case, wider pass for pre-finishing stand or narrower pass of finishing stand is to be taken. Rounded corners may also be due to worn out pass of finishing stand, pass should be changed. Worn out pass of pre-finishing stand will help to fill up the corners.

6. **One side of corner is blunt:** This may be due to above reasons and if the rolls of finishing or pre-finishing stand is cross. Top rolls should be moved towards blunt side in pre-finishing stand and opposite to blunt side in the finishing stand.

7. **Bend web:** This is due to less reduction in finishing stand or more reduction in pre-finishing stand. Adjustment should be done accordingly. When the finishing stand pass is worn out, there is a chance of bend web. Pass should be immediately changed.

8. **Inside corners are not filled up:** This is due to reasons explained in 5 and 6 above.

9. **Waviness of the web:** This is due to unequal reduction between flange and web. Sudden more reduction in one web (pressing of top roll) will also result in this.

13 Defects of Beam and its Rectification

10. **Wavy Flange:** If there is excessive work on the flanges, they try to elongate more than the web. Because the web is holding them back, the extra length has to go somewhere so the flanges become wavy at the tips. At the extreme, the flange elongation can cause migration of material from the web to the extent of pulling holes in the web. This defect is generally observed in universal rolling of beams.

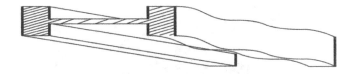

Fig. 13.38 Wavy Flange.

Remedies are:

1. Tighten rougher vertical rolls and/or
2. Loosen rougher horizontal rolls or
3. A combination of these depending on loads and finished bar dimensions
4. At finisher, tighten horizontals and loosen verticals

11. **Under filling mark on the top:** This is due to reasons explained at 5 and 6 above.

Fig. 10.39 Under Filling Mark on the Top.

12. **Worn out mark on the flange, web, and fire crack mark on inside corners:** In case of less worn out mark on pass can be rectified by grinding and pass can be used again. The bottom roll grove of all stand gets heated up very quickly as accumulation of heat takes place at that point, water cooling of bottom roll should be very efficient, If ,it is not removed, then pass should be immediately changed.

13. Scratches/Grooves

Scoring / scratching appearance on finished product. Beams are particularly prone to this problem.

Fig. 10.40 Scratches.

- Mainly mechanical marking

Remedies are:
- Check guides/tackle
- Check troughs, deflectors
- Check for scale build up in rolling line

14. Holes in Webs:
Result of web being stretched due to too much flange work

Possible causes
- Draughting ratios out of synchronisation between web and flanges
- Buckle appears due to either too much work on the flange or insufficient web work (inverse of web buckle).

Remedies
- Adjust roughing mill to allow more stock into finishing mill.
- Adjust finishing mill to balance web / flange ratio.
- Check for possible interstand tension in a continuous mill arrangement

15. Off Centre Web:
It is due to following reasons:

Off centre web is called, when top and bottom flange heights are different beyond permitted tolerance limit. Possible causes are:
- Rolling line is either too high or too low.
- Intermediate stock imbalanced on tongue and groove passes.

13 Defects of Beam and its Rectification 395

Fig. 10.41 Off Centre Web.

Remedies

- Adjust rolling line, addition of packers/running in plates or removal of stand and removing liners.
- Check roll alignment, guides and guard setting of roughing and intermediate group to get balance stock as best as possible.

❑❑❑

11

Rolling of Rail

1 INTRODUCTION

Rails provide the running surface for the movement of locomotives, wagons and coaches to move men and materials. Railways are the only means of transport, whose name is closely linked with the basic material they use *i.e.,* the steel rail. Railways came into being at the same time as the iron and steel industry and have developed together throughout the 19th century proving to be the basic factors of extraordinary economic development. Since then, the rails have remained the symbol of railway transport and the progress of railways and steel rails have followed completely parallel tracks.

The ever increasing growth of rail roads in developing countries around the world alongwith the advent of larger and faster locomotives have led to a continuous development of rail rolling practice to achieve better dimensional tolerances, surface qualities and physical properties of the steel rail. These developments have also been fuelled by increasingly stringent inspection requirements being setup by railway authorities to ensure safety on the railroads.

Railways are basically a development of the tram ways and plate ways of the 18th century. These, originally consisted of two parallel lines of slabs, stones or timber, laid flush on a road surface to facilitate haulage of heavy loads by horses. These were mainly used in mines and the early iron works. The changeover from stone or wood to iron took place alongwith the industrial revolution in the 18th century. Various shapes of rails were in use during the early years. Some of these are shown in Fig. 11.1.

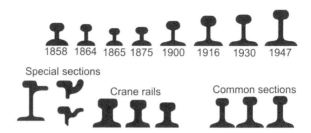

Fig. 11.1 Developments in Rail Sections.

The presently used rail shape with a wide flange base was first developed by Stevens in America in 1832. Large scale usage of rails was started around 1858 and the running weight of rails had reached to 30 kg/m starting from 12 kg/m for the earliest tracks. Steady developments leading to increased axle loads, as more powerful locomotives become available, have increased the weight of rails in use today to 60 to 70 kg per meter.

While the railways were able to achieve considerable technical progress in the first half of last century, simultaneously there were major developments were also made in road transport and aviation industry, which has caused considerable shifts in the volume of passenger and freight traffic handled by the railways to these industries. As a consequence most railways fell into a slump after the second world war. The energy crisis of the seventies has however created a new impetus for rail transportation which has the lowest specific energy consumption as compared to road transport or shipping may be seen from Fig. 11.2.

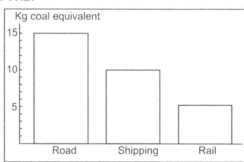

Fig. 11.2 Specific Energy Consumption vis-à-vis modes of Transport.

The advent of faster and heavier rail transport has led to requirements of better material qualities, heavier sections and stringent inspection procedures. The railways are meeting the challenges by increasing the speed and haulage capacities of trains. Notable work in this regard has been done in Japan and France. This in turn has created severe loading conditions on rails resulting in fast wear and frequent failures. Rail maintenance work and rail replacements have become more frequent. This has led to demands for rails with better properties and these are being achieved through developments in feed stock, shaping practices, heat treatment and rail finishing equipment.

2 DEVELOPMENT TRENDS

Blooms rolled from semi killed steel ingots produced by the open hearth and Bessemer process were the traditional feed stocks for rails. Improved practices resulted in killed steels also being used for production of rails from the early sixties. The basic oxygen process of steel making which also emerged in a big way around this time was soon adopted to meet the needs of rail steels.

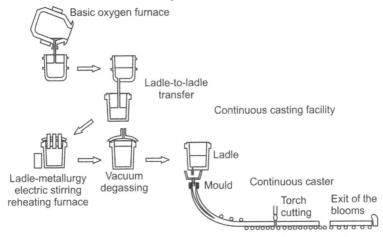

Fig. 11.3 Continuous Casting Practice for Rolling Rails.

The major development in the last few decades has been in the use of the continuous casting process to directly cast rail blooms of killed steel quality. Trials with rails produced from continuous cast blooms were held in the mid-sixties and as these rails were found acceptable, the trend towards continuous casting has gained momentum. Comparison of steel quality show that the continuous cast material to be markedly superior in surface quality and macro cleanness as compared to the ingot route material, as shown in Fig. 11.4.

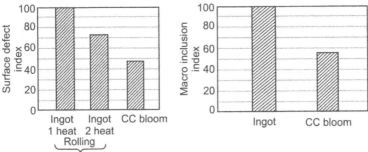

Fig. 11.4 Comparison of Quality of Continious cast vis-à-vis.

400 Chapter 11 Rolling of Rail

The tendency towards more pronounced center segregation has been alleviated by optimizing the rectangular section and using low superheat casting practice. Electromagnetic stirring is also applied for greater freedom from temperature control. Another development has been the use of vacuum degassing techniques for rail steels. Besides giving very uniform chemical compositions over large heat sizes, degassing practice considerably lowers the hydrogen content of steel so that slow cooling of the material at the bloom stage or after rolling to rails is avoided.

On account of the favourable properties, virtually all important users have amended their standards to accept continuously cast material.

3 RAIL SHAPING PRACTICE

The process of rail shaping involves rolling the ingot/bloom down to a minimum required rectangular section, followed by 5-6 intermediate shaping passes and finally a finishing pass to obtain the desired rail section. The rail section is typical of its kind having a bulky and narrow head to support the railways wheel, a wide and strong base for fixing onto the sleepers and a tall and strong web to connect the head to the base. Whereas various rail profiles have differences in the amount of metal in the head as compared to the base, the web section is always the smallest.

3.1 Conventional System

Several types of pass sequences are employed in rail rolling. This is associated with the specific features of production in each plant. Nevertheless, all of these sequences have certain features in common.

In accordance with their form, all passes applied in the rolling of rails may be divided into two types (1) hat passes and (2) edging passes.

Hat passes serve to form the base of the rail; three or four such passes are used. The first of the edging passes is of the web cutting type (with knifing). There are usually from 5 to 7 edging passes.

The rough profile of the rail is formed in the hat and first web cutting passes. These are known as the shaping passes. The rest of the edging passes impart the final form and size of the hot rail to the bar.

The mean co-efficient of elongation per pass ranges from 1.20 to 1.22 and in the finishing pass-from 1.07 to 1.09.

Typical rail pass sequences are illustrated in Fig. 11.5

3 Rail Shaping Practice

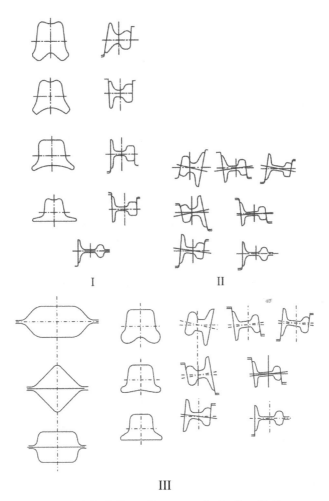

Fig. 11.5 Pass Sequence for Rolling Rails.

Sequence I is designed for rolling rails in nine passes of which four are hat passes and the rest edging passes. In this sequence the first edging pass for web cutting is positioned diagonally to enable considerable side work to be performed and to reduce roll wear.

The advantages of diagonal positioning for a web cutting pass was thus demonstrated in practice and this served as an impetus to the development of diagonal pass sequences in rail rolling.

Sequence II is a diagonal pass design system. The application of diagonal pass design for rails led to a reduction of the number of passes, less roll wear and prolonged roll life.

In the present times roll pass design aims at devising methods for rolling rails in which the base and head are formed from metal located at the corners of the bloom. This is based on the fact that the corners of the bloom are free or almost free of surface defects and the structure of the metal is fine-grained and dense. In this connection, it is proposed an interesting solution, *i.e.* to include an oval-square sequence in the rail passes.

Sequence III is based on oval-square pass design principle. Rolling in an oval and then a square pass renews the corners of the bar and as a result the metal of the corners on the initial bloom becomes the head and base of the rail.

3.2 Basics of Rail Design

The first step in rail pass design is to select the pass sequence. Next step is to determine the number of passes, which will be in accordance with the type and size of rail, available power of the mill drive motors and the capacity of the rail and structural mill. Generally, nine to eleven passes are used in most modern mills of this type.

The next step is to distribute the co-efficients of elongation among passes and the elements of the rail profile.

The co-efficient of elongation for the web in the finishing pass is usually assigned in the range from 1.05 to 1.15, the co-efficient for the head is slightly larger.

Co-efficients of elongation for the web cutting pass may be as high as 1.5 or 1.6 for the web, up to 1.30-1.45 for the flanges and upto 1.4 or 1.5 for the head.

Intermediate values of elongation, between those for the initial and finishing passes, should be assigned for the various elements of the profile in the roughing passes.

The amount of pulling down for the flange in the dead hole is taken as 5 to 7 mm for the finishing pass and 7 to 10 mm for the other passes. The increase in flange height in the live holes is taken as 0.5 to 1 mm.

The depth of the head is to be reduced in all passes: from 1 to 3 mm in the finishing pass and from 3 to 5 mm for the dead half and 1 to 2 mm for the live half in the other passes.

A spread of 2 to 3 mm is assigned in the finishing rail pass.

Since, the top of the rail head should be of convex shape, the pass walls forming the head are to be inclined for the finishing pass (Fig. 11.6).

These inclined walls are tangents to the required curved from of the rail head. They are drawn from a point on the rail axis, 1 to 1.5 mm from the highest point of the curvature of the rail head. The convex surfaces of the rail head is obtained in the finishing pass as a result of spread of the metal. A clearance of 8 to 5 mm is provided between the rolls and a radius of 6 to 8 mm where the side of the pass goes over into the roll parting.

3 Rail Shaping Practice

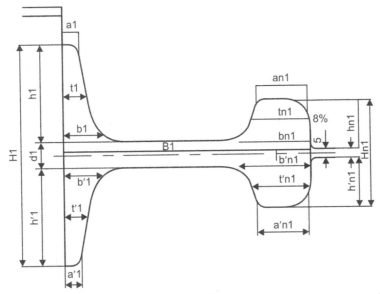

Fig. 11.6 Design of a Finishing Pass.

The upper flange is live and the lower dead to avoid the necessity of using double guides.

A finishing rail pass, used in rail rolling has the pass axis inclined to the roll axes. The break, however, was left at the center of the head to obtain convexity. Satisfactory results were obtained by this design.

The remaining passes are designed on the principles discussed previously.

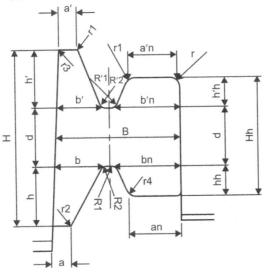

Fig. 11.7 Design of a Web-cutting rail (edger pass).

Certain features may be noted for calculating and designing the first web cutting edging pass (Fig. 11.7).

The web cutting element is designed with an angle of 50° to 60° and its vertex is rounded off. The axis of this element is not in the center of the pass width as the case in beam rolling, as the thickness of the head always more than that of flanges. The position of the web cutting element is determined from the head-to-base thickness ratio which is generally equal to 1.7 to 2 in the finished rail, it will prove more expedient to increase this ratio for the web cutting pass, as halves of the head are always reduced more than the flanges during the course of rolling. Even If the above mentioned ratio for the web cutting pass is taken in the range from 1.8 to 2.5, then also no difficulty will be encountered in finding the position for the vertical axis of the web cutting element.

The web thickness in the web cutting pass must be checked for the bite condition *i.e.* it is necessary to see whether the rolls will accept the bar.

The hat passes differ from the edging passes in that here the bar is subject to extremely non-uniform deformation. Any non-uniform deformation may become the cause of internal stresses in the metal and may lead to the formation of cracks or fissures in the finished rail. The plastic properties of the metal are much higher in the beginning of rolling than at the end. Therefore, maximum non-uniformity of deformation should be applied at initial passing so that the deformation employed at near the final passing will become as uniform as possible.

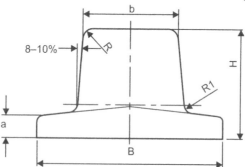

Fig. 11.8 Design of a hat pass for rolling Rails

The form of hat passes employed at present time in rail production is illustrated in Fig. 11.8. A knifing action is used to cut the stock to a depth up to 35 to 65 mm and leave two inclined protrusions which are later gradually opened out and thinned to form the flanges. This improves the working of the middle part of the base and helps to break up the initial columnar crystal structure in the metal.

In addition to direct draught, side work is also applied in hat passes. In general, the working of the metal in hat passes ensures fine-grained structure of the metal and high mechanical properties of the finished rails.

3 Rail Shaping Practice

The design of the hat passes is based on the dimensions of the web cutting (edging) pass. The height of the hat pass is easily determined from the width of the web cutting pass by the following formula .

$$H = B - \Delta b$$

Where: Δb is the spread in the web cutting pass and B is the width of web cutting pass.

The spread in the web cutting pass is usually taken from 8 to 10 mm. The width of the base in the hat pass is taken 7 to 15 percent more than the web cutting pass :

$$B = (1.07 \text{ to } 1.15) H$$

This larger base width will cause for the pulling down of the flanges in the web cutting pass. The width of the head in the hat pass is equal to or slightly larger than the head depth in the web cutting pass

$$b \geq Hh$$

The walls in the hat passes are inclined 8 to 10 percent.

The thickness at the flange tips in the hat pass equals to that of the edging pass.

After designing the hat pass, the angle of contact is to be calculated for the web cutting pass.

The other hat passes are designed to provide for both direct draught and side work, though later passes will give reduction in magnitude.

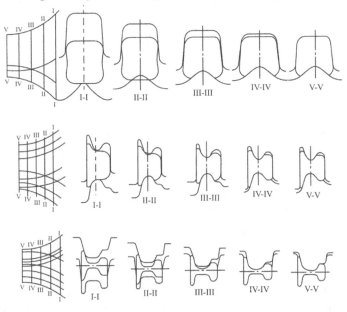

Fig. 11.9 Stages of Rolling Along The Zone of Deformation.

406   Chapter 11   Rolling of Rail

When rectangular bloom is rolled in the first hat pass, severe non-uniform deformation occurs along the width of the bar. Through the elongation of the bar as a whole will itself be very high. Certain parts of bar are subjected to considerable draught. The gradual knifing action of the lower roll and the filling of the pass is shown in [Fig. 11.9 (a)]. From the moment the bar enters the rolls, is subject to considerable side work. At the same time, the bar is knifed by the V-element of the roll.

The consecutive stages of deformation in the web cutting pass are illustrated in [Fig. 11.9(b)]. Large draught at the middle of the bar will cause induced spread and pulling down of the flanges. The deformation process in an intermediate edging pass is shown in [Fig. 11.9(c)]. Here, work on the web begins after reduction of the head and flanges.

4 SALIENT FEATURES OF RAIL PASS DESIGN

While in a few instances, minor shaping can be given in the input bloom itself, but in the majority of cases the first shaping operation in the development of a finished rail takes place in the roughing rolls. This roughing and shaping is performed in several ways.

4.1 Draughting

Draughting is one of the major considerations given in the design of all rail passes. The ultimate aim is to develop the shaping in such a manner, so as to permit proportionate draft over all the component parts of the rolled section and through the reduction schedule. In the case of rail, the only way by which proportionate draughting can be accomplished to throughout the reduction schedule would be to start with rail shaped bloom. Otherwise, when it is not possible to use shaped bloom, it is always the endeavour of roll pass designer to have proportionate draught and consequently equal elongation of three parts – head, web and flange shape should be effected as soon as conveniently possible in later part of rolling. Mill layout also dictates the choice of convenience of selecting a draughting schedule and it is because of these variations in layouts, among other reasons, that the rough shaping of rail in roll pass design is approached differently on various mills.

After years of development, the roll designer and mill operator has come out with the best possible reduction sequence, the rolling methods. Present day, pass design of rails has become fairly standardized atleast to the extent that the so called slab edging method is used almost excessively. In using this method, however, there are differences as to the total number of passes required to produce the finished rail. At times these differences can be reconciled by the

4 Salient Features of Rail Pass Design

fact that many different sizes of mills are used to produce the same size of finished rail.

4.2 Design Efficiency

Efficiency in design of roll passes is not merely based on producing a section to the required tolerances, but the design must take into account the mechanical feature of mill and use these features so as to produce the bar in minimum time with minimum of operating delays. Consequently, the number of passes used for the design for producing a rail section is not considered as criteria of best design. Reduction in total number of passes can't be considered, always a means to increase productivity of the mill. For example if a reduction schedule with lower number of passes with higher reduction has been designed in a such a way, which may require greater power consumption, which may in turn force the roll pass designer to roll the bar at minimum rpm, thereby it will be actually requiring more time for rolling than can be experienced in rolling with more number of passes with lower reduction and higher speed of rolling. In this case, the problem of difficult biting may also not arise. Cobbling of bar while entering the pass with heavy reduction and at high speed causes operating delays, which may be nullified with the anticipated increase in production with reduced number of passes.

4.3 Roughing Pass Design

To consider the design of roughing passes for producing rails, it is always desirable to perform extra work on the head in order to produce denser structure which, in turn, will result in improved wearing qualities. It is impractical to perform extra work at later passes of rolling, as proportionate draughting should be increased with the approaching the finished cross section of rail. Because of this, the great changes in shape of the bloom must undergo at roughing stage of rolling itself. There are practically no pass in the roughing schedule where work can be applied equally to all parts of section, with this work on one part of section soon may become out of balance, compares with work on other part. A flow of metal is setup from one part of the pass to another. The great difference in shape is evident due to above reasons. The control of this flow and keep its in balance is the real problem faced by the designer in designing workable passes for rolling rails.

In the roughing design, the flanges must receive most careful attention. Their development in the slab and edge method is accomplished in number of ways, but two passes are common to all approaches of roll pass design. These are used in various position and different numbers of times. First is the hat pass and second is the pass formerly referred to as the open edge pass in roughing. In the latter part of reduction schedule, the hat pass is referred as dummy.

4.4 Combination Rolls

An important feature to consider in the pass design of the roughing rolls for rails is the combination roll–that roll which is designed to produce sections of more than one size. It is a set of roughing rolls used into produce the rough shape for various sizes of finished sections. By designing the rolls to produce sections of more than one weight, the roughing roll need not have to be changed while section changeover, thus make it available more time for productive rolling. It requires proper scheduling of pass design to obtain the maximum results from such an arrangement.

The design of roll entails more than the knowledge to produce workable passes:

- Only by analyzing the frequency of rolling a section in conjunction with sales and order department, roll designer can designs it a combination rolls in roughing.

- In addition to eliminate downtime for roll changing, the ability to produce more than one section from one set of roll helps to keep the roll inventory to a minimum.

The important point where many sections are being produced *i.e.*, if without proper consideration, passes are placed in the rolls, just to make a combination, then actual loss may be encountered due to imbalance use of passes in combination roll. This not only adds to the rolling cost by virtue of the cost of dressing of rolls, but also adds to the cost by loss of roll life, since a duplicate working pass might well have been placed in the rolls instead of seldom used pass.

4.5 Proper Pass Distribution

Another factor which must receive careful consideration is proper pass distribution. It is essential for obtaining maximum production. By placing too many passes only in one set of rolls, the following roll sets may become idle for the greater part of time. It requires a careful study by roll pass designer to balance the rolling time, plus the handling time of full series of passes. Every set of rolls should produce rolling tonnage as uniform as possible. Naturally, there are limitations to it also, can be imposed by physical arrangement of stands, length of roll body, available power etc., over which designer has no control. After taking into account the various limiting factors, a suitable balance will derive the most effective design.

Now question arises "How many passes are required to produce section". In case of rails, it depends upon a number of factors, among which:

- The type of mill available for effecting the necessary reduction.

4 Salient Features of Rail Pass Design 409

- The power available, the design of table and other physical features/ equipments of mill layout.

Generally, passes to roll rail varies from 9 to 11. While there may be a difference in number of passes employed in different mill, but there will not be much major variation in rail design.

A general procedure to follow in design of rail passes is to develop the shape of section as quickly as possible, so as to approach proportional draughting as rapidly as practical in the reduction schedule. First shaping is usually performed in four to six passes, because unequal reduction in later passes will set up stresses, which may likely to produce rupture in steel.

In addition to the difference in total number of passes used, there will be slight variation in the distribution of passes in the rolls from mill to mill. The layout and physical conditions again decides the most suitable arrangement.

4.6 Size of Bloom

In the rolling of light and heavy rails from the same size of ingot, then internal structure of the finished rails will not be similar because with a decrease in temperature due to higher number of passing, the effect of draught will be different. On this basis of experience, it is generally suggested that the cross sectional area of bloom should be around 15-20 times that of finished rail.

4.7 Rolling Temperature

Unlike other sections, rolling temperature for rail production should be very closely controlled. Rolling at lower temperature improves the quality and strength of steel. The mechanical properties developed in rail must meet rigid rail's specifications. These mechanical properties are equally as important as the development of well-designed rail passes. At higher temperature, the influence of reduction on the steel is less effective. However, as the finishing end of the reduction schedule gets approaches, the effect of draught on the mechanical properties, particularly on the flange, is especially noticeable.

4.8 General Consideration

There are a number of features in rail design, which must be addressed; if rails of prime quality are to be produced, then head must have the proper contour and exact width and rail flanges must also be of proper width, in order to fit on the base of tie plate. The flange or base must be flat as well as of proper width to prevent the rail from rocking.

Cambering is a peculiar problem to rail rolling, adds another operation, known as cambering, which is performed on a straightening machine. It is well known fact that thinnest part cools at fastest rate. In case of rail, the head contains the larger mass and therefore, retains the heat to the highest. Consequently contraction at the head will be more than that of flange, with the result that in cooling, the rail would naturally sweep or pull towards the head, causing a very high bow formation in rail which would make it extremely difficult for further processing. To overcome this condition, after rail has been sawed to length. It is passed to cambering/straightening machine, which give a reverse curve in rail, so that during the process of cooling it will straighten it.

4.9 Mill Setting

There are many conditions prevailing in rail rolling, which makes it impossible for the acceptance of the finished roll products.

For example, suppose the top flange of the finished rail is running short. In an endeavour to correct this condition, while in production, the live hole in the leader pass, which fits into the bottom or dead hole of finisher should be thickened by pushing the roll sideways. This thickened live flange will then fit more tightly into the dead hole of the finisher, driving the metal into the short live flanges. The reverse procedure can be adopted in reducing the height of flange in the live hole of the finishing pass. With the live flange thinned in the leader pass (again can be accomplished by shifting the rolls sideways but in opposite direction will produce a loose fit in the dead hole of finishing pass. Then metal would be robbed from live flanges of finisher as the steel tend to flow to fill the loose dead hole.

If both flanges of the finisher are running short, it is usually found necessary to adjust the rolls a little further back in the reduction schedule–back to the former and 'dummy' passes. The 'former' is opened to the point, where head is just wide enough not to shear in the entering 'dummy' pass. Then 'dummy' will pull the metal, thereby thinning the flanges and at the same time it produces additional length of flange through spreading action. Of course, there is a limit to this action also. The flanges can be thinned only to the point where it gets into the first edger pass; they are not to be made so thin, so as to permit wire drawing from web working action. Naturally if both flanges are running long, reverse procedure of roll adjustment can be tried.

Proper fitting of the flanges in various passes is extremely important for rail rolling, with too thin flanges working in a dead hole, where a certain amount of edging work can be performed, the upsetting action of edging will produce a good toe at the end of flanges, but it is generally found that the flange will be hollow at a short distance from toe. The work cannot spread the metal enough to fill dead hole uniformity because of insufficient stock. Waving flanges also can be caused by improper design or roll setting.

4.0 Salient Features of Rail Pass Design

The principle applied to beam, channel is also applicable for rail, but with the rail design, there is an additional important factor *i.e.,* "Fishing" is to be considered. Loose fishing is usually caused by under filling of the pass under the head at the second edger or leader pass. To fill these passes the roll at the 'former' pass should be raised, thereby transmitting more stock and consequently more work to the first edger, second edger or leader passes or both. If the fishing is loose at the top of rail, and leader is properly filled. It is usually follows that the top flange is too thick and the slight sideway adjustment of rolls can correct this condition. With a loose fishing in the bottom of rail, the head usually has a low radius and tends to be flat.

To correct a flat hat, leader should be raised slightly, thereby adding more work by virtue of increased stock at the finishing pass. With a tight fishing in the bottom of the rail, the rolls at the leader pass should be shifted sideway to thin the bottom of the head in the leader pass. A ridge under the head of the rail can be caused by the rolls being out of alignment. Kinks in the rail can be caused by the lack of proper work on the web.

4.10 Diagonal Rolling

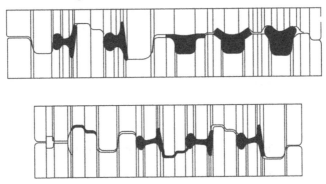

Fig. 11.10 2-high Roughing and Finishing Diagonal Design of Rail.

Difference between conventional tongue and groove and diagonal pass is that in the former the break is at the top of head and flange, so that it becomes a closed pass. The surface of head and foot must be inclined slightly outward to permit correct entry and delivery of stock, but the finishing pass must square up the section.

In the diagonal rolling of rails, the head and flange can be rolled at right-angle to the web due to inclined position of the profile and positioning of break at the opposite corners of the section.

In some instances, several of the passes in the rolling have been placed diagonally in the roll. While applied primarily for rolling rails of smaller size (industrial size) this system has been used to produce the larger T-rails. Here, the cutting in action of the first roughing pass resembles the shaping operation performed in beam. All three parts of rails (head, web, and flange) are started simultaneously. Which is quite different from standard rolling practice, where head and flange received first working and web working is started in later stages of rolling?

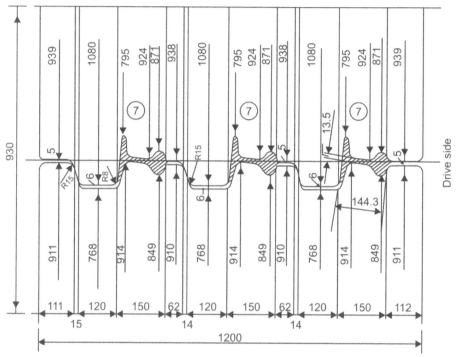

Fig. 11.11 Finishing Stand of Diagonal rolling of Rail.

The use of the diagonal system permits heavier working of flange than in case of conventional rolling, where the flanges are perpendicular to axis of roll. This placement is also conductive to the building of the flanges. By virtue of approach, a true slabbing condition, the more number of passes can be placed in rolls in the diagonal system. Higher amount of work can also be applied. There are other factors, which limits the amount of diagonal positioning.

The angle of entering stock into the pass is limited only by the amount of taper in the flanges. As stated earlier the live hole performs the slabbing action and in dead hole, only edging or reduction in height of flange can be obtained.

A slight choke is usually given in the dead hole of the flange in order to compensate for the elongation caused by slabbing action of live hole. The choking also helps in to straight delivering of rail. With the provision of diagonal pass, the live hole is usually permitted to run free. This procedure allows flange to spread freely without any fear of fin. The succeeding pass

4.0 Salient Features of Rail Pass Design

uses a dead hole for the preceding live flange; it helps in controlling the height of flange and also eliminates any tendency towards tearing because of free spreading.

No sharp corners are allowed on any of the collars, nor at any of the parting of diagonal passes. These corners should be well rounded to prevent spalling of the roll.

Finishing pass of the diagonal system may give end thrust and in this case, the strong thrust collars are necessary to resist axial movement of rolls.

4.11 Bearing Collars

The advantage of diagonal pass is the rapid reduction and easy flange development. Such positioning develops an end thrust, which force the designer to use bearing collars to be used for this purpose.

Using bearing collars creates new problem also. Since, thrust setup by the rolling causes serious wearing. These collars should be dressed to attain the original shape as per design, so as to maintain the passes to proper relationship. Worn collars will make the rolls to get shifted, thus increasing flange and head thickness of the live holes. Therefore, bearing collars are made inclined, because of wear condition, so that they can be "dressed-in" more rapidly, with a minimum reduction of roll diameter.

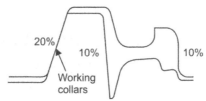

Fig. 11.12 Bearing Collars for Rail Design.

It is also important to keep these collars balanced through the pitch line in order to keep the wear to the minimum. There is considerable difference in the speed of the two surfaces at the extreme of a collar. One of the collars has a large diameter and consequently a high surface speed, while the mating roll will have low surface speed. At the pitch line of these collars, the wear is at minimum, as there is no difference in surface speed of both rolls, therefore, there will be no rubbing action.

4.12 Roll Spring

The prime importance in the design of roll passes is the spring of the rolls, or the amount of opening allowed in the rolls at the time of turning. This is the most important factor, which affects the final section of rail.

It is to be kept in mind that sufficient opening should be designed to permit bringing the rolls close enough together in actual rolling operation to compensate the spring the mill.

In other words, the scope of adjustment must be available whereby the rolls can be set at the thickness which will produce bar to the designed dimensions, because all mills have different mill spring. Mill spring of a mill depends upon the rigidity of mill proper, body length of roll, distribution of roll passes within roll, number of bars in the roll at one time etc. However, a general procedure can be formulated to know the mill spring. The amount of roll opening get decreases as finishing stand is approached. The mill spring of finishing stand is always less, because draughting get reduced, the number of bars rolled is usually only one, the length of roll body of finisher is usually shorter and roll is therefore stiffer.

4.13 Roll Diameter

The effect of roll diameter has a great influence on spreading and elongation. Larger diameter causes greater spreading and small elongation, because larger diameter rolls have a greater contact area with the steel being rolled and consequently it forces the steel to get spread than to elongation. It is to be also noted that forward slip has also a definite influence on spread. It reduces the spread.

5 CONVENTIONAL RAIL AND STRUCTURAL MILL

These mills are generally used for rolling of rails and heavy structural of varying sizes. A common productions schedule for 0.75 MT Rail & Structural mill is shown below in Table 11.1:

Table 11.1

Production Capacity Per Yar (as Finished)	750,000 T
Rails (For Indian Railways)	
52 kg/ metre	
60 kg/ metre	500,000 T
Structurals	
Beam 150 to 600 mm	
Channels 250 to 400 mm	
Angles 150 to 200 mm	
Crane Rails	
Crossing Sleepers	250,000 T
Billets	

5 Conventional Rail and Structural Mill

5.1 Rail Specification

Some of most popular rail specification used worldwide, chemical specification vis-a-vis mechanical properties is shown below in Table 11.2.

Table 11.2

S.N	Rail Standards	Specifications	Chemical Composition Percentage					Mechanical Properties	
			C	Mn	S Max	P Max	Si	Tensile Strength kg/mm²	Percent Elongation
1.	Indian Railway	IRS-T-12-96 Grade-880	0.60-0.80 *	0.80-1.30	0.035	0.035	0.10-.50	88	10
2.	British Standard	BS-11-1985	0.65-0.80	0.80-1.30	0.040	0.040	0.10-0.50	72	9
3.	Japanese Industrial Standards	JIS-E-1101	0.60-0.75	0.70-1.20	0.040	0.035	0.30 (Max.)	80	8
4.	Egyptian Railways	PW – 101	0.45-0.60	0.90-1.20	0.05	0.05	0.35 (Max.)	70	14
5.	Union of International Railway	UIC-860-0-700	0.40-0.60	0.80-1.25	0.05	0.05	0.05-0.35	70	14

5.2 Layout of Rail and Structural Mill

A typical rail & structural mill layout is shown below:

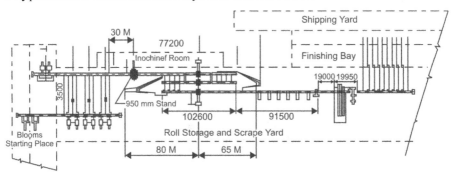

Fig. 11.13 Layout of a Conventional Rail Mill.

5.3 Details of Equipments: Mill and Furnace

Detail of major equipments used in the mill and furnace are appended in Table 11.3 below:

Table 11.3

S.N	Equipments	Details
1.	Reheating Furnace (3 Nos.)	Pusher Type Reheating Furnace with a total capacity of 200 T/Hr. Fuel used = Mixed Gas of CV.1900 KCAL/M^3 soaking zone Temperature – 1260° + 10°C (for Rails).
2.	950 MM Roughing Stand (1 no.)	Av. Dia of Rolls = 950 MM. Two High Reversible type Power Supply = One 865 V, 4000 kW, DC Motor Speed = 0-70-120 RPM.
3.	800 MM Intermediate Stands (2 Nos.)	Av. Dia of working Rolls = 800 MM Three High Non-reversible type Power Supply = 665 V, 4000 kW, DC Motor speed = 0-80-160 RPM Provision of Tilting tables and Hydraulically driven tilters and manipulators.
4.	850 MM Finishing Stand (1 No.)	Av. Dia of Working Rolls = 850 MM. Two High Non-reversible type Power Supply = 750 V, 1800 kW D.C. Motor Speed = 0-96-180 RPM Bearing Used = Roller Bearing
5.	Hot Saws (7 Nos.)	Sliding Type, adjustable for any length (Min-6 M) with two stamping machines for stamping cast nos. of Rolled Products.

5.4 Details of Major Equipments in Finishing Area

The list of major equipment's used in the finishing area is appended in Table 11.4 below

Table 11.4

S.N	Equipments	Details
1.	Slow Cooling pits	Each pit with capacity of 120 Rails of 13 M length for slow cooling to eliminate H$_2$ flakes.
2.	Rail Roller straightening machine.	Having six cantilever rollers mounted in 2 Rows at a distance of 1200 mm from each other peripheral speed of working Rollers = 0.8 to 1.6 M/sec.
3.	Vertical gag press	One for each finishing group for final straightening Straightening force = 200 T Stroke of slide = 70 mm No. of strokes per min - 30.
4.	Milling and Drilling machines	2 Milling & 2 Drilling machines for each group for end Milling & for Drilling Fish Bolt Holes.
5.	Inspection Beds	With a capacity of about 100 Rails of 13 M length per bed. Rails are tilted manually for surface inspection on all the sides.
6.	Cold Saw	For processing cut bar Rails with two tempering machines.

5.5. Welding

In this plant, rails of 65 m are welded to 130 or 260 m. The welding machine has integral stripping system and is fully automated with dedicated microcontroller for the weld process control and PLC for sequencing of operations.

Fig. 11.14 Welding of Rails.

The anti-twist clamps and automatic alignment ensure perfect profile matching of rail ends while welding. The distributed clamping ensures highly rigid and slippage free clamping of rails while welding. The profile-grinding machine grinds both the running faces and top of the rail with a very high precision and finish. Ultrasonic testing ensures flawless joints.

6 RAIL TESTING

A number of tests are carried out to ensure that the rail conforms to the products specification. Falling weight test, tensile test, hardness tests, micro-examination, chemical analysis, hydrogen content test, sulphur print residual stress, fracture toughness, fatigue strength and inclusion rating level are carried out. The details of major tests and their procedure is appended in Table 11.5 below:

Table 11.5

Tests	Testing Requirement	Testing Method
Chemical Analysis	Sampling collected during teeming of the steel. Samples of rails taken from standard position.	Spectroscopic analysis. Both spectroscopic as well as chemical method of analysis.
Falling Weight Test	Rail sample collected from position corresponding to top of ingot in case of ingot route or from any stand in case of continuous cast route.	A tup weighing 1270 kgs falls freely from a height on the full rail section of 1.5 metre length placed on two bearer 1.2 metres apart. For 60 kg. rails height of drop is 8.4 metres. The samples has to withstand the blow.

Contd...

Tensile Test	Samples taken from the standard position of rail.	The test is conducted in a standard universal tensile testing machine. The samples are prepared to : Gauge diametre -13.82 mm Gauge length – 69 mm Parallel length – 76 mm UTS should be minimum 90 kg/mm^2 for 90 UTS rails.
Hammer Test	Rail sample selected at random.	Sample of 50 to 60 mm length full rail section is tested by giving blow on side head, flange being held in fixed position. A deflection of 1 mm of rail axis is produced by hammering. The sample has to stand the blows.
Pendulum Impact Test	Rail sample collected from position corresponding to the bottom of ingot.	Length of sample is 25 mm. The blow of a swinging weight given on side head with foot held in fixed position. The sample has to stand energy of impact of 11.6 kg.M.
Micrographic Test	Rail samples are collected from positions corresponding to the top of ingot as well as to the bottom of ingot in case of ingot route or from any strand in case of concast route.	The samples are tested for sulphur print (Bauman) type for checking internal soundness and compared with album of UTC 860-0.
Hardness Test	Rail sample selected at random.	The decarburized layer is removed by filing from the rail table and subjected to Brinnel hardness test. For 90 UT rails, hardness value of 260 BHN minimum is required.

7 ROLL PASS DESIGN FOR RAIL

Roll Pass Design for a rail section is illustrated in Fig. 11.15 below:

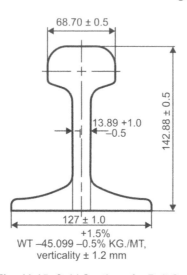

Fig. 11.15 Cold Section of a Rail Section.

7.1 Rolling Scheme of Roughing Line

Mill has the configuration of 950 mm Roughing stand as shown in mill layout (Fig. 11.13) earlier. The cast bloom 300 × 325 mm, supplied from the continuous and casting shop is first stored in the bloom yard for inspection. The ok blooms cleared by inspection department are charged into the furnace, and after proper heating and soaking, are then hot rolled in 950 mm roughing stand of the mill first. This is a 2 high reversing stand and the reduction to the bar is given in forward as well as backward directions.

Total 7 passes are given for rolling this section in the roughing stand. The roll diagram of 950 mm Roughing stand with pass diagram are shown below in Fig. 11.16.

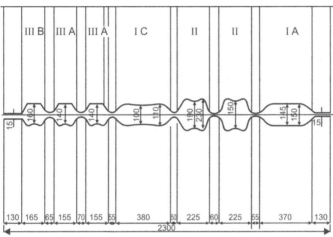

Fig. 11.16 Roll Diagram of 950 mm Roughing Stand of Rail Mill.

Rolling Scheme of Roughing Line

Table 11.6: Roughing stand, rolling scheme

No. of Pass	No. of Passings	Dail Reading	CC 300 × 335
IA	1	110	260
	2	55	205 × 345
II	3	90	280
	4	30	220 × 225
IC	5	65	175
	6	15	125 × 245
IIIA	7	20	160 × 155

The shaping of rough rail section has been performed in roughing passes of 950 mm stand. The reduction schedule for 950 mm stand is shown in Table-6, above. Firstly, the bloom is rolled in the box pass IA, wherein the con-cast bloom 300 × 335 mm are rolled forward and backward two times to get a rectangular section of size 205 × 305 mm, as shown in Fig. 11.17.

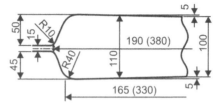

Fig. 11.17 PASS –(No.-2)- Roughing Group.

Metal, after tilting to 90°, is then transferred to pass II, whereas also two passing's makes the size to 220 × 225 mm. The second pass is also an open box pass where metal is pressed from the top and bottom roll in a box pass. Pass IC is a barrel pass, where in metal is again after tilting 90° is passed twice, to make the stock rectangular to a size of 125 × 245 mm and finally in pass IIIA (Fig. 11.18), metal is slightly pressed from the top and bottom to give a shallow shape, so that it can enter the first hat pass of finishing line.

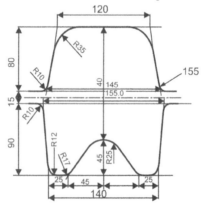

Fig. 11.18 Pass III A of 950 mm Stand.

While considering the design of roughing passes for producing rails, it is desirable to perform extra work on the head in order to produce denser structure which, in turn, would result in improved wearing qualities. The great changes in shape of the bloom must undergo at roughing stage of rolling itself, as metal is hot at this juncture and resistance to deformation will be at minimum. There are practically no pass in the roughing schedule where the work can be applied equally to all parts, with this work on one part of section soon may become out of balance, compares with work on other part. A flow of metal is setup from one part of the pass to another. The great difference in shape is evident due

7 Roll Pass Design for Rail

to above reasons. The control of this flow and its balance comprises the real problem faced by the designer in producing workable passes for rolling rails.

7.2 Rolling Scheme of Finishing Line

The 800 mm intermediate mill is a 3 high non-reversing mill. The mill has two 3 high stands coupled together and driven by a single drive motor. The scheme employs 3 passes in each stand. Lifting tables lift the rolling bar upto the upper passes for reversal of direction. The distance between the pinions is 800 mm and a barrel length of rolls is 1900 mm.

The final passing is given in the finishing pass of the finishing stand, which is a 2 high finishing stand; provide the final shape and dimensional accuracy to the rolled profile. Rails are branded with month, year, profile, grade of steel and manufacturer's name at the finishing stand. The working rolls are mounted on anti-friction bearings and have hydraulic roll balancing and a closed top housing to minimize mill spring and roll gap variations. The distance between the pinions is 850 mm and barrel length of rolls is 1200 mm.

Total 7 passes are designed for rolling of this rail section, three passes each in 800 mm intermediate stand and 1 final passing for the finishing 850 mm stand. The shape and major dimensions of these 7 passes of diagonal design are rails are shown in Fig. 11.19 below:

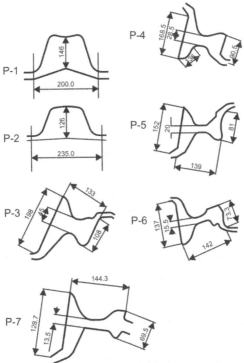

Fig. 11.19 The shape, position and dimensions of passes of Finishing line.

7.3 Pass Design of Individual Pass in Finishing Line

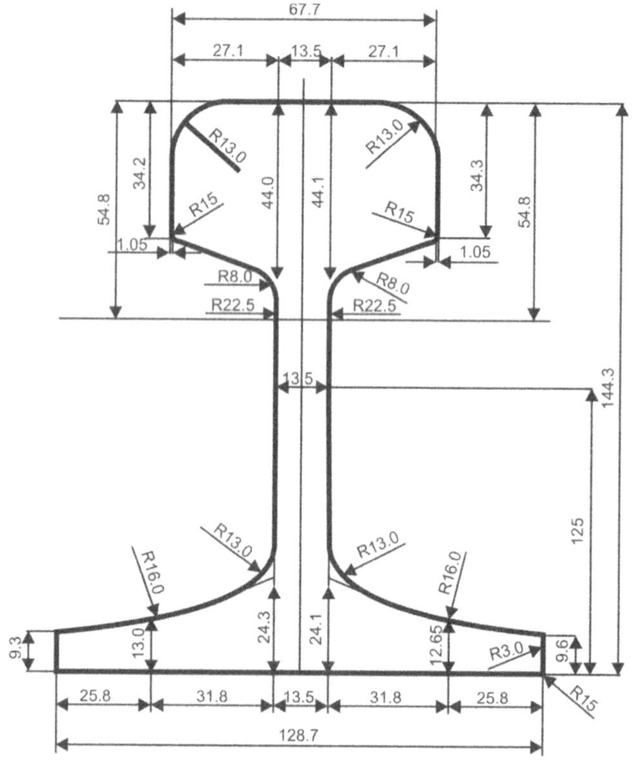

Fig. 11.20 Finishing Pass –(No. 7).

The design of the finishing pass is very important, in point of view of getting the final section of the profile. There is a lot of difference between the rail design, in compare with the beam and channel design, due to the strict tolerance of rolling of rail profile.

The most important elements of rail design are web thickness and of flange height. The tolerance on web is given as +1.0 mm and -0.5 mm. During design, the dimension for web thickness should always be taken slightly on –ve side, because in actual rolling, web thickness can get increase by the mill spring. In roll pass design of this section, it is taken 13.5 mm, against the 13.89 mm of cold section. During rolling; the web thickness can be increased or decreased by adjusting the collar gap.

The flange height of rail section is given in cold section as 127±1.0 mm. The hot section of profile on negative side comes to 128.5 mm, (127 × 1.012). It is recommended to take slightly more the computed dimension by 0.2-0.4 mm, as there will be slight increment in live side of the flange and to avoid

7 Roll Pass Design for Rail

the fin at the parting. Here, it is selected as 128.7 mm, 0.2 mm more than the computed dimensions.

The thickness of flange on live side is to be generally kept 0.3-0.5 mm less than the dead side, because of greater wear on the live side, in compare to the dead side of the flange. In this design it is taken as 9.3 mm and 9.6 mm for the live and dead sides respectively.

To avoid the fin formation at live side of flange, the flange is made of double taper type, which helps in to form good tip and height of the pass.

The pull down in dead flange is generally taken 5-7 mm and height increment of flange in live hole is assumed as 0.5-1.0 mm.

Head is the most sensitive part of the rail profile. Principle applied for the formation of flange, may not be worked in case of head. The finishing head works as a closed pass, because parting is at the center of the roll body in finishing pass, so both part of the head work as a close one. The pull down of head in finishing pass is around 1-2 mm, while for other passes; it varies from 2-3 mm.

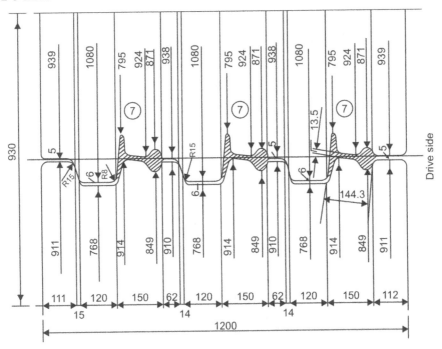

Fig. 11.21 Roll Diagram-Finising–(No.-7).

424 Chapter 11 Rolling of Rail

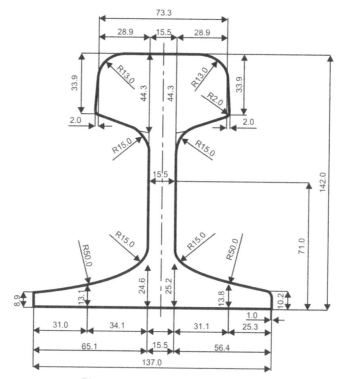

Fig. 11.22 Pre-Finishing Pass–(No.-6).

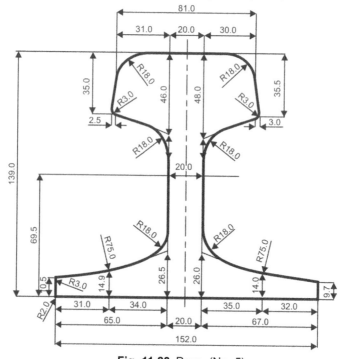

Fig. 11.23 Pass–(No.-5).

7 Roll Pass Design for Rail

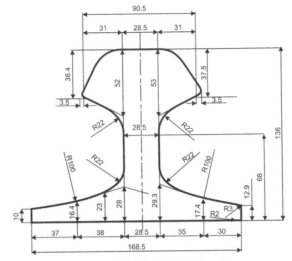

Fig. 11.24 PASS–(No.-4).

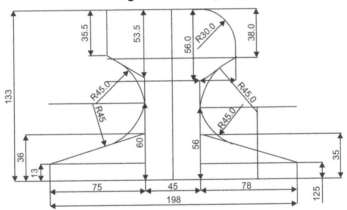

Fig. 11.25 PASS–(No.-3).

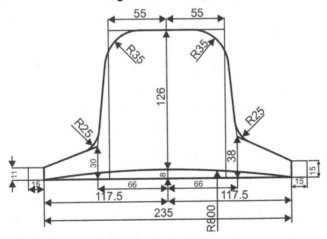

Fig. 11.26 PASS–(No.-2).

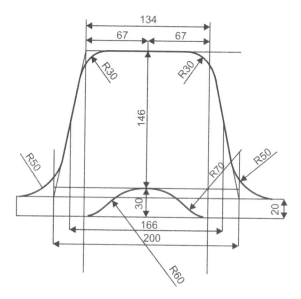

Fig. 11.27 PASS–(No.-1).

7.4 Computation Sheet for Rail Design

Table 11.7

Pass		Flange	Flange Height	Root Thick-ness	Tip Thick-ness	μ_{Root}	μ_{Tip}	Pull Down /Ht. Increment	Web Width	Web Thi-ckness	Web Area	μ_{Web}	Spre-ad
7th	Dead	57.65	24.3	9.6	1.01		7.45/	144.3	13.5	1920	1.15	2.3
	Live	57.65	24.1	9.3	1.05	1.10	+1.2					
6th	Dead	56.4	25.2	10.2	1.03		10.6/	142	15.5	2200	1.26	3.0
	Live	65.1	24.6	8.9	1.08	1.18	+0.1					
5th	Dead	65.0	26.5	10.5	1.05		10.0/	139	20.0	2780	1.39	3.0
	Live	67.0	26.0	9.7	1.13	1.33	+2.0					
4th	Dead	65.0	29.3	12.9	1.19		13.0/	136	28.5	3870	1.54	3.0
	Live	75.0	28.0	10.0	1.29	1.29	0					
3rd	Dead	75.0	36.0	13.0				133	45.0	5980		
	Live	78.0	35.0	12.5								

7.5 Graph for computation of Co-efficient of Reduction

As shown in Table 11.7 and 11.8, the roll pass designer has to make a balance among the co-efficient of reduction of following elements of rails:

- Tip thickness (μTL) of live and dead side (μTD)
- Root Thickness (μRL) of live and dead side (μRD)
- Web thickness (μweb)

In initial passings, μweb should be kept at minimum; otherwise it will lead to shortening of the flanges or may cause waviness of the web. It will get increases further in pre-finishing passes and will be 1.54, at the pass-4.

7 Roll Pass Design for Rail

Table 11.8

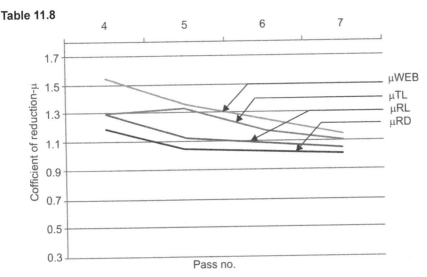

Likewise the µTL will always be kept more than the µRL of liveside. This will help in the formation of flange; otherwise pulling down action will take place. µTL and µRL$_t$ of finishing pass are 1.10 and 1.05, while for pre-finishing pass; it is 1.18 and 1.08 respectively.

The pulling down or height reduction takes place in the dead hole. It is 7.5 mm in the finishing pass, while 13 mm in the pass-4. The height increment is negligible in the live hole (<1 mm).

The spread in the web like beam and channel rolling is assumed to a value. Here, it is taken as 3 mm.

Due to the uneven metal distribution, rolling of rail sections needs utmost care and skill in operation. The conventional process of rail rolling through calibers between two rolls however has certain disadvantages which are listed below:

1. Lower forging effect is obtained in the rail head and the rail base which are the main load bearing surfaces.
2. Since, deformation is only at one direction, the dendrite residual network is not got broken and thereby it becomes the potential sources for crack propagation.
3. Due to asymmetrical rolling, residual internal stresses will remain in the rail.

These drawbacks led to considerable thought and eventually after a series of tests in the 1960's, rolling of rails in universal passes was established and put into commercial use in the early seventies. This has rapidly gained acceptable and latest rail mills are coming up with universal design for rolling with universal stands.

8 UNIVERSAL ROLL PASS DESIGN

The universal rolling stand as is well known has two driven horizontal rolls and two idle vertical rolls with axes in the same vertical plane. This permits mechanical working on all four surfaces at the sametime. The principle has been widely applied for rolling of beams. Fig. 11.28 shows the arrangement of rolls and stock in universal stand.

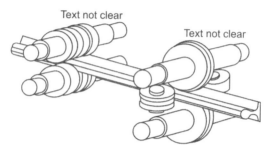

Fig. 11.28 Universal Rolling of Rails.

The rough rail section produced by the roughing stand is fed to the universal stand wherein the web is worked on by the horizontal rolls whereas the head and basis worked on by the vertical rolls.

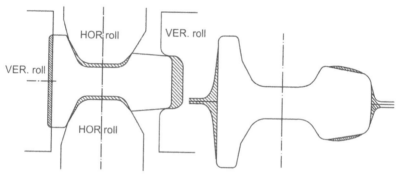

(a)- Universal Pass (b) –Edging Pass

Fig. 11.29 Types of Passes used in universal Rail Rolling.

As may be seen from [Fig. 11.29(a)], a certain amount of free spread takes place at the rail foot ends and the flanges of the rail head. A comparatively high elongation (from 1.25 to 1.40) in the universal pass guarantees a good 'forging' of the entire rail cross section.

The universal pass is normally followed by an edging pass with two horizontal rolls [Fig. 11.29(b)]. The edger pass is not intended to reduce the cross section of the stock but to improve the shape of those parts not worked on in the universal pass *i.e.,* the width of the rail foot and sides of the rail head.

8 Universal Roll Pass Design

The rail finishing pass can also be an universal stand. In one of the processes followed in Europe, the finishing is done in a half universal stand having only one vertical roll (Fig. 11.30).

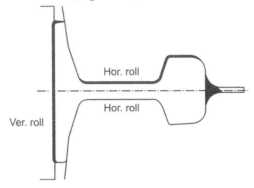

Fig. 11.30 Half Universal Finishing Pass.

The basic purpose of finishing pass is to work the stock applying direct pressure on profile with very slight reduction in order to control and adjust the finished profile.

The peculiar nature of the rail shape has led to certain developments in universal rolling stands which are described below:

(*i*) **Universal stand with small vertical roll**

This stand uses small diameter vertical rolls used with two large diameters back up rolls on each vertical roll.

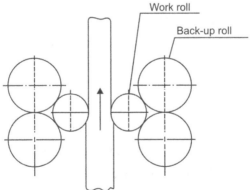

Fig. 11.31 Vertical Work Roll and Back-up Roll.

With this arrangement (Fig. 11.31), the spread of head and foot is minimized leading to better dimensional accuracy.

(*ii*) **Universal stand with vertical rolls on different diameters**

The cross sectional area of the head and base of the rail are different and when equal reduction is applied the relative drafts applied by the two vertical rolls differ. In addition the area of contact for the head and base are also different. Due to differential rolling loads on the vertical rolls there is a tendency for the rolling position to shift and this may result in warping of the rail blank. In order to counteract this situation the vertical roll on the rail base is made larger in diameter as compared to the roll on the head side as shown schematically in (Fig. 11.32).

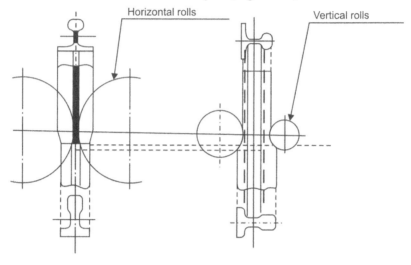

Fig. 11.32 Vertical Rolls of Different Diameters.

8.1 Advantages of Shaping in Universal Pass

The following advantages are offered by shaping of rails in universal stands.

1. Greater forging effects are obtained in the rail head and rail base.
2. The dimensional accuracy of the rolled cross section is superior.
3. The rail surface condition is excellent.
4. The internal strain generated during rolling process is minimized due to symmetrical formation of the rail from all sides.
5. The specific roll consumption is significantly lower.
6. Longer as rolled rails can be produced.
7. Greater freedom of roll adjustment is possible during the rolling process thereby making roll adjustment easier.

8.2 Mill Configuration with Universal Rolling Practice

The typical mill configuration for rolling of rails has two reversible 2-hi breakdown stands followed by one or two reversible universal stands each followed by an edger and a universal finishing stand. Some typical mill configurations are shown in Fig. 11.33.

8 Universal Roll Pass Design 431

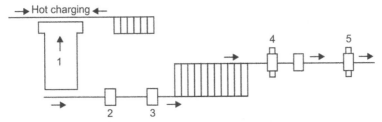

1-Furnace, 2-Break-down mill #1, 3-Break-down mill #2 ,4- Universal Rougher w/edger,5-Universal Finishing

Fig. 11.33(a) Mill Layout-1.

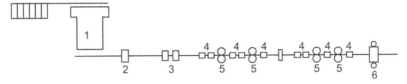

1-Furnace, 2- Break Down Stands, 3- Horizontal Stands 4-Horizontal/Universal Stands ,5- Vertical Stands, 6- Universal Finishing.

Fig. 11.33(b) Mill Layout-2.

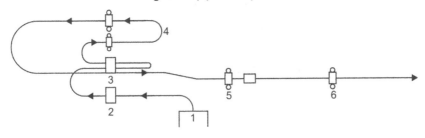

1-Furnace, 2-Break-down mill #1,3-Break-down mill #2,,4-Double Universal Stands, 5-Unit Rougher Edger, 6-Half Universal Finishing

Fig. 11.33(c) Mill Layout -3

8.3 Layout of Finishing Line

Layout of equipment's required for the finishing operations on rails of typical modern mills are shown in Fig. 11.34.

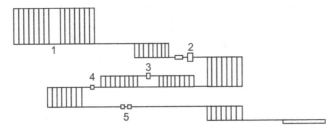

1-Cooling Beds, 2- Hor. & Ver. Straightening Machine 3-Testing Unit 4-Charging Rack, 5-Sawing & Drilling m/c-5 nos., 6-Chamfering m/c.

Fig. 11.34(a). Finishing Layout-1

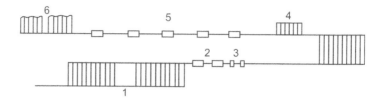

1.-Cooling Bed, 2-Hor. and Ver. Straightening m/c, 3.-Testing Unit, 4.-Charging Rack, 5.Sawing and Drilling M/c, 6.-Chamfering m/c

Fig. 11.34(b). Finishing Layout-2.

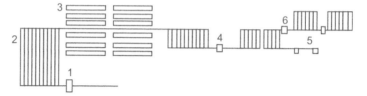

1,-Hot Saw, 2-Cooling Bed, 3.-Cooling Boxes, 4.-Straightner, 5.- Sawing and Drilling m/c 6.Hydraulic Straightener.

Fig. 11.34(c). Finishing Layout-3.

9 DEFECTS AND ITS RECTIFICATIONS

Defects in rail can be classified as:

1-Steel Defects

2-Mill Defects

3-Finishing Defects

9.1 Steel Defects

9.1.1 Non-Metallic Inclusions

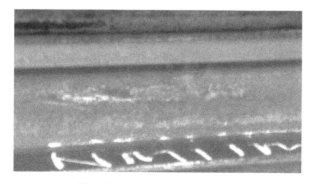

Fig. 11.35 Non-Metallic Inclusions.

9.1.2 Casting Powder Entrapment

Fig. 11.36 Casting Powder Entrapment

9.1.2 **Inclusions** arises as de-oxidation product/Re-oxidation product/from entrapped slag/from unstable refractory lining and from mould powder compound during casting

Fig. 11.37 Inclusions.

9.2 Mill Defects

9.2.1 Lap

Rail is a complex section. During rolling, it is subjected to non-uniform deformation. Lap appears on the surface of rails caused by excessive material in a given pass, which are rolled back into the stock, without being welded.

The formation of lap in rail rolling can be linked with deep gouges in hot stock from titles finger, loose manipulator liner, skid mark from pusher type furnace. Use of excessive wear out passes. All the above mentioned factors contribute to genesis of lap in rails.

Fig. 11.38 Lap.

9.2.2 M.D.M.

Defects generated during rolling from mechanical devices such as dogs of chain transfer, rollers of roll table, from guide rollers, sharp edges on roll table, from dogs in cooling bed, from piling guide etc., is called as defects generated due to mechanical devices (M.D.M.).

Fig. 11.39 MDM in Rails.

9.2.3 Sticker Mark

These are the cutting of edges of stock from sticker in the pass cutting occurs at equal distances. Smaller stickers are more dangerous than the the bigger one.

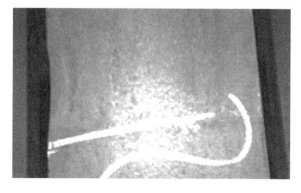

Fig. 11.40 Sticker Mark.

9 Defects and Its Rectifications

9.2.4 Split End Due to Bad Bloom End

Fig. 11.41 Split End Due to Bad Bloom End.

9.2.5 Pit Mark

Fig. 14.42 Pit Mark.

It is a protruded (extra) metal in the finished product. Pit mark arises due either

- Over-used roll.
- Pits on the pass.
- Excessive wear out of pass.
- Improper guiding, when stock hits the roll.

9.2.6 Under Filling

Under–filling is due to incomplete filling of the metal both on head table and either inside and outside surfaces of the flange.

Fig. 14.43 Under-filling.

9.2.7 Guide Mark

It is small shallow longitudinal mark/grooves of varying depth in the direction of rolling.

Fig. 14.44 Guide Mark.

9.2.8 Quenching Crack

Appears in the form of transverse cracks, mainly at the side head and at flanges. It occurs due to metal jamming in the pass and at subsequent rolling stages.

Fig. 14.45 Quenching Crack.

9.3 Finishing Defects

9.3.1 Camber / Reverse Camber

The arc or curvature occurred in the arch or curvature occurred in the rails during cooling process at cooling beds.

9 Defects and Its Rectifications

Fig. 14.46 Camber in Rail while transferring in cooling bed.

Cause: Since the rail head contains a larger mass of metal than the base *i.e.* flange, which causes flanges get faster cooling than the head, which will have more temperature due to higher mass. Rail will first begin to bend downwards towards flange *i.e.,* reverse camber and then it start upward bend towards head, as head cools afterwards.

Solution: The straightening of cambered rail leads to appearances of additional stresses, which lower the quality of rails. This should be avoided as much possible by as to make the top surface of head complex from end to end. A properly cambered rail will become almost straight after cooling. In long rail, the provision of pre-cambering m/c is a must.

9.3.2 KINKY

This is sharp/knotty bend present in any length or part of the rails, caused due to improper alignment of pushing dogs, while shifting the rails cooling beds and improper handled by cranes, both in hot or in cold condition

Fig. 14. 47 Kinky.

9.3.3 MDF (Mechanical Defects During Finishing)

Mechanical defects arises during finishing of rails are called as MDF. Rubbing mark of rolls of straightening machine and bending of flange are two important causes of this defect.

Fig. 14.48 MDF (Mechanical Defects during Finishing).

9.3.4 TWIST

A condition where rail have been forced to rotate in relatively opposite directions about its longitudinal axis.

9.3.5 Down Sweep / Up Sweep

It is the deviation from a straight line in plane of heads of the head of rail. In case of up-sweep, the head is curved upward and in down sweep, the head is curved downwards.

In case of upward sweep, the gap is at center. The gap is allowed to 0.5 mm at the centre. The gap is at the end, in case of downward sweep, gap is towards end of rail head.

9.3.6 Bent End

The rail may have short bend on either ends within 2 meters length.

Fig. 14.49 Bent End

Causes for Bend end/up-sweep/down sweep

- Blunt hot saw
- Grill plate level
- Stopper hitting
- Piling Guide Setting
- Improper Stacking
- Improper Pushing of Hot Rails in Cooling Bed.

❏❏❏

12

Operation, Safety and Quality Management

1 OPERATION MANAGEMENT

1.1 Principal Elements of the Operation Management

The following are the principal elements of the operation management in a rolling shop:

1. Keeping a steady eye on the stability of the rolling process parameters : optimum metal heating cycle, rolling temperature range, rolling rhythm and speed, standard conditions of deformation (reductions and tensions in rolling), working surface condition of the rolls and roll fittings, roll cooling (or lubrication) conditions and conditions of descaling on the metal surface.

2. The stability of the rolling process is controlled by means of pyrometers, ammeters (load indicators), dynamometers and tachometers. Hourly rolling schedules and continuous short time testing of final intermediate samples are used to regulate the rolling process including pass and roll changes.

3. Preparatory operations for the production, *viz.*, preparing billets for rolling in accordance with monthly, weekly, daily and hourly schedules and getting the mill ready for the operation (preparing sets of rolls for roll changes, as well as roll fittings, straightening rolls, blades and special equipment).

4. Repairs of furnace installations and of processing and auxiliary equipment of rolling mills (regular schedule, duration, characteristics of repairs and methods of their organization). Long interval repairs called general

440 Chapter 12 Operation, Safety and Quality Management

overhauls or capital repairs. The periodicity of overhauls may be varied. Lengthy shut downs of rolling mills once a year, adopted habitual generally by most of the rolling mills, cannot be considered as a good practice. A method of fractional overhauls with large scale restoration including modernization of the equipment carried out during running repairs and preventive maintenance should be considered more expedient. Running repairs are made one to five times every month, depending on the characteristic features of rolling mills and conditions of the repair service management at the works. A method of repairs involving the replacement of whole units or assemblies is a progressive one. Preventive maintenance between shifts and during roll changes includes, in addition to minor repairs, tightening of screwed joints, lubrication and checking of the equipment.

It is expedient that the rolling mill repair crews be integrated so as to form the so-called specialized and centralized repair groups, especially at large works.

5. Introduction of the system of efficiency engineering in rolling mill shops, to provides for proper allocation of jobs, multi-specialization of workers, no duplication of functions, gradual elimination of labour-consuming operations, introduction of industrial closed-circuit television systems, automation of production processes, automation of the production management and control including job rating, wage accounting etc.

6. Planning and recording of expenditures on basic and auxiliary materials, recording of fuel, power and water consumption, of replacement equipment, rolls, repair services, of semi-finished products, work pieces in production, waste and finished products, alongwith the day-to-day calculation of the actual rolling and final costs.

In a well-managed rolling mill shops the routine production conditions are determined not only by the technological instructions, but also by flow charts drawn upon the basis of optimum process parameters for each type of rolled product. With the improvement of the production techniques these charts are subject to revision. The preparatory stage for the actual production, as well as primary records of the process parameters are regulated by the programmed sequence of rolling operations, hourly rolling schedules and time tables for roll changes, production changeovers and replacements of roll fittings, bearings and shearing blades.

1 Operation Management 441

1.2 Technological Instruction on Preparation of Rolls and Rolling in Light Structural Mill

(a) Planning & Preparation during planned roll changing

These include :

1. Preparation of rolls.

2. Preparation of roll tackles, tilter inserts ,guards for each stand and for each profile, entry and exit boxes, guides for the boxes, inserts for finishing stand, other tools and devices.

3. Preparation of rollers of straightening machines, rollers and tools for the same.

4. Preparation of cold shears, shear blades for different profiles.

5. Roll changing is to done as per the rolling programme. Preparation of rolls and roll tackles and rolling programmes. Preparation of rolls and roll tackles and roll changing programme is prepared by I/c (Operation).

In charge (operation) should have following Information:

1. The stock of rolls for different profiles, placement of orders for new set of rolls. In charge (operations) should have the list of wornout rolls to be sent to roll turning shop for redressing and schedule for getting them back for further use.

2. The stock of roll tackles, develop new roll tackles for new profiles, get the wornout roll tackles replaced/modify the existing roll tackles, if necessary.

3. The preparation of roll changing programme, as per profile, which should see that the combination of passes in different sets of rolls so that the roll changing of different stands can be reduced and rolls can be economically used.

4. In charge (operation) should see that the rolling scheme of the profile to be rolled should reach to Mill Foreman (operations) in time. Any change in the rolling scheme or any modification of passes for better result should be intimated to Mill Foreman.

5. In charge (operation) should select sets of rolls for rolling. And give these rolls to bearing shop for the assembly of bearings. Spare rolls are also to be given to meet the emergency.

6. In charge (operation) should see the condition of passes of different rolls before bearings are fitted.

442 Chapter 12 Operation, Safety and Quality Management

7. In charge (Mech.) should see that these rolls are ready with bearings fitted at the time of requirement.

8. In charge (operation) should see that the furnaces are to be filled with the required quality and size of billets before the start of Mill for profile changing. This job will be done by Shift Managers as per the instruction of In charge (operation).

In charge (Finishing & Shipping), should have following information for the preparation of straightening machine rollers and shear blades:

1. The stock of rollers for each profile. He should order for new rollers when old rollers became under-sized, it become soft. He should send worn out rollers to the roll turning shop for redressing and get them back in time for further use.

2. He should know the stock of Shear Blades for different profiles. He should arrange the replacement of old, broken chipped off blades by new ones.

3. He should earmarked rollers and shear blades, which will replace the changing rollers in straightening machine and changing of blades of shears.

4. In charge (Mech. - F&S) should see that condition of chain transfers, rope transfers and other equipments which will be in use for the profile should be in good condition.

5. In charge (Elect. - F&S) should see that the condition of Electrical equipments which will be in use for the profile should be upto the mark.

(b) 1. Planning and Preparation during breakdown

1. In case of roll breakage or urgent roll changing, the command for roll assembling can be undertaken by the Shift Manager of that particular shift, with the consent of the In charge (operation). Before the start of roll changing, passes of the assembled rolls should checked carefully to see the condition of pass and whether the correct pass is marked, especially in case of finishing stands rolls for beams and channels. If the pass is found defective. Foreman (operation) should see that these passes are not used. He should inform to the in charge (operation).

2. Shift Manager should co-ordinate the roll changing and other jobs in such a way, so that all help and co-operation is available from all sides (Mech. Elec. oprn.) to finish the work satisfactorily in time.

3. Furnace staff should be intimated, about the probable time the job of roll changing will be taken for change over. Foreman (operation) should be made responsible for carrying out this job with the consultation to Shift Manager.

1 Operation Management

1.3 Pre-cautions to be Taken During Roll Changing

(a) Horizontal Stands

1. The outgoing Shift Foreman (operation) should inform the Shift Foreman (Mech.) before hand about the Roll changing programme during the next shift and keep the assembled rolls on the Mill floor.

2. The Shift Foreman (Mech.) should ensure that the rolls are kept ready fitted with spindle heads and lubrication fittings.

3. While fitting spindle heads the key-way on the rolls, and end should be checked. The slide blocks should be checked and the worn-out slide blocks are to be replaced and loose spindle heads should not be used.

4. Before shifting the stand it is to be ensured that all the hydraulic wedges are declamped fully.

5. While disengaging and engaging the coupling, care should be taken so that the pinion stand should not move to the extreme end, as it may cause breakage of lubrication pipelines and jamming of spindle shaft in the coupling.

6. At the time of starting of the roll changing, the bottom screw-down mechanism should be operated both ways to ensure proper working of the mechanism.

7. After the stand and pinion stand got declamped, the displacement mechanism should be checked both ways to ensure for its proper working of the mechanism.

8. After leveling of the bottom roll, the ratchet handle should be kept vertical by means of providing locking arrangements. It should be left resting on the displacement mechanism screw. It should never be left resting on sides as in the position; there is a possibility of it getting stuck up and bend, while stand is displaced.

9. The stand base channel should be thoroughly cleaned with water when the stand is in disengaged position as due to the accumulated scale in the base, the free movement of wedge bolts may get obstructed, resulting in breakdown in displacement mechanism reducer.

10. While engaging the stand, grease should be put thoroughly in slide blocks.

11. While putting top roll in stand, when top roll sits on the bottom roll, the pressing of top roll should not be made, unless a gap has been made between two rolls.

444 Chapter 12 Operation, Safety and Quality Management

12. While pressing the top roll, it is to be ensured that the spring is not compressed fully as it may cause breakage of it in the housing itself.

13. Inclination of the spindles should be equal and opposite for both the spindles.

(b) Vertical stand:

1. While roll house lifting mechanism is operated to take the same in upward or downward direction, care should be taken so that there should not be any obstruction on the way of movement.

2. At the extreme down position, the housing should sit on the groove of the apron plate. When the housing is sitting on the groove, the lifting mechanism should be operated carefully to see that the screw of the mechanism is well above the extreme limit.

3. When raising the housing, care should be taken to see that the housing does not cross the extreme upper limit marked on the main housing.

4. While taking the housing in or out of the main housing, care should be taken to see that the housing does not get any obstruction from both sides *i.e.,* from bottom and top.

5. The inclination of spindle, when coupling is engaged on the rolls, should be equal.

1.4 Points/areas to be Checked after Roll Changing

(a) Mechanical:

1. All side coupling guards should be put in position properly.

2. Any crow bar, side guard etc., should not touch any rotating parts.

3. Stands should be clamped properly and only after that starting of main drive may be allowed to rotate.

4. Before giving lubrication permission, all inlet and outlet pipes for oil film bearing should be checked thoroughly.

5. Oil flow to the reducer, pinion stand, bearing etc., should be checked.

6. All the floor plates should be put in position properly.

7. Adequate cooling arrangement for the tilter *i.e.,* used for tilting the work piece by 90° in between two horizontal stands must be made.

8. After the main drives are allowed to rotate, the following points should be checked from time to time.

    (a) Bearing temperature.

    (b) Whether spindle heads are getting heated up.

1 Operation Management

(c) Vibration of the stands or any other part of main equipment.

(d) Oil flow to the main equipment.

9. Checking of all the equipments and their proper working is a must.

10. Preparation and checking of speed points required for different profiles is to be ensured.

11. All mechanisms, fittings changed/ repaired during roll changing are to be checked during trial or initial running for proper performance.

(b) Electrical

1. During roll changing time, the connections and plugs of screw down mechanism should be checked and defects should be immediately rectified.

2. When the vertical stand housing is required to be in or out, then the trailing cable should be taken care of, so that it may not get stuck up somewhere and then, subjected to tension or any other pressure.

3. The cleaning of photo cell and testing of auto for tilters is to be ensured.

4. While mill is started the temperature of main motors should be checked at least 3 to 4 times in a shift.

5. Those motors, having ventilation system, it is to be checked and to ensure that they are in working condition.

6. The temperature of the bearings of main drives and main M.G. sets should be checked every hour of the running shift.

7. Leakage from main hose pipes and connections should be checked atleast once in every shift.

8. Loads of all the main drive should be checked atleast four times in a shift.

9. Sparking of the commutater should be checked atleast once in a shift.

10. Major mill equipment panels working on relay contactors etc. should be checked atleast 3 times in every shift.

11. Checking and rectification of cooling bed roll table motors should be done regularly in every shift.

12. Survey of roll table motors and auxiliary fixing should be done in every shift.

13. Checking of all equipment's and their proper working is a must.

14. Preparation and checking of special points required for different profiles is to be ensured.

446 Chapter 12 Operation, Safety and Quality Management

1.5 Starting the Mill After Major Roll Changing

General rules for mill adjustment after roll changing are as follows:

(a) Before the start of mill:

1. Rolling scheme for the section should be as per roll pass Design and should be confirmed by in charge (operation)

2. Passes of all stands should correspond to the roll pass design. Any deviation is only allowed with the permission with In charge (operation).

3. Fixing up of rolling tackles of entry and exit sides of rolls should correspond to the ones confirmed by In charge (operation). It is forbidden to use modified rolling tackles without consulting the In charge (operation).

4. During the mill adjustments, twisting of the metal around its axis should be removed for all sections.

5. Bending of pieces side wise, up and down, bending of front end and tail ends should be completely eliminated.

6. Distortion of section in all stands should be removed during mill adjustments.

7. Finished products should be free from all rolling defects.

8. The final speed adjustments should be done only after all above operations are completed; after which the mill is considered to be adjusted.

(b) After the start of mill

1. Mill adjustment is carried out by Foreman (operation).

2. During mill adjustment, all changes in the reduction scheme and adjustment of rolling tackles can be made with the permission or by order of Foreman (operation). Slight changes adjustment in reduction in case of finishing and pre-finishing stands within the tolerance of finished product can be done by roller.

3. Only the Shift Foreman (operation) will be allowed to start the mill or any individual stands after he has checked the same.

4. Unauthorized persons should not be allowed in the Mill. Area mill, especially while Mill adjustment is being done. All Mill personnel should watch carefully the movement of bars in stands and outside the stands.

5. Signal for pushing out the billets from the furnaces is to be given by Shift F/M (operation). He should see the condition of the billet, before

1 Operation Management

the start of pushing of billet from furnace. Only one billet at a time is pushed out for Mill adjustment.

For the purpose of mill adjustment, return billets without heat numbers are to be charged into the furnaces. For this purpose, 3 to 4 billets depending upon the section to be rolled and are charged into the furnaces.

6. For pass burning, sample pieces are to be kept in the furnaces for heating, which are taken out during pass burning. This piece should be selected by F/M (operation).

7. Checking of dimension of pieces after the roughing, intermediate and finishing group should be carried out by Shift F/M (operation). The Foreman should also check loads on different stands for all sections.

   Corrections during Mill adjustment should be done in consecutive in order of rolling.

8. When the bar is passing through pre-finishing and finishing stands, the roller should see the following:

   (*a*) Proper gripping of metal by rolls.

   (*b*) Direction of twist of the bar.

   (*c*) The nature of bending of the bar.

   (*d*) Over filling and under filling of pass.

   The roller should correct these inaccuracies and take a sample to examine the dimensions. Then again, necessary correction should be made in the mill to have correct dimension of the bar. The Foreman should see the profile and satisfy himself and give a permission to start the normal rolling.

9. Sample to be collected from sample saw with the permission from Foreman or roller.

1.6 Examination of the Finished Products

These are general rules for taking samples and examination of metal on the cooling beds.

1. Final estimation of the section can only be given after double check of the finished product *i.e.,* checking of technological sample at sample saw and examination and checking of dimensions of finished products on the cooling bed. It is forbidden to start normal rolling, if metal on the cooling bed is not checked and if there are no defects observed on the sample.

2. Defective bars rolled during the mill setting should be separated from the rolled metal and afterwards cut separately. This separation should be carried out by recorder of the cooling beds on the instruction of the roller.

448 Chapter 12 Operation, Safety and Quality Management

3. The roller should examine the bar, out of which the sample is taken and check the dimensions at different distances from the back end.

4. Samples from the saw are taken to check the dimensions as per the desired tolerances of given standard and also examine the surface of the product. In case of any rolling defect detected, the rolling should be stopped to take necessary action to rectify the defect.

5. When examining the bar on the cooling bed, special attention should be paid to the defect. Periodicity of defects (roughness, scabs, sticking bar etc.) on any side of the section is the result of defect in the pass in one of the stand. In such case, rolling should be stopped to rectify the defect.

6. In the process of steady rolling after every 100 to 200 pieces, sample should be taken at the hot saw. The last sample taken during the running shift should remain on the control table for the next shift.

7. The roller should periodically examine the pieces on the straightening machine approach roll table when rolling is going on.

1.7 Technical and Economical Characteristics of Rolled Metal Production

Each ton of the annual output of rolled products, including that of the cogging division, requires capital investments 3-4 times greater than those for the blast furnace (excluding ore mining and coke production) and steel making processes. The unit capital investment in rolling increase parallel with the improvement of the rolled product quality.

Taking as an ordinary example, most popular structure of a hot rolling shop and comparing an approximate correlation of percent costs among the main stages of rolling, that the cost of the rolling mill, including shearing and strengthening facilities, will be around as high as 70%, that of the heating facilities to 16% and of the finishing installations and store rooms upto 14%.

It is customary to hold the rolling mill proper to be the main unit heading all other divisions of a rolling shop. Such an approach is also justified by the average upkeep expenditures of individual rolling divisions. These amounts to 20-30% for heating, 50-60% for rolling and 15-30% for finishing.

The following general formula may be used for determining the actual hourly output of a rolling mill :

$$P = Q.q \frac{3600\,k}{t}$$

Where
P = net yield of the mill, t/h

Q = mass of billet or ingot

q = net yield factor

1 Operation Management

$$t = \text{rolling rhythm}$$

$$k = \text{rolling mill use factor}$$

The rolling rhythm is determined in a number of ways, depending on the rolling mill arrangement and rolling scheme. For a single stand blooming mill this is the total time of rolling in the stand, including intervals between passing and successive ingots. For multi stand rolling mills this is the time of passing in the finished stand, including intervals between successive billets.

In general terms the rolling rhythm for a stand in which several passing are performed is determined from the formula :

$$t = t_p + t_i$$

Where t_p = rolling time for several passing or total productive time

t_i = time of intervals between passing.

If rolling speed v is constant (disregarding speed gain) and the length of the work piece after passing is L, then the productive time is determined from the formula

$$t_p = \frac{L}{v} = \frac{L.60}{D \times n \times \pi}$$

Where D = effective diameter of rolls

n = roll revolutions per minute

For reversing rolling mills, where the peripheral roll speed is variable, the productive time is expressed as

$$t_p = t_1 + t_2 + t_3$$

Where t_1 = speeding-up period

t_2 = period of rolling at a constant speed

t_3 = slowing down period

The duration of the speeding up and slowing down periods is determined proceeding from the different between rotational speed n_1, n_2 and n_3 and from acceleration a or deceleration b rates (rev. sec/min.)

Speeding up time $t_{s.u} = \dfrac{n_2 - n_1}{a}$

Slowing down time $t_{sl.d} = \dfrac{n_2 - n_3}{b}$

It should be borne in mind that in most cases the beginning and the end of a passing do not coincide with the zero speed values in the speed diagram of a reversible motor and they are determined best with reference to the load diagrams of the motor or to oscillograms.

2 SAFETY MANAGEMENT OF ROLLING MILL

Fig. 12.1 Cobbles in a Structural Mill.

2.1 Accident Prevention Regulations in the Rolled Metal Production

Potential hazards from rolling come from heat radiation, high travel speeds of the worked metal, concentration of moving and rotating mechanisms in the operating area. It is also due to scattering of scales and shavings during shearing, sawing and conditioning the metal.

Safety regulations for the operation of crane, railway transport, for repairs, servicing of gas, electrical and oxygen facilities are the same as for the entire iron and steel industry. Great importance is attached to protective garments for workers of different trades, including means of individual protection (protective goggles, deflectors, nets, helmets, asbestos cover plates, special gloves, shoes and so on). Heat radiation is controlled with the help of water net screens, curtains, safety glasses. Much attention is paid to different ventilating, blowing-off and air conditioning systems.

Moving and rotating mechanisms are guarded with special shrouds.

Finishing and high speed working stands are reliably protected on the sides and at the top by means of removable wire guards.

In rolling mill shops the access to standing mechanisms, rolling trains and working stands should be made absolutely safe. To this end the so-called tag system is mainly used, *i.e.,* handing to the electricians of a division a tag taken

3 Quality Management in Rolling Mills

off a corresponding starting device and serving as a pass, this confirming that the electrical circuit of the given mechanism or of the entire division has been disconnected. Special interlocks, signaling devices and mechanical shutters serve as duplicating safety arrangements. Various safety devices (guard posts, ledges, deflectors) are used to protect the tandem rolling mill operators.

In rolling mill shops safe, clearly definable passages are made in the form of overhead bridges with railings or of tunnels.

Care should be taken to ensure good lighting, proper metal stacking and timely removal of rejects.

Varying operating conditions of rolling mills require working out of particular additional safety regulations in the form of production instructions which should be strictly adhered to.

3 QUALITY MANAGEMENT IN ROLLING MILLS

3.1 Quality concept of Products

Quality existed in every walk of life, many decades ago; but the concept upon which this discipline exists has assumed enormous importance in the recent years. What is this 'Quality' of Products and What does it mean? The simple answer is of all the concepts in Quality or Quality function; none is so far reaching or important as "Fitness for use". All industries are managed by human beings with one objective of providing quality products or quality services to human being. This becomes positive and constructive only if the goods and services respond to the overall needs of the user in price, delivery time and fitness for use. If the goods and services do respond to these overall needs, the products are said to possess "Marketability or Salability". But against the overall needs, the extent to which the product services the purposes of the user during usage is known as "Fitness for use". The concept of fitness for use is properly called by such name as "Quality" is a universal concept applicable to all goods and services. "Fitness for use" of the product is mainly determined by the features of the product which the user can recognize as beneficial to him. The 'Fitness for use' is the resultant of some well-known parameters.

A customer is one who buys a product from another. But the customer is also the user. User is one who receives the intended benefit of the product. The buyer of services is often called a client. Consumers are individuals or groups who consume goods for personal purposes. To the economist these are products and services and the user is interested only in services. To the Technologist, the responsibility is to produce products 'Fit for use' economically with the maximum.

3.2 Reliability, Availability and Maintainability of Products

In true sense, the product should never fail and its availability should be 100%. But in actual practice products do fail, so that an essential sub parameter of availability, *i.e.*, freedom from failure is formed as Reliability. In other words reliability is the probability of performing without failure, a specified function under given conditions for a specific period of time. Let us take hot rolled products like rails for railways, structures for bridges, rods for buildings etc. the quality involves, use for long time or in other words their products should have maximum reliability, which means they should have the quality properties of materials. Another term 'Operation Reliability' is used often to mean the attained reliability. Reliability is not to be confused with conformance to product specifications but it is also the actual use of the product over a period of time plus collection and interpretation of data on performance and failures during that period.

Hence, the important factors which are to be considered, are the (*a*) Availability and (*b*) Maintainability. A product is said to be available, when it is in service or in an operative state. The total time in operative state called as "up time" is the sum of time spent in active use, similarly the total time in the non-operative state called as "down time" is the sum of the time spent under active Repair (diagnosis and Repair) and waiting for Spare parts etc.

$$\text{Availability is } \frac{\text{Up time}}{\text{Up time} + \text{down time}}$$

$$\text{Or availability Ratio is } \frac{\text{Mean time between failures (MTBF)}}{\text{MTBF} + \text{mean time to repair (MTTR)}}$$

Maintainability is the need for continuity of service which is the effort to improve the maintenance of long life products.

Inspection: Work of check the product for size, height, quality and any other specifications.

Accuracy of inspection = % of defect correctly identified,

$$i.e. \qquad \frac{d-k}{d-k+b}$$

d is the number of defects reported by Inspection, k is number of good units rejected by inspectors, as determined by check inspection. $d - k$ true defects found by Inspector; b is the defects missed by the inspector, as determined by check inspection. $d - k + b$ will be true defects originally in the product.

3 Quality Management in Rolling Mills

Suppose, the number of defects reported by inspection d is 45, and out of this 5 were found to be good by check inspector $i.e.$, $k = 5$. Hence, $d - k = 40$, which is the true defects found by Inspector $i.e.$, $d - k + b = 50$ ($i.e.$ the 40 found by the Inspector $+$ 10 he missed).

$$\% \text{ accuracy} = \frac{d - k}{d - k + b} = \frac{45 - 5}{45 - 5 + 10} = 80\%$$

Standards Specifications: All products produced, manufactured and sold to customers for whole end use, requirements are well understood, studied and basic quality requirements are published in codes known as 'Standards, or Specifications either by Government agencies or various trade agencies.

3.3 Rolled Product Quality Control

The quality of rolled products is determined primarily by that of the steel melting and casting and by appropriate use of each heat conformable to the data on the actual conditions of melting and casting, as well as to the chemical analysis for each specific order.

With modern mass production, the pre-requisites for obtaining high quality are steady conditions of processing and well organized control at all stages of the metallurgical cycle.

All the processing data from melting to shipping are documented by a through certificate of a heat or a ladle, if the metal of the given heat is tapped into more than one ladle. The heat certificate allows one to follow-up the routing of each ingot mould buggy, stool or even of each metal zone along the height of an ingot.

The heat follow-up makes it possible to precisely determine the actual metal consumption factors for given steel and to analyze all losses in processing.

One of the main functions of the process control is checking on the proper observance of the existing regulations for the routing of heats within the process flow and for separating reliably the heats and sometimes, the ingots as well as billets and bars with markings denoting the metal zones along the height of the ingot. At works producing high grade steels the metal is checked by employing the spark test, but this should be done only as an additional safeguard against misidentification of the metal.

Lastly, the conformity of the rolled products to the order specifications is determined on the basis of the procedure control information and on that of the data subsequent to the control over the dimensions an surface condition of the products. It carried out in the course of production as well as on the ground of the final data on finishing, sorting, packing and marking, and on that of the findings obtained as a result of the final mechanical tests. For some types of

454 Chapter 12 Operation, Safety and Quality Management

the rolled products the control involves an analysis of the macro and micro structure.

Stepped turning, as well as X-ray, ultrasonic and other modern nondestructive tests are used for detection of internal flaws.

Works producing high grade steels practice, the so-called heat control as the basis for consigning a heat to fill important orders. This, a sort of a tentative method of control, consists in that the entire set of the rolled product tests is carried out effectively on products made of individual ingots, this being followed by deciding as to the consignment of the entire heat or outlining a special type of processing.

Now-a-days high quality of the rolled products is attained by carefully investigating the properties of given steel, selecting most favourable conditions for its production and optimum ingot structure, as well as by ensuring standard processing conditions and a rational production control pattern.

There importance is for the properly selected system of the procedure control. It is closely linked with the filling in of the heat certificate. In the course of rolling quick upsetting, torsion and pickling tests are made concurrently with the intermittent and continuous control over the regularity of the shape and dimensions of the products. These tests make it possible to quickly eliminate rolling defects, such as back fins, folds, seams, surface tears, scratches and rolled-in scale.

The final control of finished rolled products is accomplished in accordance with the requirements of corresponding standards or specifications, by using specimens taken after rolling. This is often carried out in the course of production, the samples being cooled under conditions similar to those prevalent in the production. A check on flakes is done after expiration of a predetermined metal storage period (incubation period). In addition to the chemical analysis mechanism and processing tests, internal flaws, if required, are detected by viewing the metal fracture and macrostructure.

The proportion of non-metallic inclusions, the degree of decarburization, the presence of a carbide network etc., are determined by analyzing the microstructure of the metal.

The processing tests include the cupping (Ericson), stamping and bending ones, as well as flanging and expansion tests (for tubular products) and pendulum impact tests (for rails). Repeated tests are envisaged with an increased number of specimens. If these tests give unsatisfactory results, then rolled products of these heats or batch are to be rejected or their grade may be changed to suit less important consignments.

3 Quality Management in Rolling Mills

3.4 Quality Control in the Production of Rolling Products

Role and responsibility of Quality Control Section is to monitor system adherence, report and control deviations, ensure traceability of material, confirm and certify product as per specification.

(*a*) Control of input material:

- Receiving, inspection and stack verification of billets in the yards to avoid/minimize supply of defective billets for rolling. Monitoring and follow up of all the special steel billets rolled to avoid mix ups by giving gap/separation blooms between casts and grades.
- Hot chalk marking of cast number and demarcate end of previous cast and start of next cast immediately after stacking.
- Color codes for different grades.
- Quality separation while charging into the furnaces.
- Inspection of billets in the stack for all the visible sides, end cross section and rejection of the defective billets.
- Red painting of all rejected billets.
- During supply of billets from stack, removal of defective billets marked on the stack.

(*b*) Controlling and monitoring the technological process parameters:

- Control of heating regimes and rolling parameters.
- Monitoring of heating regimes followed in furnaces for different grades of steel.
- Monitoring and control of process parameters like soaking zone temperature of furnace, temperature before finishing stand qualization temperature during TMT rolling in bar Mill laying temperature in Wire Rod Mill, water pressure used for quenching in both mills. Quality change with atleast three hook gaps on hook-conveyors and atleast two pallets gap on pallet conveyors for Wire Rod Mill and 1 meter gap on cooling beds in bar mill

(*c*) Controlling profile during rolling:

- Profile checking at the beginning of each shift or after pass/roll changing.
- Regular profile checking at an interval of 20 minutes.
- Regular surface checking on cooling beds at an interval of 20 minutes.
- While rolling of special steels cooling bed/pallet conveyor separation shall be ensured.

(*d*) Inspection of finished products

- Confirmation to technical delivery condition (TDC).
- Identification and segregation of non-conforming products.
- Rejection of bundles in rolling mill if they contain more of the defective pieces, and piece by piece inspection and also if they contain partially defective pieces.
- Random Bundle inspection with respect to shearing and surface defects at cradles. Inspection of all coils on hook conveyors /coil hangers of wire rod mill.
- Red painting of defective material.
- Proper identification of the coils with respect to cast number, section, grade, rolling date and shift.
- Checking of products as per specification tolerances.

(*e*) Sampling and testing

- Collection of samples as per scheme and size required.
- Samples are collected at an interval of ½ to 1 hr.
- In case of special steels, collection of test sample and its identification is done in presence Quality Control Department.
- Number of samples and cast number shall be verified for sending it for testing.

  Colour coding of sample pieces and verification of details marked.

(*f*) Colour coding

All bundles and coils shall be colour coded as per specified colour code. In case any heat is diverted on testing, verification of the change of colour code.

(*g*) Bundling and packeting

Proper bundling in bar mill and binding/strapping in wire rod mill. Strapping of required number of straps in coils and bundles.

(*h*) Pre dispatch inspection

Random checking of loaded wagons for defective pieces.

(*i*) Testing of material and Issue of test certificates.

Testing of products (Mechanical, chemical) to confirm the specification Issuing of test certificates.

3 Quality Management in Rolling Mills 457

3.5 Requirement of TMT Rolling

(*a*) **Input material:**

- Length control of input material at Billet Mill and also at cooling beds of Billet yard.
- If base length is falling out of this range, give immediate feedback to Billet Mill.
- Checking of location of the stack.
- Checking of cast number written on the billets and cast separation marking.
- If there is a quality change, keep a higher size billet like charging of 150 × 150 billets between two qualities.
- Tallying the number of billets stacked with production to find out mix-up.
- Checking of separation gap between heats on cooling beds.
- Checking of colour code on billets:

    Fe 415-NO COLOR

    Fe 500-ONE YELLOW DOT

    Fe 550-TWO YELLOW DOTS

    Fe 415-A ONE BLUE STRIP

    Fe 415-M ONE BLUE DOT

    Fe 500-A TWO GREEN DOTS

    Fe 500-M TWO BLUE STRIPS

Stack inspection for visual defects and marking on end face with chalk
Checking of removal of defective billets on feeders.

(*b*) **Process parameters:**

(*i*) **Furnace Stage:**

Checking of soaking zone temperature 1200 ± 50°C

Checking of temperature after finishing stand : 950 ± 50°C

Checking of equalizing temperature: 610 °C (Depending of grade getting rolled).

(*ii*) **Rolling stage:**

Checking of samples at stage inspection bench for D_{tr} (ht of lug), S_{tr} (spacing of lug), L_{tr} (Length of lug), weight tolerance, length tolerance of cut lengths and quality of surface and calculation of projected area.

(*iii*) **Sampling of product:**

Checking of Samples for size, marking on the samples, number of samples as per sampling scheme.

(*iv*) **Inspection of final product:**

(*a*) Checking of colour codes given on bundles:

Fe-500 HCR TMT (Gr. M-500)   Two Blue bands

Fe-500 HCR TMT (Gr. A-500)   Two Yellow bands

Fe-415 HCR TMT (Gr. A-415)   One Yellow band

Fe-415 HCR TMT (Gr. M-415)   One Blue band

(*b*) Inspection of finished product:

Each bundle is checked for visual defects. If majority of the bundle is found with defects, bundle is rejected. Otherwise marked for full inspection.

(*c*) Pre-dispatch Inspection:

When wagons are getting loaded bundles are checked.

(*d*) Testing:

Test for confirming yield strength, ultimate tensile strength, percentage elongation, chemical analysis, bend property. Issue of test certificate after dispatch.

13

Packaging of Steel Products

1 INTRODUCTION

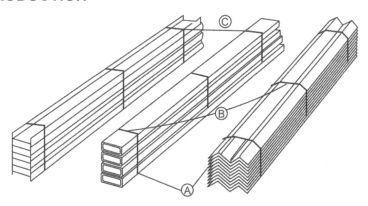

Fig. 13.1 Method of Piling/Bundling of Long Products.

2 OBJECTIVE

2.1 The steel products from the steel plant are dispatched to the stockyards or directly to actual consumers in railway wagons or by road transport. Though the steel plants have been equipped with facilities for lifting/handling/storage to achieve the dispatches of finished material at rated capacity. In most of the units the facilities for on-line/off-line packaging of products are not available except for some portable pneumatic tensioners and sealers.

2.2 These products after production are transferred to the stacking bay, tied/packed manually to whatever possible extent, stacked in dispatch area, and loaded to wagons as and when required. Many times, the packaging is damaged in the plant itself while handling/loading. The materials suffer damage during loading/ transit/unloading at customer's premise/ stockyard.

2.3 Over the years, the customer's expectations have gone up and packaging has emerged as an important element of quality products. This coupled with competition in the market due to liberalization has necessitated a relook at our packaging standards. In the context of the changing business scenario and with stiff competition, the condition in which material finally reaches to the customer has become a paramount importance. A steel consumer today has more demanding customers wants:

- A convenient lift size.
- Adequate packaging.
- No damage during handling, transit or storage (both at plants/ stockyards) either to the package or to the material.

3 CONCEPTS OF PACKAGING

3.1 Role of Packaging

Packaging is a significant tool in the modern competitive market system. Today, almost everything ranging from daily need like bread to sophisticated goods like business machines and computers are funneled into the market in attractive convenient packages.

Packaging is also one of the priorities for better customer's satisfaction. Packaging operation is looked upon as a necessary, vital and integrated part of production operation. Packaging is also closely interconnected with the process of identification, labeling, handling, transfer, transportation, loading storage and stacking of the product.

The purpose of packing is as follows :

1. Protect against ingress of moisture and water.
2. Protect from damage at the point of loading.
3. Take care of stress generated during transit.
4. Protect against damage during unloading.

All the prime materials should be shipped in packed condition to the customer unless otherwise specified by the customer. However, arising can be shipped in only strapped condition unless otherwise asked by the customer.

3.2 Theoretical Aspects of Packaging

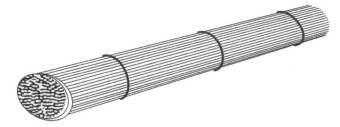

Fig. 13.2 Packaging of Rounds.

(*a*) **Definitions**

There are many definitions of packaging; two most widely quoted are:

1. Packaging is the art, science and technology of preparing goods for transport and sale.
2. Packaging may be defined as means of ensuring the safe deliver of a product to the ultimate consumer in sound condition, at the minimum overall cost.

There is another definition which sets out to explain what packaging is by saying what it does. It is: 'Packaging must protect what it sells, and sell what it protects'. This adds to the first two definitions, the important subject of sales appeal. Packaging must be considered at the design or formulation stage of the product/processing line for the production itself.

(*b*) **Packaging criteria**

The final form of any package can be influenced by many factors but logical packaging development can be achieved by considering various packaging criteria. They are five in number, namely:

  (*i*) Appearance
  (*ii*) Protection
  (*iii*) Function
  (*iv*) Cost
  (*v*) Disposability.

The relative emphasis placed on these criteria depends on the product and its marketing requirements.

(*c*) **Appearance**

The appearance of a package has to perform following important functions:

1. It should ensure proper identity of the product throughout the distribution chain and especially when it reaches the final consumer. The requisites are:

(a) Identification must be clear and positive (no ambiguity).

(b) The identification must be long lasting under all prevalent climatic conditions.

2. It is expected to carry instruction for use.
3. It must carry information about the contents in order to abide by the statutory requirements (as with poisons, most foodstuffs and breakable items).
4. It will usually carry the brand name or the name of the manufacturer, or both.
5. It can act as an important sales aid.

(*d*) Protection

Although protection may not be the most important criterion for a particular packaging situation, but its requirement cannot be ruled out completely. For some products it may be a major requirement of packaging. The protection required by the product varies with the nature of the product, the final destination, the distribution system and the total time of protection requirement. Protection is normally required against two main hazards; chemical and physical.

Chemical hazards

Chemical hazards generally include :

- Product/packaging material compatibility
- Ingress of liquids and vapors
- Loss of liquid vapors
- Micro-organisms.

(*e*) Physical hazard

Physical hazards in distribution may be static or dynamic and can summarized as below:

- Impact,
- Puncturing,
- Vibration,
- Effect of temperature,
- Effect of light,
- Macro-organisms,
- Pilferage.

(*f*) Function

The functions which a package is required to perform depend on:

1. End use of the package.
2. Type of packaging/filling line.

3.0 Concepts of Packaging

End use performance

End use performance of package is significant as faulty performance will lead to dis-satisfaction with the product and reduction in sales. End-use package functions include :

1. Display.
2. Ease of opening.
3. Convenience.
4. Dispensing.

(g) Cost

Factors which contribute to the overall cost of packaging a particular product are:

1. Packaging material cost (carriage inward to factory).
2. Storage and handling costs of the empty package.
3. Filling cost (including quality control and handling of filled packages).
4. Storage costs of the filled package.
5. Transport costs of delivering filled packages (carriage outward cost).
6. Insurance costs involved in transport.
7. Losses due to breakage or spoilage of the product (including loss of goodwill).
8. Effect of the package on sales.

(h) Disposability

The term disposability includes methods in which the packaging material can be eliminated or converted at the end of its useful life as a package. These are :

- Recycling
- Non-packing applications
- Waste disposal
  - Land fill
  - Incineration.

3.3 Approach to Packaging Design

Packaging is no longer in the hands of packers neither it is the last operation on production lines. It is an integral part of :

- Production
- Distribution

464 Chapter 13 Packaging of Steel Products

- Marketing,
- Buying,
- Consuming,
- Waste disposal,
- Recycling,
- Total economy factors,

All the above said need not be factors of the final packing selection but certainly are considered before final decision is made on type of packaging. Selection of packaging method influences/affects requirements of various parties in different manner as shown below :

Table 13.1

Parties involved	Needs
Producer	Productivity, protection, profit
Packer	Ease, simplicity, safety, incentive
Transporter	Ease, economy, identification, sturdiness
Marketer	Sales turnover, space utilization, cleanliness and elegance
Consumer	Risk free buying, consuming convenience, value realization
Government	Consumer protection, resource conservation, environment protection

In a nutshell, packaging is a total activity that helps to achieve all the above needs at minimum overall costs, balancing the conflicting needs for the final goal of consumption. Thus, final selection of packaging depends on:

1. Product characteristics

   Weight, dimension, surface finish, shelf life of material.

2. Marketing techniques:

   Retail outlets, consumer needs, buying capacity, want, competitor's methods, family size, and consumer education.

3. Available packing media, alternatives, materials, design:

   If there are no complaints on packaging, it means the product is given over packaging and too many complaints means under packaging.

3.4 Importance of Steel Product Packaging

The basic factors to be considered in designing an effective packaging system are:

1. Customer needs.

2. Product characteristics.

3. Rigors of handling and transportation.

4. Weather and storage conditions.

4 Packaging of Long Products   465

If packaging is not good, even the best of material may reach to the customer in damaged condition. Some foreign steel producers are using mechanized packaging lines to avoid the damage to the product and reduction in the manual labour.

3.5 Advantages of Packaging of Steel Products

The main advantages of piling/bundling/binding/packing can be summarized as follows:

1. Minimization in space requirement.
2. Avoidance of heavy physical labour.
3. Lowering of noise level.
4. No damage to product while handling/storing/transporting.
5. Lowering of noise level.
6. No damage to product while handling/storing/transporting.
7. No pilferage in transportation.
8. Higher net sales realization.
9. Better appearance.
10. Better customer satisfaction.
11. Lower rejections.

4 PACKAGING OF LONG PRODUCTS

Packaging of long products involves piling, bundling, binding and packing.

4.1 Piling

- It is also called stacking/nesting,
- It can be done manually or automatically.

Advantages of automatic systems are :

- Compact packs
- High capacity performance
- Elimination of hard physical labour
- Can be employed on-line or off-line
- Optimization of production and handling costs.

Proven designs are available for piling of virtually any section *e.g.*, angles, channels, I-section, H-section, rails, flats, double tees etc. Sketch-1 and show the method of piling of different sections.

Fig. 13.3 Piling of stock by chain Transfer.

Salient features of piling and bundling system of a Light and Medium Merchant Mill.

1. 4 cross transfers are provided after cold shear for piling and bundling equipment. One is for 12 m and 3 are for 24 m width; out of these, one is for structural products.
2. 2 lines for bundling of bar products. The bar bundling is provided with counting device.
3. Automatic sorting of under length.
4. Automatic tying of bars with 2.0 to 10 t bundle weight.
5. Structural products are automatically piled to form square packets in nesting arrangement for tying.
6. Flats and squares are also piled and tied. The tying machine uses 6 mm wire.

4.2 Bundling

It is also a type of piling, generally used for bars/round products. Bundling can be done in stages *i.e.*, in first stage small bundles are formed and then these are converted to big bundles. Advantages of bundling are:

- Strict customer requirement regarding lengths can be fulfilled (sorting by length is possible).
- Better handling, even magnets can be used.
- No damage during handling/transportation.
- Lower operating costs.
- Higher sales realization.

4 Packaging of Long Products

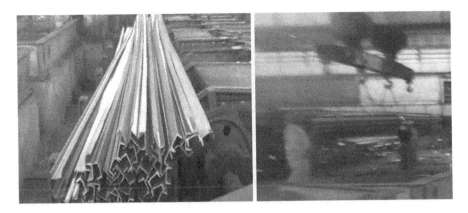

Fig. 13.4 Bundling of Channel.

- Bar counting possible
- Automatic weighting of bundle possible.
- Ease in handling at customers end (within his limited means. Fig. 13.5 Shows the bundling of bars.

Fig. 13.5 Strapping of Rounds bundles.

4.3 Binding

Binding is generally done by :

Strapping   : Using manual portable tools or automatic machines.

Bar binding   : Manual or automatic

468　　　　　　　　　　　　　　　Chapter 13　Packaging of Steel Products

Fig. 13.6 Straps for Strapping of Bundle.

Different designs of manual portable machines (starting from complete manually to completely automatic) are available for strapping. These is commonly used methods for binding the wide variety of steel. Fig. 13.6 above show the binding of bars and binding tool.

❑❑❑

14

Energy Efficient Practices and Technologies

1 INTRODUCTION

Iron and steel industries is one of largest consumer of energy. Due to steady increase in the input cost, the cost of production has increased many folds. Cost of energy accounts nearly one-third of production cost of steel in India. Hence, it is imperative to upgrade the design of equipment/process and the operational parameters in rolling of steels for optimal uses of energy and better quality of product, to survive in present competitive scenario of steel industry. Most attention is required in reheating furnace and mill technology of the rolling mill. A typical energy consumption of a rolling mill is shown below (Fig. 14.1).

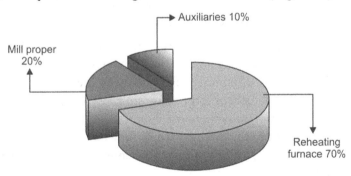

Fig. 14.1 A Typical Energy Consumption on a Rolling Mill.

It can be seen from above that maximum energy is consumed in rolling *i.e.*, about 70% of total consumption is consumed by the furnace, which is followed by mill, about 20% and rest 10% is consumed by auxiliaries.

470 Chapter 14 Energy Efficient Practices and Technologies

Auxiliaries includes:

- Drives for Roller Tables, Fans, Pumps and Shears etc.,
- Operational and Maintenance spares reclaiming facilities etc.,
- Cranes,
- Lighting.

The position of Indian re-rolling units in the field of energy consumption against the international benchmarking is shown below:

Table 14.1

Sl.No.	Energy Parameter	Type of Mill	Indian Re-Rolling Units	International Benchmark
1	Heat (Kcal / t)	Bar & Rod Mills	400-500	300-350
		Section Mills	500-620	320-350
2	Power (kWh / t)	Bar & Rod Mills	95-120	90-95
		Section Mills	65 - 75	45-50
3	Mill Yield % of Input Material	Bar & Rod Mills	90-92	96-98
		Section Mills	85-90	95-96
4	Mill Utilization % of available Hours	Bar & Rod Mills	65-70	80-82
		Section Mills	55-60	75-80

2 ROLE OF REHEATING FURNACE

Reheating furnace plays an essential role in the hot rolling mill production plant. The purpose of a reheating furnace is to provide properly heated billets at the discharge end of the furnace, before they are further processed in the mill. It is therefore, desirable to improve the furnace efficiency for saving more energy and have more yields of possible way of improvement is by controlling the reheating process precisely.

The energy optimization in a reheating furnace in rolling becomes essential to reduce the cost of product and to be price competitive. It enhance the profitability and also improves the quality of the product. Optimization of the process parameters for improving productivity and thus reduction of specific heat consumption is the most important and least cost approach for energy saving. The present walking beam furnace has been developed with state of art technology with low specific fuel consumption of 280 kcal/ton and scale loss limited to 0.6%.

2.1 Type of Furnace

Both the basic types of furnace *viz.*,

(*a*) **Batch furnace:** As the name indicates, in batch furnaces, the entire batch of materials (billet/bloom/slabs) is charged and heated to the desired temperature at a time.

(*b*) **Continuous furnace**: Material is charged from one end and the heated material is discharged from another end.
  – Pusher type.
  – Walking beam furnace.

2.2 Purpose of Heating Metal for Rolling

- Softening of metal suitable for rolling.
- Providing a sufficiently high initial temperature so that rolling process is completed in fully austenitic temperature region.
- Surface scaling for removal of surface defects.

For the energy efficient rolling practice, the reheating furnace plays a vital role. It not only improves the yield of the product, but improves the mechanical properties of the final product. The modern furnaces are fully computerized to avoid over/under heating for achieving fuel efficiency.

It is recommodated to use top and bottom fired reheating furnace to make it more energy efficient and to have less temperature gradient in billet to have consistent rolling with less cobble and homogenization of properties.

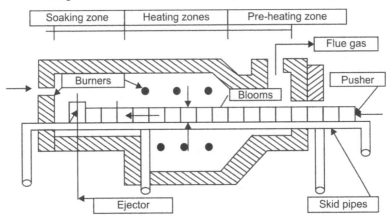

Fig. 14.2 Top and bottom zone fired reheating Furnace

2.3 Problem Associated with Heating

- Achieving the desired minimum temperature consistent with achieving the correct temperature and metallurgical properties at the finishing stand of the mill.
- Minimizing temperature difference between surface and center to a desired level as low as 15°C.
- Minimizing local cold spot (skid mark) due to water cooled skids.
- Avoiding over heating and burning of metal.

- Elimination of scratches on the bottom surface of boom/billet/slab in pusher type furnace.
- Avoiding thermal stresses and cracks.
- Minimizing scale formation, decarburization.
- Effect of sulphur in fuel, which cause severe scaling.

3 OPERATIONAL ENERGY EFFICIENT IMPROVEMENT PRACTICES IN FURNACE AND MILL

Operational improvement are the most vital factors for fuel saving and improvement in the quality of heating. Any improvement or the modification in the design of furnaces has to be appropriately adopted in the operational practice to gain the benefits. Successful implementation of most of operational improvement depends upon the operating personnel. Such important practices are:

3.1 Optimal Operation of Combustion System

The basic requirement for good combustion of fuel in reheating furnace are :
- Complete combustion of fuel,
- Desired flame configuration,
- Minimum oxygen in the outgoing product of combustion,
- No occurrence of overheating of furnace elements,
- Minimum pollution.

The excess air in flue gas should be kept at minimum level. The excess air should be set in such a way that there is no unburnt fuel in the flue gas. To achieve complete combustion of fuel with minimum excess air, factors such as type of burner, fuel, combustion air pressure. It's preheat temperature is also very important factor. The walking beam furnaces have oxygen analyzer to trim and control the oxygen in flue gas. The heat loss is higher at higher excess air resulting in higher fuel consumption. The scale loss is also high at higher excess air. Fig. 14.3 shows the rate of fuel saving against excess air-fuel ratio (m).

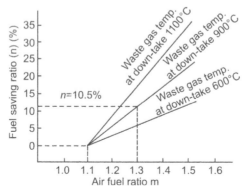

Fig. 14.3 Rate of Fuel saving by reducing Air-Fuel Ratio.

3.2 Optimization of Thermal Regimes

The basic requirement of heating of stock in the reheating furnaces is that the metal should reach the desired level of temperature within the permissible tolerances. It is also necessary to make the metal ready only when it is to be discharged for rolling. Heating the metal much before it reaches discharge end leads to high fuel consumption and scale loss, as the stock remains in high temperature zone for a longer period. In case of continuous pusher type or walking beam furnaces, it is necessary to maintain required temperature profiles along the length of the furnace in such a way that the metal will get ready when it reaches to the discharge end. In case of batch-operated furnaces, the stock is heated at prescribed rate of rise of temperature (known as RAMP heating mode or step heating mode) with the help of programmable temperature controllers.

3.3 Furnace Pressure Regimes

The pressure inside the reheating furnace is normally maintained at 0.5 to 1.0 mm WC, to avoid ingression of atmosphere air. This is achieved by operating the chimney damper in auto/computer mode in walking beam. The pressure inside the reheating furnaces is normally maintained positive at skid level in soaking zone. When the furnace pressure exceeds the normal limits, flame shoots out of the doors causing heat loss and may damage to the furnace walls. More detrimental effect is the loss of combustion air. Thus, both high and low furnace pressure will result in increased fuel consumption.

3.4 Hot Charging of Material

Hot charging of the material is one of the most efficient fuel saving measures and now being used worldwide. However, this requires installation of suitable facilities in the mill and also requires the good synchronization of rolling with the feeding mill. In hot charging, the available sensible heat of the stock can reduce the heat requirement in the furnace considerably, thus improving the fuel saving. The walking beams furnace has facilities to choose the charging temperature of slabs/billets/blooms and accordingly, then computer choose the heating curves and set zones temperatures for optimum heating. Fig. 14.4 shows how hot charging makes furnace energy efficient.

It is important to note that the benefits of hot charging are tangible only when the thermal regime is adjusted suitably, otherwise overheating/melting of stock takes place. The regimes can be adjusted only when a considerably number of blooms/slabs are charged continuously. It is practically impossible to work out regimes for hot charging where it is intermittent and temperature of input stock is widely varying. In some plants to supply consistently uniform temperature of stock, hot boxes are used.

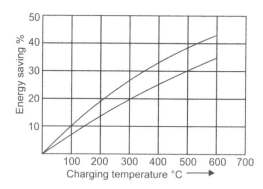

Fig. 14.4 Potential for Fuel Saving by Hot Charging of Slabs/Blooms.

3.5 Furnace Productivity

Furnace productivity is one of the important factors, which affects the specific fuel consumption in the furnace. There is an optimal level of productivity at which the furnace should be operated to derive maximum thermal efficiency. The under loading of the furnace results in high fuel consumption. To avoid the under loading, it is essential to run minimum numbers of furnaces.

3.6 Hearth Coverage

Optimization of rolling schedule can increase the hearth coverage of a furnace, leading to increased furnace throughput and reduced energy consumption. However, the potential reduction in energy consumption will depend significantly on design of furnace and the feasibility of improving hearth coverage. Fig. 14.5 shows the effect of hearth coverage on furnace productivity and specific fuel consumption.

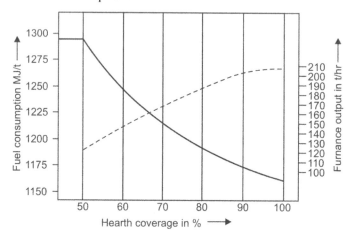

Fig. 14.5 Effect of Hearth Coverage on Furnace Productivity and Specific Fuel Consumption.

3.7 Use of Coil Box

During rolling in the finishing stands of the Hot Strip Mill (HSM), the tail end of the transfer bar will experience a temperature drop. Normally, it is necessary to compensate for this drop of temperature by increasing the speed of mill during rolling, which results in increased energy use. The coil box reduces temperature loss and reverses the transfer bar between the roughing and finishing mills, thus eliminating the temperature rundown and reducing the electrical energy requirement of mill by ~10%. The use of a coil box also allows lower drop out temperature from the furnace, which would result in a ~5% reduction in fuel saving.

3.8 Unfired Preheat Zone

The length of modern walking beam reheating furnaces is in the range of 40-50 m and there is 18-20 m unfired preheating zone. One of the biggest influences on furnace efficiency is the length of unfired preheat zone. Within this zone, excess energy in the waste gas is transferred to the slab/blooms/billets, while still retaining sufficient energy to allow economic preheating of combustion air. In a conventional furnace operating with cold charge slabs/blooms, the waste gas temperature on the entry to the recuperator is in the range of 850-1000°C, while for a furnace with 20 m unfired preheat zone this is reduced to the order of 700°C. It means that energy released to the waste gases can be reduced by ~30%, leading to an overall reduction in energy use of ~10%. The benefits of a long unfired preheating zone decreases when the temperature of charged slabs/blooms increases.

3.9 Recovery of Waste Energy

After the transfer of sensible heat into the slabs/blooms/billets being heated in unfired zone of walking beam, the next largest energy recovery can be from the sensible heat in cooling water and in waste gas. Efficient furnaces will operate such as to minimize the losses to cooling water and the waste gas encompasses conventional energy conservation technologies, such as increased levels of skid pipe insulation, as well as standard combustion improvement such as excess air control and combustion air and fuel pre-heating *via* recuperator. Of these latter techniques, combustion air preheating is most commonly practiced. The advantages of preheating of combustion air are saving in fuel, increase in flame temperature. Fig. 14.6 shows the rate of fuel saving by preheating of combustion air. Energy recovery from skid system can be used to heat water or to produce steam depending on the necessities of the mill. It also shows the fuel saving of 20% at the combustion temperature 400°C and the waste gas temperature of 900°C.

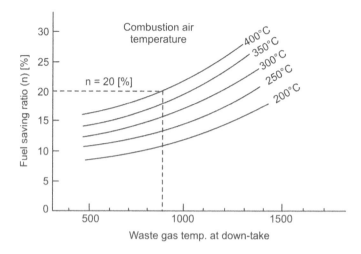

Fig. 14.6 Rate of Fuel Saving by Preheating Combustion Air.

3.10 Stack Efficiency

The stack efficiency is a measure of the amount of energy that is carried away with the exhaust gases. The higher process temperature, the higher exhaust temperature and therefore, the lower the stack efficiency. The stack efficiency is approximately the difference between the flame temperature and the exhaust temperature; divided by the flame temperature.

3.11 Thermal Cover of Roll Table in the Mill

Thermal cover along the tables between the roughing and finishing mills will save ~70% of the transfer bar temperature drop, considering a transfer bar thickness of 25 mm, a roughing mill exit temperature of 1060°C and a finishing mill entry temperature of 1030°C. Thermal cover may also allow lower furnace drop out temperatures in association with coil box.

3.12 Computer/combustion Control Model

Computer/combustion control model optimize the sensible heat that slabs/blooms/billets absorbs and can be applied along the full length of a furnace to achieve the desire drop out temperature with minimal fuel use. When firing zones are not fully isolated and, therefore, not subject to individual control, this leads to inefficient fuel use due to uncontrolled flow of waste gas within the furnace atmosphere. Computer models can combine information concerning fuel CV, excess air in furnace atmosphere and firing zone temperatures as well as other parameters such as slab/bloom/billets entry temperature, maximum hearth coverage and mill status etc. The main advantage of process control system is their ability to optimize the ramping of furnace set point temperatures

3 Operational Energy Efficient Improvement Practices in Furnace and Mill 477

over time during unscheduled delays on mill. There will be reduction in scale build up during period of low production, as facilities will be available for lowering the furnace set point temperature. Energy consumption will also be reduced by 3-7%.

3.13 Discharge Temperature

The fuel consumption in a reheating furnace depends not only on the mean discharge temperature, which itself dependant on the size of the slabs/blooms as well as the quality of the steel but also on production time, and the schedule downtime and the delay time. Thus, higher discharge temperature will increase fuel consumption, scale formation and reduce furnace efficiency. Adherence of optimum thermal regimes and delay strategy will ensure optimal discharge temperature resulting in reduction in fuel consumption and scale formation.

Table 14.2 Burning and Reheating temperature for different carbon content in steel.

Description	Temperature for different carbon content						
Carbon content in percentage	0.1	0.2	0.3	0.5	0.7	0.9	1.1
Burning Temperature in °C	1490	1470	1410	1350	1280	1220	1180
Reheating Temperature in °C	1350	1320	1280	1250	1180	1120	1110

3.14 Using Higher Mill Speed

To save the power consumption, the spare capacity of motors can be utilized to increase the mill speed of rolling. This also saves the specific heat consumption as time taken for rolling cycle will be less and rolling can be started with less discharge temperature, due to higher finishing temperature of the finished product.

3.15 Using Bigger Diameter Roll

Roll housing always has a cushion, which can be utilized for bigger diameter roll.

Table 14.3 The selection of max./min. diameter vis-a-vis stand size.

Type of stand	Dia of pinion mm	Stand opening Max. mm	Stand opening Min. mm	Dia of roll Max. mm	Dia of roll Min. mm	Max. Take-off mm
Horizontal	500	600	480	535	480	55
Vertical	450	520	380	420	380	40
Horizontal	400	500	380	420	380	40
Horizontal	350	470	340	370	340	30
Vertical	350	460	340	370	340	30

478   Chapter 14   Energy Efficient Practices and Technologies

3.16 Running Mill Concept

Generally, it is seen that at the start of the shift, mill is abruptly stopped at the beginning of the shift by operational personnel of take over shift to check the pass/tackles condition, for tightening of the tackles and to know the roll gap. In addition, mechanical staff also would like to know about the equipment condition during mill check-up, without knowing that the furnace is ready to deliver the metal and stoppage at this juncture will not only affect the furnace health, but will also cause wastage of the energy. Running mill concept is evolved to avoid such practice and these routine and avoidable activities can be clubbed and taken up during roll/pass changing time.

3.17 Establishing Communication System Between Furnace and Mill Proper

It is very important for an energy efficient mill to establish a well effective communication system between mill and furnace. Any planned/emergency stoppage should be communicated to furnace staff by mill personnel, atleast 30 minutes before the planned stoppage of mill to take the appropriate action in time to reduce the intake of gas to the furnace. Like-wise delays from finishing, cause stoppage of mill to furnace should also be informed in advance. The advance communication system will help in this regard to achieve the establishment of the effective system.

3.18 Idle Running of the Mill

Idle running of mill for longer time should be avoided to save the power consumption of the mill.

3.19 Automatic Stoppage of Mill Drives/Roll Tables

Provision is to be made in the mill for automatic stoppage of the mill drive and roll table in case of interruption in the rolling.

3.20 Use of VVVF Drives with Air Blowers and Auxiliaries

Energy loss is quite high in case of AC motors running at fixed speed. Thus, for optimum utilization of power (low-output) required output for loads which are of varying nature such as for pumps, roll table etc.

Fast and accurate control of torque and speed of drives is required under all operating conditions of rolling process.

The use of VVVF drives control system has a direct impact on the quality of end products since mill drive motors have high dynamic response and have high efficiency. It also reduces spare part inventory. It gives a fuel saving upto

4 Use of Energy Efficient Operating Technologies 479

1.5 – 2 % and saving in scale reduction upto 0.5% and power reduction of 1.2 kWh per ton of steel production.

Variable voltage variable frequency drives are combination of latest microprocessor based digital technology, static inverter equipment as power switching devices with other hardware platform, software and high speed signal processing. These VVVF drives use the following latest state-of-art technology

- Modern Integrated Gate Bipolar transistor (IGBT) technology resulting in low loss operation.
- Integrated Gate Commutative Thyristor (IGCT) for higher drive ratings with direct torque control.
- Vector control, v/f control and servo controls as closed loop controls depending upon application.

4 USE OF ENERGY EFFICIENT OPERATING TECHNOLOGIES

The main consideration in modern mills is for the superior quality of product and low cost of production. In recent past, many technological developments have been taken place in the mills area and main focus are on improving energy conservation, mill availability/utilization, rolling rate, yield, dimensional accuracy, surface finish, mechanical properties, inspection procedures, packaging etc. Reheating furnaces are being designed with computer control for reducing heat consumption and uniform heating of stock. Hot charging and hot direct rolling is being practiced to utilized sensible heat of concast bloom/billet. Low temperature heating and rolling reduces fuel cost and at the same time, increases product yield because of less scale formation. New mills are being designed with about 800°C dropout finishing temperature.

The quest for higher input billet and high finishing speed always remain with the modern mill designer. Concast rounds/squares of 200 mm diameter are being used to produce wire rods and to rebars of different sizes. Modern bar mills of speed upto 40 m/sec are in use. The salient features of high speed mill are rigid horizontal/vertical cantilever stands in roughing stands, housing less stands in intermediate and precision sizing block before finishing stands for size free rolling arrangement, bar counter and automatic bar bundling and binding facilities.

The compact housing, better guides, quick roll change device, tension free rolling in roughing and intermediate stands, loop control, modern cooling beds and size free rolling have helped in achieving higher mill utilization. Rolling failures have reduced to a large extent by proper mill monitoring

480 Chapter 14 Energy Efficient Practices and Technologies

through PLC, computerization, CCTV and other fault finding devices. Mill utilization has increased to over 90% in modern mills. Control in the furnace and lower dropout temperature has reduced the scale formation to less then 1.0%. Endless rolling, lower cobble generation and less scale formation have helped in achieving about 98% yield.

On the accelerated cooling by Thermax, Tempcore and Temprimar processes have helped in achieving designed strength level of TMT bars by optimizing cooling parameters. Desired yield strength has been achieved by controlling depth of tempered martensitic layer and self-tempering temperature of the core by mathematical modelling and micro structural engineering.

5 IMPROVEMENT IN THE ROLLING OF REBAR TECHNOLOGY

5.1 Reheating

The major developments relate to stock inspection, reheating and conditioning. Surface and internal defect detection is important to ensure appropriate feedstock quality. Several techniques like the ultrasonic, electrical discharge, induction heating and laser based ones are used for this purpose.

The modern reheat furnaces are of the walking beam type. They offer the advantages of uniform heating, low decarburization, scale losses (less then 1%) and energy consumption (0.28 Gcal/t). The use of computerized combustion control, which can facilitate real time dynamic temperature monitoring and control, has led to a reduction in energy consumption by almost 10%. Some of the other energy conservation measures include lengthening of heating zone, decrease of exhaust gas temperature, furnace insulation, optimization of regime and waste heat recovery.

5.2 Low Temperature Rolling

Low temperature rolling is another technique to reduce energy consumption. The lowest rolling temperature can be determined through mathematical modelling, which takes into consideration factors like ductility of steel, load on the roughing stand and finish rolling temperature and metallurgical property requirements. The mills are now being designed for a drop out temperature of about 800°C.

The new trends is to employ hot charging of stock (HCR) and thereby save on energy consumption by about 30-50%. This practice, developed by NKK, requires the upstream concast practice to be extremely good so that internally sound blooms with no surface defects are available. HDR (hot direct rolling)

5 Improvement in The Rolling of Rebar Technology

is also being done in some mills. HDR requires synchronization with casters and places the same quality requirement on the concast practice as the HCR. In order to minimize the incidence of rolled in scale, high pressure rescaling of the stock on exit from the reheating furnace is done.

5.3 Increased Capacity

There has been a trend towards increased inputs stock sizes and high finishing speeds. Concast rounds/squares of 200 mm size are used to produce rounds of sizes 15 to 75 mm. In order to satisfy the varying needs of the customers, four different layouts of bar mills have come to be widely used.

(*i*) High speed bar mill.

(*ii*) High productivity bar mill with slitting.

(*iii*) Flexible bar mill.

(*iv*) Bar mill for quality and special steels.

5.3.1 High Speed Bar Mills

High speed bar mills have finishing speeds of upto 40 m/s and the product sizes range from 8 mm to 40 mm in diameter. A typical high-speed bar mill comprises the following:

- Roughing mill with horizontal/vertical cantilever stands. The cantilever stands are cheap and require less space and have less maintenance requirement.

These stands features:

1. Universal spindles with spindle spotting.
2. Chock spindle carriers for off-line setup of stand assemblies, better support and alignment of roll end and coupling box.
3. Quick disconnect service couplings for water and air/oil lubrication.
4. Minimum unit piping on stand assembly for better maintenance and operation.
5. Universal container for both horizontal and vertical stands to minimize off-line stand equipment.
6. Stand positioning through hydraulic cylinder.
7. Duplex roller thrust bearings.
8. Axial stiffness locking bar on work side entry and exit chock assemblies to eliminate axial movement.
9. Axial adjustment of ± 3.0 mm.
10. Mechanical disc springs for increased stiffness, ease of maintenance and elimination of hydraulic unit piping.

11. Dual roll neck seal arrangement for prevention of ingress of water and scale.
12. Four (4) row cylindrical roll neck bearings for increased separating force capacity.
13. Air/oil lubrication of the roll neck bearings for better lubrication.
14. Symmetrical roll parting adjustment using a hydraulic motor mounted on the spindle carrier.
15. Spring loaded, hydraulic release bed clamps, two (2) entry and two (2) exit for increased stand rigidity.
16. Common designed on-line beds for both the vertical and horizontal stands minimizes spare parts

Intermediate mill with PRD (pre-stressed rigid design) cartridge type housing-less stand. The PRD system has low mill spring (about 0.2 mm) and offers quick roll change facility, requiring no adjustment after roll change. As a result, the mill availability is increased. These stands have low maintenance requirement and adaptable to various configurations. The H/V finishing blocks have speeds of about 36-40 m/s and use tungsten carbide rolls.

H/V finishing block: In addition to advantages shown above PRD stands, the H/V finishing blocks have speeds of about 36-40 m/sec and it use tungsten carbide rolls, which provide high tonnage/pass and very good surface finish of the finished products.

Fig. 14.7 H/V Finishing Block.

- The modular and versatile design features of the No-Twist Mill allows the mill to be supplied in 4, 6, 8 or 10 stand configurations with pass reductions from 10 to 25%. It allows a wide range of steel grades to be rolled from carbon steel to stainless steels and even tough to roll heat resistant alloy steels.
- The No-Twist Mill can be configured with 250 mm, 230 mm or 160 mm. Cantilevered roll housings depending on the processing requirements

and product size range. Due to the unique design of the pinion and bevel gear housings the roll housings are interchangeable allowing the mill configuration to be changed to increase the mills product size range or enhance the processing requirements through lower temperature rolling.

- The roll housings are generally equipped with symmetrical roll parting adjustment with either manual or remote powered screw downs. They are capable of adjustment under load adjustment if required (not provided in base offer). Provision for roll parting position feedback and bearing temperature monitoring are provided on all roll housings. The roll housings use carbide roll rings for improved surface quality and extended roll life. Dual pressure hydraulic or mechanical roll mounting can be provided depending on rolling demands; either system is interchangeable to cater to the changes in rolling regime.

Reducing/Sizing Mill

Reducing/sizing mill is provided in wire rod line to produce rods in the size range from 5.5 mm to 26 mm.

A four (4) stand Reducing/Sizing Mill is installed between the No-Twist Mill and laying head to provide the following features:

- Single family rolling throughout the complete mill, simplifying the rolling process and increasing mill efficiency by reducing roll change times. With the high annual production specified the Reducing/Sizing Mill becomes a critical feature of mill layout to provide added capacity and rolling flexibility for the future.
- Provides increased maximum finishing speed for wire rods 5.5 mm to 110 m/s.
- Improved rod tolerance to ± 0.1 mm and ovality to 0.12 mm over the complete size range (5.5 to 26.0 mm).
- Provides thermo-mechanical rolling capability over the complete size range depending on roll housing selection.
- Provides improved surface quality through patented round-oval-round-round-round pass sequence.
- Controlled temperature rolling, with thermo-mechanical rolling capability down to 750°C. Entry temperature depending on size and steel grade.
- Increased size range 5.0 to 26.0 mm with low temperature rolling through to 26.0 mm depending on steel grade.

- The 150, 230 or 250 mm roll housings are interchangeable within the reducing mill roll unit for increased capacity which depending on size range and rolling process.
- Powered screw down under load is available for all roll housing sizes.

Rake type cooling bed with HSD (High Speed Bar Delivery) system.

Fig. 14.8 High Speed Cooling Bed.

- **Short bar recovery**
- **Bar counting system**
- **Automatic bar bundling and binding facility**

Fig. 14.9 Typical Bundling Station with Strapping Machine.

Inclined Pinch Roll and Laying Head

The latest inclined intelligent pinch roll and laying head arrangement is specifically designed to provide the optimum setup to cover the maximum size range at the highest operating speeds while providing a consistent ring pattern on to the conveyor for uniform cooling.

5 Improvement in The Rolling of Rebar Technology

Fig. 14.10 Inclined Pinch Roll & Laying Head.

To improve the ring pattern of large diameter product (maximum 26 mm) at lower laying temperatures the laying head is inclined at 20° to the horizontal, with the pinch roll inclined at 10°. It minimize the wear on the pinch roll entry and delivery guide pipes.

5.3.2 High Productivity Rebar mill with Slitting Facility

Fig. 14.11 4-Strand Slit Rolling.

4-strand slitting for 8 to 12 mm diameter rods and 2-strand slitting for 16 to 20 mm diameter rods are being used and yearly production of upto 700,000 T has been achieved. The maximum speeds on the cooling bed are 18 m/s and 15 m/s for 2-strand and 4-strand slit rolling respectively.

5.3.3 Flexible Bar Mill

Such mills produce small to medium section and comprise rigid roughing stands: PRD stands in intermediate and finishing stands, cooling bed, multi-stand in-line straightening machine, cold flying shear and automatic stacker.

5.3.4 Bar Mill for Quality and Special Steels

Some of the important features of these mills include high reduction stand coupled with two 3-roll blocks, equipment for controlled temperature rolling and in line heat treatment. It has been possible to produce cold-upsetting steel, ball bearing steel and valve steel, through optimization of process technology. The mill has features like intermediate, finishing and sizing blocks of three-roll type and has facilities like speed control system, walking beam, hot bed and controlled cooling.

5.4 Dimensional Tolerances

Several developments have contributed to produce rolled products with improved dimensional tolerances of rebars. Single strand rolling is being used in preference to multi strand rolling because of lower roll deflection in the former. Stands, with high rigidity and H-V configuration in the roughing and intermediate stands together, with no-twist blocks have helped in achieving improved tolerance. Tension free rolling is now possible through loop control in the finishing stands. Inter-stand tension causes size variation along the bar. It results in oversized head and tail portions that have to be removed before dispatch. In the roughing stands tension control is achieved through measurement of torque at successive stands and adjusting the speeds of the stands wrt to a pre-determined value. Dimensional tolerances are affected by the quality of the rolls. Tungsten carbide sleeve rolls are being used in the finishing blocks as well as in the intermediate stands. Improved roll cooling systems have also helped in retaining pass contours over longer durations. Pass designs are being done with the help of computers and CNC machines are now being employed for roll turning. Improved heating practices have also helped in minimizing stock dimension variations along the length.

5.5 Endless Welding and Rolling (EWR)

Endless welding and rolling (EWR) (Fig. 14.12) is a process where billets are welded on-line at the delivery of the reheating furnace. This facilitates uninterrupted production by eliminating the gap between two billets.

The welding process consists of the following:

- Billet descaling.
- Feeding the head end to the tail end of the billet being rolled.
- Start up and speed-up of the EWR unit to billet speed.
- Clamping of two billets.
- Being rolled position adjustments of the incoming billets to line-up with the centerline of the billet.
- Preheating.
- Flash-melting.

5 Improvement in The Rolling of Rebar Technology

- Pressing for welding.
- Cooling and debarring.
- Unclamping.
- Slowdown to stop the EWR unit
- Return stroke to the starting position.

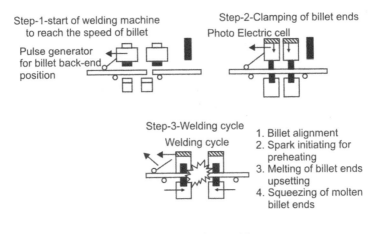

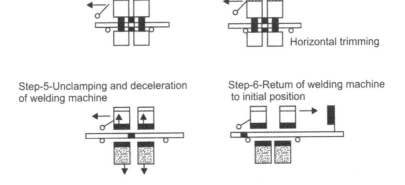

Fig. 14.12 End less welding of Rod (EWR) Process.

The weldable range for square billet is 100 × 100 mm to 200 × 200 mm and 100 to 200 mm diameter for round billets.

The EWR process yields the following benefits:

(*i*) Improved mill efficiency and productivity through elimination of the gap time between two billets.

(*ii*) Elimination of head and tail cropping resulting in improved yield.

(*iii*) Reduced costs due to lower specific energy consumption.

5.6 Improved Mill Utilization

Mill utilization in bar mills has increased on account of several factors. Among the more important ones are compact housings, better guides, and quick roll changing device, tension-free rolling, loop control, modernized cooling beds and on-line straightening machines.

5.7 Size-free Rolling

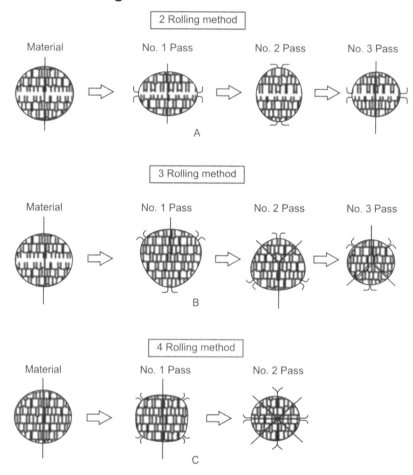

Fig. 14.13 Size Free Rolling.

Size-free rolling is a new technology (Fig. 14.13), developed by Kawasaki Steel, which, use a four-roll technique to enable unlimited ranges of sizes to be rolled with few roll changes. The advent of tungsten carbide rolls has helped in improving roll life significantly while rolling failure have been reduced by the use of CCTV's, PLC's and several fault finding devices.

5.8 Thermo-Mechanical Treatment (TMT) Process

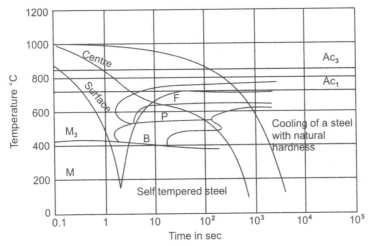

Fig. 14.14 Thermo-mechanical treatment (TMT) Process.

The thermo-mechanical treatment process (Fig. 14.14), known by various names such as Tempcore or Thermax, has become popular in the production of rebars. Rebars following hot rolling are made to pass through specially designed cooling boxes. Cooling in these boxes is aimed at affecting a predetermined thickness of martensitic layer on the surface of the bar while the core remains austenitic. After the rebar emerges from the cooling boxes, the heat of the core tempers and the martensitic layer while the austenitic core itself transforms to ferrite and pearlite. The process thus comprises three stages.

(*i*) Quenching of the surface layer.

(*ii*) Tempering of the martensite.

(*iii*) Transformation of the core.

The cooling process results in significant improvement (150–250 mpa) in yield strength and bar with 415, 500, 550 and 600 mpa are the popular grades available in the market.

6 RECENT DEVELOPMENTS IN PRODUCTION OF STRUCTURALS

The impetus for developments in regard to the rolling of structural sections includes the need to enhance the range of product sizes and production capacity, add value to the rolled product and improve dimensional tolerances. Some of the major developments related to the production of structural sections have been outlined below.

6.1 Size-Free Rolling of H-shapes

Size-Free Rolling technology for H-shapes, developed by Kawasaki Steel Corporation, has been used to overcome the limitations of conventional rolling of these sections. In conventional rolling, the inner width of the web is determined by the width of the horizontal rolls of the roughing and finishing universal mills while the flange depth depends on the depth of caliber of the edger mill rolls. The thickness of the flange and the web are controlled by roll gap adjustment. It is not possible here to vary the web and flange thickness while keeping the same web depth and flange width. Also, on account of internal stresses, which may result in shape deformations, there is a limitation on the flange to web thickness ratio as well as on the web thickness itself.

The size-free rolling technique comprises the following (Fig. 14.15):

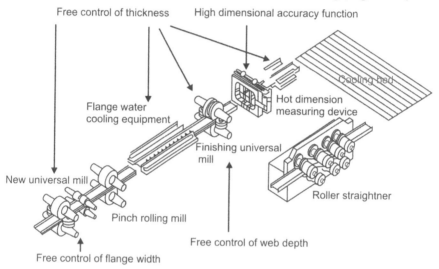

Fig. 14.15 Size-free Rolling Technique.

Free control of web-depth

The vertical rolls of the universal mill are used to effect web inner-width reduction while the horizontal rolls restrain the web. An essential component of these techniques is the use of horizontal rolls having adjustable width. It is possible to roll H-shapes with uniform web depth but varying flange thickness.

Free control of flange width

In the new compact type universal mill, the flange width control and web off-center control operations are separated from each other. The edging mill, which is attached to the universal mill, has drum shaped, groove less rolls. The inner surfaces of the flanges are supported during rolling by guides having

6 Recent Developments in Production of Structurals

adjustable widths. In order to control web off-center, the centre of the flange is aligned with the pass-line of the mill.

Free control of thickness

Through the use of size-free rolling technology it has become possible to have higher flange/web thickness ratios than before. However, special consideration is to be given to the fact that there is, as a result, an increased asymmetry of mass distribution across these sections. Because of the differential cooling of the web and the flanges, the web tends to develop compressive stresses while the flanges experience tensile stresses. Since, this stress pattern may result in buckling of the web, control of the temperature profile across the H-section is essential. A cooling strategy is therefore, required to overcome this problem. Since, cooling is intimately linked with the material properties, simulation of internal stresses and material properties is an important component of cooling strategies for achieving thickness control.

The cooling itself comprises two parts:

(*a*) Localized cooling for the external surfaces of profile.

(*b*) Uniform cooling over the entire flange width (Fig. 14.16).

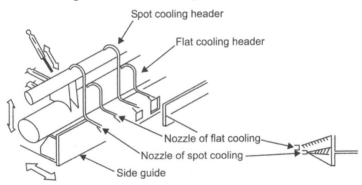

Fig. 14.16 Uniform cooling over the Profile.

6.2 Quenching and Self Tempering (QST) Process for Structural Steel

The QST process has been jointly developed by Arbed, Luxembourg, CRM, Belgium, and British Steel for producing superior quality beams. This technology seeks to address the following limitations of the conventional thermo-mechanical (TM) rolling of beams:

- In the conventional rolling of beams, high thickness cannot be rolled without sacrificing on yield strength (YS). One way of achieving high YS values in thermo-mechanical control rolling is by increasing the

Carbon equivalent (CE). However, high CE values have a deleterious effect on weldability and toughness, besides entailing increased costs.
- Thermo-mechanical TM rolling requires high deformation rates at low temperatures, which have implications on the load bearing capacity of the mills.

The QST process comprises the following:

(*i*) Selective cooling of the flange-web intersection at the roughing and intermediate groups to ensure uniformity of temperature across the beam section when it enters the main cooling bank located after the finishing stand.

(*ii*) Uniform quenching of the beams in the cooling banks, from a temperature of about 850°C to a self-temperature of about 600°C. The idea here is to stop the quenching process before the core gets affected (Fig. 14.17).

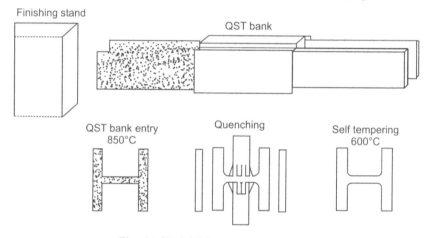

Fig. 14.17 QST Process for Beam Profile

The QST process uses a mathematical model to achieve computerized control over the key process parameters. Using temperature data during rolling, necessary modulations in the cooling intensity during selective cooling can be made. Also, water distribution and stock speed in the main cooling banks can be modulated through the computerized process control.

Some of the advantages of the QST technology include:

- Ductile-brittle transition temperature is reduced and low temperature toughness improved.
- Property variation in the QST process is less than that in the TM process on account of computerized process control.
- Oxy-cutting of QST beams is easier as no pre-heating is required to avoid cracking.
- Overall fabrication costs can be reduced.

6.3 Endless Welding and Rolling (EWR) for Structural Steel

The EWR technology. It is used to weld billets on-line. Prior to welding, on-line de-scaling of billets is done so as to ensure proper passage of current for the flash welding process. While in one billet is being bitten by the roughing stand rolls, the head of succeeding billet is welded to its tail end. Subsequently, cooling and deburring operations are done. All of these operations are done at billet speed, employing clamps for centering and holding the ends of the billets during the operations. In the final sequence of a cycle of operations, unclamping is done and the unit slows down, stops and then traverses back on its return stroke to reposition itself for the next cycle of operations.

The advantages of EWR process include higher yield because of reduced cropping and cobbles and greater production capacity on account of elimination of the time-gap between successive billets.

6.4 High Precision Rolling (HPR)

The HPR system is used to achieve high tolerance levels on the structural rolled product. The HPR system is located after the finishing stand and comprises a vertical and a horizontal stand. The horizontal stand is axially adjustable while the vertical stand is both axially and horizontally adjustable. Both the stands are housing less have rings of high alloy steel fitted onto the rolls. The horizontal stand has pre-stressed rolls; the force applied for pre-stressing is 20% higher than the expected roll separating force. Pre-stressing of rolls ensures that the effect of thermal expansion of the rolls is nullified. In order to minimize spread, small roll sizes and reduction are used.

HPR can be used for rolling both for rods and bars as well as for sections.

6.5 In-line Head Hardening of Rails

Because of the limitations of off-line head hardening systems for rails in regard to the case-depth achievable as well as the production costs, in-line head hardening or rails has assumed significance. The main objective of the process is to affect a certain thickness of hardened layer on the head of the rails through in-line cooling (Fig. 14.18). The web and foot are also cooled; however, the aim here is to ensure that thermal stresses are minimized so that stock remains straight. The cooling unit incorporates roller guides for guiding the rail, while horizontal rollers transport the rail through the cooling system. Temperature of the rail before entry into the cooling system, as well as temperature of water, is measured and the data fed to a computer which then determines the dwell time and the water flow rates for the cooling of rail. The temperature of the rail on exit from the cooling system is also measured and serves as an important feedback for process control.

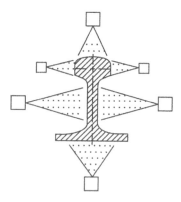

Fig. 14.18 Head Hardening System for Rails.

6.6 Vanadium Steels

Vanadium containing steels (V-steels), originally developed by Bethlehem Steel, USA, have been used in the production of structural sections and plates. In these steels precipitation strengthening is brought about by the precipitation of Vanadium Carbo-nitirides V (C, N) in ferrite. In order to ensure that there is no free Nitrogen, which reduces toughness, the V: N ratio in such steels needs to be maintained at the stoichiometric level of 3.64:1 or greater. Under this condition, the yield strength (YS) of steel has been found to increase with increasing Nitrogen content as the V (C, N) particles get more finely dispersed.

Some of the V-steel sections include parallel flange channels and forklift mast sections.

Index

A

A cluster mill 82
Accelerated cooling 68
Alignment 312
Alpha iron 62
Angle design method 325
Angle of bite 18
Angle of contact 15
Angular acceleration 54
Angular Contact 132
Appearance of a package 461
Asymmetrical rolling 427
Austenite 63
Automatic sorting 466
automatic stoppage 478
Automatic sub-merged arc welding 123
Automatic tying 466
automation 3
Availability 452
Average diameter 27
Axial load 171

B

Backlash 333
Bad apex 316
Bakhtinov formula 9
Balanced draughting 326
Ball bearing 131
Bar mills 481
Barrel pass 156
Batch furnace: 470
Beam blank 337
Beam blank method 324
Beam method 323
Bearing collars 413
Bearing temperature 444
Bellying of bloom 159
Bend web 392
Bifurcates 336
Billet 165
Billet descaling 486
Binding 465
Bite 364
Blooming mills 138
Blooms 135

Blunt apex 316
Boron 67
Bottom pressure 33
Box pass 151
Breakdown passes 23
Brittle Fracture 120
Buckling of the web 369
Bullhead pass 150
Bundling 465
Burnt Blooms 161
Bush 134
Butterfly design 284

C

Cambering 409
Capital repairs 440
Carbon equivalent 234
Cast bloom 419
Cast irons 63
Cast number 457
Cast steel 103
Cemented carbide roll 104
Cementite 63
Changing programme 441
Checking of dimension 447
Chromium 67
Clear chill rolls 96
Close design 261
Closed Pass 24
Coefficient of linear expansion 48
Coil box 475
Cold dimension 49
Cold working 59
Collar gap 250
Collar taper 25
Collaring of rolls 314
Color codes 455
Communication system 478
Compact housing 479
Compact housings 488
Composite sleeve rolls 105
Composition of the flux 127
Concave edges 270
Conformity of the rolled products 453
Contact area 15

Index

Continuous casting process 399
Continuous furnace 471
Continuous mill 87
Continuous rolling 50
Continuous sheet bar mill 84
Conventional cooling 113
Cooling beds 173
Copper 67
Cost of energy 469
Counter 323
Counter flange method 323
Crack 160
Cracks 388
Critical temperature 66
Cross country 86
Cross country mill 168
Cross Rolling 77
Crystalline features 121
Crystallization 137
Crystallization temperature 60
Cupping 454
Curly marks 333
Customer's satisfaction 460
Cutting of metal 314
Cyclic temperature variation 111

D

DC motors 172
Dead flange 46
Dead hole 44
Defective material 456
Delivery guides 352
Depth of pass 156
Depth of Pass 26
Design Efficiency 407
Design of fillet 327
Diagonal Difference 237
Diagonal method of rolling 361
Diagonal pass design system 401
Diamond-diamond 176
Diamond–square system 177
Differential cooling 491
Direct compression method 75
Direct draught 5
Discharge temperature 477
Distribution of draught 34
Dog bone pass 249
Double pour rolls 102
Double tapers 367
Double-duo mill 88

Downtime 408
Draught 4
Driving coupling 109
Dummy 407
Dynamic torque 54
Dynamometers 439

E

Eedge oval 226
Edging passes 263
Effective roll radius 19
Efficiency of roll 52
Efficiency of roll drive 52
Efficient furnaces 475
Efficient roll cooling techniques 111
Ekelund formula 18
Electrode 124
Elongation factor 5
Elongation percentage 5
Elongation Percentage 5
End Collars 29
End thrust 20
Endless welding 486
Energy consumption 469
Energy optimization 470
Energy recovery 475
Entry box 315
Entry inserts 315
Equal angles 283
Eutectic reaction point 65
Eutectoid point 65
Eutectoid steel 64
EWR process 487
Excess air 472
Excessive loading 108
Excessive side ribs 255
External friction 43

F

Fabric bearing 130
Fatigue 109
Fatigue Fracture 121
Ferrite 63
Filling of corners 390
Fin 237
Finishing pass 23
Finishing section 290
Finishing speed 483
Fish tail 318
Flakes 162
Flange method 323

Index

497

Flat 259
Flat rolling 2
Flux 126
Flying shear 173
Flywheel 55
Forged steel 103
Former pass 411
Forward slip 18
Four high 82
Free Widening 286
Friction hill 56
Frictional power 52
Fully continuous 87
Furnace pressure 473
Furnace productivity 474
Fused fluxes 127

G
Galling 369
Grain roll 96
Grinding action 41
Grishkov formula 10
Groove or Roll Groove 2
Grooved rolls 2
Group drive 169
Guard 390
Guide mark 436

H
Hard Wearing 93
Head hardening systems 493
Hearth coverage 474
Heat certificate 453
Heat extraction 113
Heat radiation 450
Heat treatment 68
Height reduction 427
High finishing speed 479
Hook formation 245
Horizontal wave 274
Hot charging 473
Housing-less stands 21
Housings 140
HPR system 493
Hydrogen 67
Hyper-eutectoid steels 64

I
Idle running of mill 478
Increase outlet 324

Indefinite chilled cast iron roll 97
Indirect draught 5
Individual drive 170
Induced spread 47
Ingot weight 137
Inner Collars 29
Inspection 458
Internal stress 3
Investigation of roll failure 122
Iron carbide 62
Iron-carbon phase diagram 62

J
Joints or Parting 24

K
Killed steel 136
Kinetic energy 55
Kinks 256
Knurling 17

L
Lap 216
Laying head 484
Leg angles 307
Leonardo de Vinci 76
Lesser slope 321
Light section mills 84
Line of rolling 31
Liquid flux 124
Live flange 42, 46
Localized cooling 491
Longitudinal rolling 92
Loop heights 251
Loop scanner 251
Low temperature rolling 480
Lubricant 118

M
Main drive motor 145
Maintainability 452
Manganese 66
Martensitic layer 489
Maximum draught 35
Maximum shear stress 60
Maximum width 31
Mechanical defects 437
Mechanical working 59
Mechanism of slip 60
Mechanized welding 125
Metallurgical investigation 122
Methods of Rolling 77

498 Index

Mill adjustments 446
Mill Module 22
Mill size 89
Mill spring 21
Mill utilization 488
Minimum tolerance 49
Molten core 162
Molybdenum 67
Moment of gyration 55
Moment of inertia 54

N

Negative allowances 49
Negative collar 30
Neutral line 26
Non uniform deformation 36
Non-metallic inclusions 454
Non-metallic lining for roll 130
Non-uniform deformation 38
Non-working collars 21

O

Off centre web 394
Oil film bearing 133
One sided rib 255
Open design 259
Open journal bearing 130
Open pass 24
Open train billet mill 167
Open train mill 85
Operation management 439
Optimum process parameters 440
Outlet 25
Ovality 236
Oval-square system 178
Over draught 33
Overfilling 313
Overloading 108
Oxygen 67
Oxygen analyzer 472

P

Packaging 460
Pass 24
Pass burning 312, 447
Pass convexity 153
Pearlite 65
Pendulum shear 172
Peripheral speed 19
Peritectic point 64
Peritectic transformation 64

Pheripheral speed 117
Phosphorus 67
Piling 465
Pinion stands 169
Pipe formation 239
Pit mark 435
Pitch line 31
Plain carbon steels 65
Plastic deformation 59
Plum ovals 224
Positive collar 30
Potential hazards 450
PRD system 482
Pre dispatch inspection 456
Pre finishing pass 23
Pre-cambering m/c 437
Pre-finishing pass 23
Pre-heating of rolls 125
Pre-stressed rolls 493
Preventive maintenance 440
Process control 453, 476
Process parameters 457
Processing operations 76
Product characteristics 464
Product mix 174
Profile checking 455
Profile radii 329
Proper pass distribution 408
Pulling-down 37
Pyrometers 439

Q

QST process 491
Quality 451
Quality of water 116
Quality separation 455
Quenching crack 436

R

Radial Load 132
Ragging 17
Ragging 364
Rail roads 397
Rail shaped bloom 406
Rail Testing 417
Rails 397
Railways 397
Reducing /sizing mill 483
Reheating furnace 470
Relative draught 4
Reliability 452

Index 499

Required power 52
Resistance to deformation 34
Restricted spread 7
Rhomboid 271
Rimming steel 137
Robbing of the flanges 324
Rock bolts 217
Roll bearings 129
Roll breakage 107
Roll changing 441
Roll cooling system 107
Roll lubrication system 118
Roll management 118, 119, 121
Roll pass design 2
Roll pass designer 1
Roll stand 139
Roll tables 146
Roll temperature 117
Roller bearings 131
Roller twist guides 170
Rolling constant 50
Rolling Diameter 30
Rolling load 51
Rolling rhythm 439, 449
Rolling schemes 244
Rolling speed 171
Rolling temperature 365
Rolling torque 52
Rolls 91
Rotating mechanisms 450
Roughing passes 23
Round edges 270
Rounds 217
Running mill concept 478
Running repairs 440

S

Safety regulations 450
Sampling 458
Scab 160
Scabs 388
Scale disposal 148
Scarfing 149
Scratches and scoring 161
Seams 160
Selection of diameter 27
Self tempering 71
Semi closed pass 332
Semi killed steel 136
Semi-continuous 87

Sensible heat 479
Separation 126
SGCI (Acicular) 100
SGCI (Pearlitic) 99
Shape deformations 490
Shearing processes 75
Sheets and Strips 80
Shelly Toes 320
Shift foreman (Mech.) 443
Shift foreman (opr.) 443
Shift manager 442
Shorter leg 308
Shrinkage 222
Silicon 67
Single strand rolling 486
Size-free rolling 488
Skid mark 471
Slabbing 139
Slabbing action 364
Slabs 135
Sleeve 134
Sleeve rolls 104
Sliding block 144
Slipping of bars 243
Slipping of metal 303
Slit rolling 247
Slitting delivery guide box 254
Slivers / Shelling 320
Softening 60
Special SGCI rolls 101
Spheroidal Graphite Rolls 99
Split of pass 26
Spread 5
Spread in channel 330
Spring steel 272
Square pass 227
Stack efficiency 476
Stamping 173
Stamping 454
Static load 54
Steel ingot 136
Steel rolls 102
Stepped rolls 260
Sticker mark 391
Straight delivery 313
Straightening machine 409
Strain hardening 59
Strapping 456
Strength of roll 364

Index

Stress relieving 125
Submerged arc 124
Sulphur 67
Surface hardening 141

T

Tachometers 439
Tag 450
Tag system 450
Taper 343
Taper of the pass 42
Tearing 320
Technical delivery condition 456
Tempcore 480
Tempered martensite 70
Templates 351
Temprimar 480
Tension 19
Textolite bearings 142
Thermal cover 476
Thermal fatigue 111
Thermax 480
Thermo-mechanical processing 68
Thick and narrow flats 269
Thick angle passes 312
Thickness ratio 153
Thin and wide flats 269
Thin flats 269
Thinner angle passes 312
Three high 81
Thrust 370
Tight widening 285
Tilting rolls 170
Tip thickness 329
TMT bars 233
TMT quenching system 73
Tongue and groove pass 262
Top guard 242
Top pressure 33
Torque arm 53
Torsional Stress 121
Transverse rib 255
Trapezoidal flats 269
Tube round mills 84
Tungsten 67
Tungsten carbide 104
Twist guides 170
Twisted blooms 159
Two high 81

U

Under filling 313
Under-draught 33
Unequal angle 307
Uniform cooling 491
Uniform quenching 492
Uniform web depth 490
Unit capital investment 448
Universal joint spindles 144
Universal mill 84
Universal pass 428
Universal pass sequence 285
Universal rolling 326
Universal rolling mill 86
Universal spindle 139
Unrestricted spread 7
Upper critical temperature line 65, 66
Use of additives 117

V

V" pass 323
Vanadium 67
Vanadium Steels 494
Vertical stands 168
VVVF drives 478

W

Walking beam type 480
Water boxes 74
Waviness of the web 392
Waving flanges 410
Wavy Flange 393
Web cutting pass 44
Web draft 46
Web formation 339
Web thickness 341
Web-cutting elements 47
Weld deposit 127
Welding 417
Wide flanged beam 361
Wire drawing 367
Work hardening 56
Work hardening 59
Working bearing collars 20
Worm reducing gears 143
Worn out pass 392

Z

Zero spread 368
Zone of deformation 15